THE ILLUSTRATED DIRECTORY OF NORTH AMERICAN LOCOMOTIVES

9000
CANADIAN NATIONAL
9000
C.N.R.
C.N.R

THE ILLUSTRATED DIRECTORY OF NORTH AMERICAN LOCOMOTIVES

The Story and Progression of Railroads from The Early Days to The Electric Powered Present

Skyhorse Publishing

Originally published by Chartwell Books, 2015
First Skyhorse Publishing edition, 2023

10 9 8 7 6 5 4 3 2 1

Library of Congress Cataloging-in-Publication Data is available on file.

Print ISBN: 978-1-5107-5658-8
Ebook ISBN: 978-1-5107-5683-0

Printed in China

Contents

Foreword by J.P. Bell

In the 1850s cries of "Manifest Destiny" rang out across the United States as the young nation stretched its muscles and expanded into the western interior. Gold was discovered in California in 1848. Adventurers and fortune seekers became desperate to reach the Pacific shore of North America. Travel to the West involved either an arduous four-month sailing trip around South America, a dangerous trip across the Isthmus of Panama, or an even more difficult overland journey through the wilderness of the intermountain west. Railroads already blanketed the East with a growing web of steel rails during the mid-1800s. As the gold rush of 1849 ensued, the need for a transcontinental railroad linking the United States became self-evident.

Secretary of the Army, Jefferson Davis, headed the group looking for possible eastern starting points of the proposed transcontinental railroad. Davis favored a southern city for this eastern terminus. The southern path across the country offered a more hospitable climate and a lower crossing of the Continental Divide. All hopes of a southern route for the first transcontinental railroad evaporated with the onset of the Civil War in 1861. Jefferson Davis from Mississippi became president of the breakaway southern confederacy. Plans for the eastern terminus of the railroad moved north.

In spite of the uncertainties of war, President Abraham Lincoln, signed the Pacific Railway Act into law on July 1 1862. The push to complete the vital transcontinental route uniting the country started from Council Bluffs, Iowa and Sacramento, California. The line was completed at Promontory Summit, Utah in 1869.

American railroads became the lifeblood of new communities in the West. Whole towns sometimes picked up and moved to new locations on the tracks of an approaching rail line. Cities across North America came into existence because of the railroad. Place names on the map still bear the mark of their railroad origins from the period of 1870 into the early twentieth century.

Railroads brought new products to rural communities. Farm crops and livestock were carried to market. Passengers came west by rail, and trains carried sons and daughters off to college and careers in the big cities. Railroads became the shining thread that tied together the tapestry of American life into a land of plenty.

Without the development of steam engines, American railroads would have remained in the era of horse drawn trolleys in the early 1800s. Steam locomotives enabled people to travel faster and farther than they could ever imagine possible. The internal combustion engine and diesel locomotives further refined motive power for the great trains that spanned the nation.

The Illustrated Directory of North American Locomotives explores the story of railroads and their motive power. Giant beasts of iron and steel once roamed the land. Their descendants still race across the country.

J. P. Bell lives and works in Fayetteville, Arkansas. His early years were spent on a farm along Clear Creek in the Ozark Mountains of northwestern Arkansas.

In 1957, as a child, J.P. and his family traveled by train from Arkansas to northern California on Santa Fe's *San Francisco Chief.* From that time forward, railroads and the American landscape through which these trains passed captured his imagination. His love of wilderness grew as he hiked in the intermountain West and canoed rivers of the Ozarks.

Bell has photographed professionally since 1987. His photographic inventory includes images from the Ozark Mountains, western United States, Africa, the Middle East, South America and Europe. His special interests include transportation, landscapes, and whitewater sports.

J.P.'s black and white prints and color murals are sold through Art Form Gallery in Van Buren, Arkansas and Cantrell Gallery in Little Rock, Arkansas. His photography has been included in fine art collections of banks, hospitals, corporate offices, public installations, and private collections throughout the country. Since 1996, J.P. has photographed extensively for Bank of the Ozarks. His black and white prints are in the offices and lobbies not only of the headquarters of Bank of the Ozarks in Little Rock, but in branch offices throughout the bank's service area.

Bell's photography and articles have appeared in *Locomotive and Railway Preservation, Railfan and Railroad,* the *Arkansas Gazette, Arkansas Times, Entertainment Fort Smith, River* magazine, *National Geographic* maps, and the *Whitefish Review.*

J. P. Bell's most recent book, published in 2014 is *Ghost Trains: Images from America's Railroad Heritage*. This 192-page book of 23 photo-essays gives the viewer a broad canvas of railroading across the U.S. and Canada. Bell's first book was *Steam Trains: A Modern View of Yesterday's Railroads.* He has also photographed for books for Border's Books, Stackpole Press, and Pepperbox Press.

Bell's photographs may be explored at jpbellphotography.com

Introduction

In the 1950s, the American railroad exerted a huge fascination on many children. It would also have been a big part of many of their lives, because many vacations and visits to relatives would still have involved a trip by rail. At this time, the railroads were also hugely important to American industry. Every day, they hauled heavy goods alongside the highway network. In many towns, working sidings connected the main line to a warren of factories. Many Americans would have dropped off to sleep to the soothing sound of the night express horn or the eerie whistle of the night freight train hauling steel and coal along moonlit rail tracks. Many ran alongside the major rivers, like the wine dark Ohio and formed part of the vital transportation links of our mighty country. This romantic experience is now virtually unknown to the modern generation as the rise of the car and truck has wiped it out. But it has left behind a generation of rail fans who are active in preserving and promoting the idea of a lasting rail heritage for the nation's future generations.

Paradoxically, America's railroad history began in October 1828 at Rainhill, nine miles east of Liverpool, England. This was where the Liverpool & Manchester Railway held eight days of locomotive trials. Several American railroad promoters and engineers, including E.L. Miller and Horatio Allen of the South Carolina Railroad, and George Brown and Ross Winans of the Baltimore & Ohio Railroad, were among the spectators. Allen had already visited England earlier that year with the purpose of learning about the most recent

Right: The steam locomotive, Savannah and Atlanta No. 750, is at Chester, Arkansas at night for the filming of the movie, Biloxi Blues in June 1987.This evocative night scene recreates the atmosphere of much railroad activity in the 1940s and 50s.

Below: The Rainhill Trials were an important competition in the early days of steam locomotive railways, run in October 1829 for the nearly completed Liverpool and Manchester Railway. Five engines competed, running back and forth along a mile length of level track at Rainhill, Lancashire. Stephenson's Rocket was the only locomotive to complete the trials, and was declared the winner. The Stephensons were accordingly given the contract to produce locomotives for the railway.

750

Left: Locomotion No. 1 (originally named Active) was the first steam locomotive to carry passengers on a public rail line, the Stockton and Darlington Railway. Built by George and Robert Stephenson's company in 1825. It is preserved at the Darlington Railway Centre and Museum.

Right: A replica of Stephenson's Rocket.

developments in railroad technology at their source; three engineers from the B&O made a follow-up visit a year later.

Both groups toured the Liverpool & Manchester railway with its chief engineer, George Stephenson. They observed the deep cuts, huge embankments, and strong stone viaducts that were designed to minimize curves and grades that Stephenson had laid out to make it suitable for his locomotives. The American visitors discussed these engineering works with George Stephenson and his son, Robert. The Stephensons also owned the engine works that had built the *Locomotion,* the first steam engine to pull a passenger train on England's Stockton & Darlington Railway in 1825.

Prior to Rainhill, the American engineers had reviewed every aspect of railroad operation with their British counterparts, and had listened to the arguments that were raging in England over the most suitable types of motive power.

To bring these arguments to a conclusive head the proprietors of the Liverpool & Manchester Railway offered a cash prize for the best locomotive design. They established rules, selected judges, and set a date. The strange and wonderful machines that ran at Rainhill included a horse treadmill, a small car that used Ross Winans's quirky "friction wheels" that two passengers operated by turning winches, and several "running coaches." One of these, George Stephenson's *Rocket*, won the Rainhill trials. The engine was the product of the untutored but brilliant coalfield engineer, and his son (Robert), an educated steam engine entrepreneur. The *Rocket* had a horizontal, multi-tubular boiler and steam cylinders that were directly connected to the driving wheels. Effectively, the machine established the prototype of the modern steam locomotive, and proved that locomotives were preferable to horses or stationary engines for providing fast, safe, and efficient railroad power.

In fact, the earliest American locomotives had gotten up steam even before the Rainhill trials.

The *Stourbridge Lion* was the first commercial, non-experimental locomotive to run n the American railroad. Built by the firm of John Rastrick, of Stourbridge, England, the *Lion* was a "walking beam" engine, which was similar to the *Locomotion.* Horatio Allen had ordered it for the Delaware & Hudson Canal Company during his 1828 inspection tour of British railways. The company had constructed a sixteen-mile railroad between Carbondale and Honesdale, Pennsylvania as this area was too mountainous for a canal. The trial was not a success. The locomotive weighed seven tons, but the line had been built for lighter traffic. Allen took the engine down the tracks in August 1829, but it was a failure. The timber viaducts sagged so much that the owners, fearing the loss of their engine placed the *Stourbridge Lion* in storage, and left it there for the duration.

Right: A replica of the Stourbridge Lion is kept at the Smithsonian Museum of Science & Technology at Washington, DC.

Far right: The Tom Thumb replica on display at the B&O roundhouse.

ROCKET
On tour from
National Railway Museum

PETER COOPER'S "TOM THUMB" 1830 BALTIMORE & OHIO R.R

Above: A working replica seen here, built in 1928, is currently on display at the Atlanta regional headquarters of Norfolk Southern Railway.

In 1830, Ross Winans was one of the passengers on Peter Cooper's *Tom Thumb* when it made a twenty-six mile round trip on the B&O Railroad. Although the engine was to pull some regularly scheduled passenger trains, Cooper's locomotive was designed as an experiment. It has been built by a prosperous New York businessman and semi-literate mechanic who had based his speculative purchase of several thousand acres in Baltimore on the future success of the railroad. Following the Rainhill trials, the B&O had ordered a Stephenson locomotive. But the engine was involved in an accident while being loaded aboard ship, and never actually reached the United States. Meanwhile, there was concern among investors that the B&O railroad's sharp curves would prevent the use of steam power.

Cooper set out to prove them wrong. His engine was assembled by a team of Baltimore mechanics and weighed just a ton - hardly any larger than a railroad handcar. The locomotive boasted a vertical, multi-tubular boiler, a single cylinder, drive gears, and rode on the "friction wheels" designed by Ross Winans, who compared the engine's power to that of *Rocket*. With Cooper at the controls, the engine managed to achieve the heady speed of 18 miles an hour. Some of the passengers pulled out notebooks and wrote down their thoughts to prove that human beings could function normally at such high velocities. But just a month later, Cooper's engine lost its famous race with the "dappled gray" horse of Stockton and Stokes, due to a mechanical failure. Stockton and Stokes had provided the B&O's first motive power, but *Tom Thumb's* "triumphant demonstration" removed all doubts about the feasibility of using steam locomotives on the B&O Railroad. *Tom Thumb* also established a tradition of idiosyncratic design that Ross Winans extended during his long career of engine building, mostly for the B&O.

In November 1830, the *Best Friend of Charleston* was America's first locomotive built for general railroad service. The *Best Friend of Charleston* made its initial trip on the South Carolina Railroad. Horatio Allen was the line's engineer and E.L. Miller helped him to design the engine, which was manufactured by New York City's West Point Foundry. The *Best Friend* had a vertical boiler, dual-angled cylinders, and direct inside connections to the wheels via axle cranks.

Unfortunately, the engine exploded just five months later when the fireman fastened down the safety valve. The fireman himself dies in the accident. The company's second engine, the West Point, was bought from the same foundry. It had a horizontal, Bury-type boiler. When it went into service, the engine was equipped with a "barrier car," stacked with cotton bales, designed to protect the passengers in the event of a second mishap.

The significance of Rainhill was not fully realized until long after the event, but an anonymous correspondent for the *Scotsman* (the leading Scottish newspaper of the day) recognized the promise of the railroad. After chiding his London colleagues for ignoring the trials in favor of their usual fare of politics and murders, he wrote: "the experiments at Liverpool have established principles which will give a greater impulse to civilization than it has ever received from any single cause since the Press first opened the gates of knowledge . . . They may be said to have furnished man with wings, to have supplied him with faculties of locomotion, of which the most sanguine could not have dreamed a few years ago. Even steam navigation gives but a faint idea of the wondrous powers which this new engine has put into our hands. It is no exaggeration to say, that the introduction of steam carriages on railways places us on the verge of a new era, of a social revolution of which imagination cannot picture the ultimate effects."

Nowhere were the railroad's effects to be more dramatic than in the United States. Edward Pease, the financial backer of England's Stockton & Darlington Railway, could have had the United States in mind when he said, "Let the country but make the railroads, and the railroads will make the country!" Certainly, there was a nation to be made. In 1830, only a few dozen miles of track had been laid, mostly for coalmine tramways and for the two general-purpose railroads that had made tentative forays from the eastern seaboard. The United States' population of thirteen million people was mainly concentrated in New England and the mid-Atlantic region, and distributed among just twenty-four different states. Just two of these (Missouri and Louisiana) lay to the west of the Mississippi River. As Frederick Jackson Turner commented in *The Frontier in American History,* "Prior to the railroad, the Mississippi Valley was potentially the basis for an independent empire."

No one understood this better than George Washington, the first "commercial American" who realized the need to bind the disparate sections of the country together with the "cement of interest," - ties of trade and commerce. As the result of his personal explorations and military campaigns, Washington anticipated in his diaries and letters the future routes of the Erie Canal, the Pennsylvania Main Line of Internal Improvements, the National Road and, along his beloved Potomac route to the West, the Chesapeake & Ohio Canal. The B&O Railroad also ultimately benefited from Washington's visionary thinking, for it inherited the C&O Canal's western alignments. When the B&O was incorporated in February 1827, western migration had advanced to a point halfway between St. Louis and Kansas City and was moving forward at the rate of about thirty miles per year. But two-thirds of the future United States to the west of the Mississippi River was still an unexplored region.

These ambitions set a tremendous challenge to indigenous U.S. locomotive technology in the years 1830-60 to improve in order to live up to the challenge of the ambitious plans of its politicians and railroad engineers. By the mid 1850s the 4-4-0 or American type had emerged as the front runner-typified by The *General* built by Thomas Rogers of Paterson, New Jersey in 1855.

Rogers was responsible for introducing most of the features which made the *General* the success it was. The most significant development, so far was the general introduction of Stephenson's link motion, which permitted the expansive use of steam. This was in place of the "gab" or

Below: William Mason is one of the oldest operable examples of the American Standard locomotive design 4-4-0.

Above: By the mid 1850s the 4-4-0 or American type had emerged as the front runner- typified by The *General* built by Thomas Rogers of Paterson, New Jersey in 1855.

"hook" reversing gears that had been used until then, which permitted only "full forward" and "full backward" positions.

In other aspects of design Rogers gained his success by good proportions and good detail rather than innovation. An example was the provision of adequate space between the cylinders and the driving wheels, which reduced the maximum angularity of the connecting rods and hence the up-and-down forces on the slide bars. A long wheelbase leading truck allowed the cylinders to be horizontal and still clear the wheels. This permitted direct attachment to the bar frames, which raised inclined cylinders did not. To allow flexibility on curves early examples of the 4-4-0 inherited flangeless leading driving wheels from their progenitors but by the late 1850s the leading trucks were being given side movement to produce the same effect. Naturally the compensating spring suspension system giving three-point support to the locomotive also proved to be a success on the sometimes rough track as the railroads took on more rugged terrain. Wood burning was also nearly universal in these early years of the type and the need to catch the spark led to many wonderful shapes in the way of spark-arresting smokestacks.

Within two or three years other makers such as Baldwin, Grant, Brooks Mason, Danforth and Hinkley began offering similar locomotives. To buy one of these locomotives one did not need to be a great engineer steeped in the theory of design—it was rather like ordering a car today. The customer simply filled in a form on which certain options could be specified and very soon an adequate and reliable machine was delivered.

The railroads made a significant impact on the Civil War. For the first time during a major conflict it was possible for

Above: U.S. Military Engine No156, built in 1864. In early 1862, the U.S. War Department established a bureau responsible for the construction and operation of military railroads. At its head was native Philadelphian and West Point graduate Herman Haupt, chief engineer of the Pennsylvania Railroad before the war. Haupt's pioneering work allowed the Union army to use the railroads to their advantage, transporting men and supplies wherever and whenever they were needed. Rejecting a promotion to brigadier general in 1862, Haupt left the service the following year.

both sides to move vast quantities of troops, munitions, supplies, and other raw materials by rail.

As a result both sides strategized tactics that could cut off or subvert the other's railroad communications. The most famous example was the Great Locomotive Chase which featured the famous 4-4-0 General locomotive, now preserved at Kennesaw, Georgia. The Great Locomotive Chase or Andrews' Raid occurred on April 12, 1862, during the Civil War in northern Georgia. Volunteers from the Union

Above: The Civil War utilized the railroads to transport weapons and munitions like this rail-mounted mortar.

Army, led by civilian scout James J. Andrews, commandeered a train from Big Shanty depot and took it northward toward Chattanooga, Tennessee, doing as much damage as possible to the vital Western and Atlantic Railroad line (the W&A). This ran from Atlanta to Chattanooga. They were pursued by the Confederate forces, first on foot, and later in a succession of locomotives. Because the Union men had cut the telegraph wires, the Confederates could not send warnings ahead to forces along the railway. Confederates eventually captured the raiders and executed some quickly as spies, including Andrews; some others were able to flee. For the first time rolling stock was adapted for martial use such as armored cabooses.The Southern Museum of Civil War and Locomotive History at Kennesaw has an interesting collection of artifacts that pertain to both Civil War and railroad history.

The building of a transcontinental railway to unite the nation was first proposed early in the nineteenth century. Sadly, this became a reality just as the nation was being torn apart by the Civil War. Abraham Lincoln signed the Pacific Railroad Act in 1862. This set out both the route of the line and how this huge enterprise was to be financed. Theodore Judah, the chief engineer of the Central Pacific Railroad, explained the long and complicated route to the President on a ninety-foot long map. Back in 1856, Judah had written a 13,000-word proposal to build the Pacific Railroad and became a lobbyist for the Pacific Railroad Convention.

The new railroad would have a huge impact on life in the West, opening it up to settlers. A dangerous trek that would have taken at least six months in the days of the wagon trains could now be accomplished in less than a week. But the obverse of this was the decimation of the bison, along with the

Above: The celebrated Great Locomotive Chase epitomized the importance of the railroads during the Civil War.

Left: Union Pacific locomotive Number 82 and its crew. Photographed in 1872, between Echo, Utah and Evanston.

complete loss of the unique Native American culture of the Great Plains.

The course of the transcontinental line followed the earlier trail routes and Pony Express trails. It ran between Sacramento, California in the West and Council Bluffs, Iowa in the East, and was to pass through Nevada, Utah, Wyoming, and Nebraska en route. The railway did not reach the Pacific until 1869, when a new stretch of line was opened up to Oakland Point in San Francisco Bay. The line integrated into the Eastern railway system until 1872, with the opening of the Union Pacific Missouri River Bridge. Its construction required tremendous engineering feats to overcome the obstacles of the route. The line crossed several rivers (including the Platte in Nebraska), the Rockies (at Wyoming's Great Divide Basin), and the Sierra Mountains. Spur lines were to be built to service the two great cities of the Plains: Denver, Colorado, and Salt Lake City, Utah.

Above: A banner celebrating reaching Cozad, Nebraska, 247 miles from Omaha.

Unfortunately, another intrinsic characteristic of the development was to be corruption. The government legislated to award the constructors with 6,400 acres of trackside land, and a tiered payment per mile of track: $16,000 per mile for level track, $32,000 per mile for plateau track, and $48,000 for the most demanding stages. Within

two years, these rates had been doubled. The investors were careful to ensure that as much track as possible was graded into the more expensive categories. The major made a fortune in the Civil War by smuggling and stock speculation. Durant deviously tinkered with the route to ensure that it ran through his own property. Surveyor Peter Day said that "if the geography was a little larger, I think (Durant) would order a survey round the moon and a few of the fixed stars, to see if he could not get some depot grounds."

Other investors like Oakes Ames were drawn into the Credit Mobilier scandal, where dummy contracts were awarded to Durant's own company. The scandal was to ruin their reputations and those of many other investors. Lincoln himself encouraged Ames to become involved in the enterprise, but it was to become his ruin. As railroad executive Charles Francis said, "It is very easy to speak of these men as thieves and speculators. But there was no human being, when the Union Pacific railroad was proposed, who regarded it as other than a wild-cat venture."

Union Pacific's corrupt investors became synonymous with the worst excesses of the so-called "Gilded Age." The term was coined by Mark Twain to describe the post-Civil War extravaganza of industrial-scale corruption when massive fortunes were made and lost. Many of the most magnificent San Francisco mansions were built with railroad money.

But the great panic of 1893 ended the Gilded Age abruptly. The financier Jay Gould replaced the discredited Durant at the head of Union Pacific and continued to steer the project. The Central Pacific broke ground in January 1863 in Sacramento, California, while the Union Pacific waited until December that year to start work at Omaha, Nebraska. The groundbreaking ceremonies began a monumental task that was to take six years and involved the construction of 1,780 miles of track. The varied difficulties and problems of the route meant that innovative engineering solutions were required. The trains of the day could not handle either

sharp curves or an incline of more than two per cent, and the mountain ranges and canyons along the route were equally difficult to overcome.

This enormous challenge required a massive workforce of over 100,000 men, who came from a wide variety of backgrounds. The majority were Irish-American veterans from both sides of the Civil War, together with Chinese immigrants, Mexicans, Englishmen, Germans, and ex-slaves from the South. Brigham Young also provided Mormon workers for the Utah sector of the line. These men were excellent, conscientious workers who ended each day of work with prayer and song rather than women and drink.

The project also required a wide array of tradesmen, surveyors, engineers, carpenters, masons, teamsters, tracklayers, telegraphers, spikers, bolters, and cooks. The work could be very dangerous. The use of early, unstable nitro-glycerine was particularly hazardous, and it resulted in many deaths and injuries. The crews from the two railroad companies were under strong competitive pressure

Above: Corinne, Utah was the final tent town of a whole string of colorful and lawless places along the construction route.

Below: Multiple storey sleeping cars provided accommodation for the thousands of men employed in the project.

Above: Currier & Ives lithographs popularized the heroic image of trains crossing the great prairies in the 1870s.

to complete as many miles of track as possible, and their work often became sub-standard. The railway companies were paid per mile of track, not for the durability of their construction, so their priority was to get the job done as quickly as possible. Slick track laying teams laid as many as four rails per minute.

Ultimately, the Union Pacific was to build about two-thirds of the track. Anxious not to lose a minute of working time, the railroad companies housed thousands of workers in enormous work-trains. These trains had sleeping cars outfitted with three-tier bunk beds, kitchens, and eating cars. Life for these men was extremely hard, and the pay was meager. There were several strikes, particularly among the less well-paid Chinese workers, but the companies were ruthless employers.

But despite the difficulties, the two ends of the Pacific line slowly moved together and went further into the wilderness. The workforce was spread out over several miles and was accommodated in mobile tent towns that followed the route.

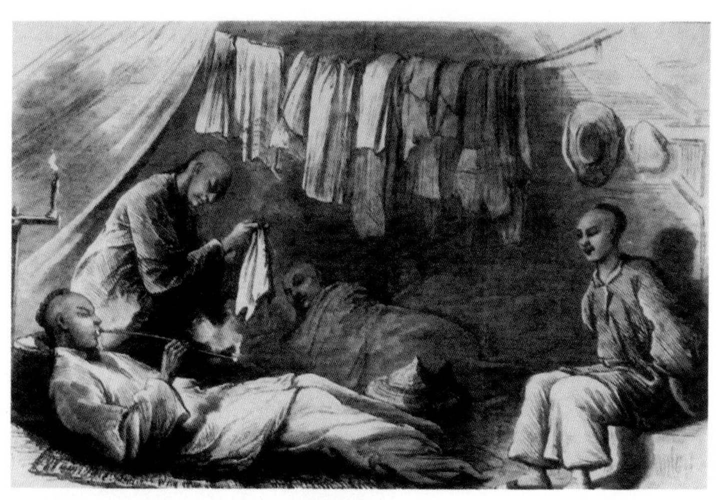

Above: Chinese workers were mainly responsible for constructing the Central Pacific Track. They lived in tents along the line.

The end-of-line boomtowns that were created were both colorful and lawless. These towns included North Platte, Julesburg, Abilene, Bear River, Wichita, and Dodge. The final tent town, Corinne, Utah, was founded in January 1869. These camps became known as "Hell on Wheels," as they were full of vice and criminality and were rough, bawdy, and brutal. Newspaper editor, Samuel Bowie coined the term and unflatteringly described their inhabitants as "vile: men and women… (a) congregation of scum and wickedness… by day disgusting, by night dangerous. Almost everybody dirty, many filthy, and with the marks of lowest vice; averaging a murder a day; gambling, drinking, hurdy-gurdy dancing and the vilest of sexual commerce."

In reality, the tent towns were conurbations of saloons, gambling houses, dance halls, and brothels. Almost all the women living in these settlements were prostitutes. Murder, arson, and violent crime were rife. Without any real law enforcement, frontier justice was the only control, and lynching was common. John Ford captured the decadent atmosphere of Hell on Wheels in his silent film of 1924, *The Iron Horse*. Although the film was not entirely accurate, Ford succeeded in showing the spirit of fervent nationalism that drove the project. Despite their inauspicious beginnings, many of these tent towns became permanent settlements. Mark Twain described the gold rush and end-of-the-line rail town at Sacramento as being no more than a "city of saloons," but the town was soon to become the state capital of California.

The railroad companies also actively encouraged immigration, from both China and Europe to swell their workforces. The Chinese population grew exponentially, from less than a hundred people in 1870, to over 140,000 men and women by 1880. The railroad companies employed agents to scout for immigrants, who were paid per head of workers delivered. C. B. Schmidt was the champion scout, responsible for settling over 60,000 German immigrants along the route of the Santa Fe Railroad.

Settlement of the prairie also led to a massive increase in American farming. The two million working farms that existed in 1860 had grown to six million by the end of the century. But this colossal increase in white settlement was a source of great anger to the Native American peoples of the Plains. The other was the decimation of the American Bison, or buffalo. This animal was unique to the Plains, and before

Above: The Governor Stanford, the first of Central Pacific's twenty-three locomotives, on its way to the joining of the rails celebrations.

the railroad came, it was estimated that as many as sixty million animals roamed the prairie in massive herds. The buffalo was crucial to the existence of the Plains Indians and also had a special spiritual significance to them. "Everything the Kiowas had had come from the buffalo," said Kiowa tribe member Old Lady Horse, "Their tipis were made of buffalo hides, so were their clothes and moccasins. They ate buffalo meat."

The other Plains tribes, including the Cheyenne, Lakota, and Apache, were equally dependant on the buffalo for their survival. In complete contrast, the railroad companies saw the ancient bison herds as a nuisance, useful only for feeding their voracious workforce. The companies hired buffalo hunters to wipe them out. The most famous of these was Buffalo Bill, who rode the Plains on his horse, Buckskin, and his gun, Lucretia. He alone shot over four thousand animals and organized many hunting expeditions. Later, the railroad encouraged "hunters" to shoot buffalo from specially adapted railcars to minimize any risk or inconvenience. Elisabeth Custer described how "the rush to the windows, and the reckless discharge of rifle; and pistols put every passenger's life in jeopardy."

This trend became so widespread that the Kansas Pacific Railroad ran it own taxidermy service to mount trophies for their customers The upshot of this dreadful slaughter was that, by the end of the end of the nineteenth century, only a pathetic remnant of fewer than a thousand animals remained from the majestic herds that had dominated Plains life for centuries.

Seeing their way of life being destroyed before their eyes, some of the more warlike Plains Indian tribes began to organize scouting parties to vandalize trains and attack surveyors and other railway workers. This gave the rail companies the excuse they needed to strike back. According to General Grenville Mellen Dodge, the chief engineer of the Union Pacific, "We've got to clean the damn Indians out, or give up building the Union Pacific Railroad." The Sand Creek Massacre of November 1864 was one of the most appalling incidents that took place. Men of the Colorado Territory Militia destroyed a village of Cheyenne and Arapaho and killed over two hundred elderly men, women, and children.

Although the massacre was widely condemned, no one was ever brought to justice. Sand Creek led to a series of revenge killings in the Platte Valley, and over two hundred innocent white settlers were murdered.

This increasing spiral of violence made it progressively more difficult for an accommodation to be found between the Plains natives and the railroad companies. The regular U.S. Cavalry was deployed to protect the security of the trains, and Dodge ordered the Powder Ridge Expedition of 1865, in which his forces rode against the Lakota, Cheyenne, and Arapaho. Although this was partly successful, hostilities soon escalated into the Reds Clouds War, which was fought against the Lakota tribe in 1866. The Lakota braves inflicted heavy casualties, and it was the worst defeat that the U.S. Cavalry was to suffer until the battle of Little Big Horn, ten years later.

Their resistance to the railroad led to the Plains tribes being confined to reservations, where they were powerless to protect their ancestral hunting grounds, or the buffalo.

On May 10, 1869, the Central Pacific and Union Pacific tracks finally met at Promontory Summit, Utah. Leland Stanford, the Governor of California and one of the "big four" investors in the Central Pacific, drove home the final, golden spike that joined the two lines. This was one of the world"s first global media events, as both the hammer and spike were wired to the telegraph line, and Stanford's ringing blows were broadcast simultaneously to both the East and West Coasts of America.

The railroad line had a great impact on the whole country, but its effects were most directly felt in the West. It proved to be a major stimulus to immigration and trade. Soon, other railroads criss-crossed the Plains. These included the Kansas Pacific, North Pacific, Denver Pacific, Texas and Pacific, Burlington and Missouri River, Denver and Rio Grande, Atchison, Topeka, and Santa Fe railroads. By 1876, it was possible to travel between New York and San Francisco in 83 hours 39 minutes.

This extraordinary achievement went on to be celebrated as an iconic element of Western culture. The railroad is familiar from any number of movies, as the great iron horses drive over the monumental Western landscape. In 1936, Cecil B. DeMille released *Union Pacific,* which explored the corruption that surrounded the building of the line. 1960s's epic movie, *How the West Was Won,* also dealt with the dramatic construction of the Union Pacific line, especially with how the railroad bosses drew the rage of the Native

Right: The Great Event poster announces the staggering achievement of a railroad from the Atlantic to the Pacific, serving "Travelers for Pleasure, Health, or Business."

American tribes on their workers, and how the Cavalry attempted to protect them. One of the film's most famous scenes is of a herd of buffalo stampeding across the railroad.

The Union Pacific left a permanent mark on American life in both the East and West. The line itself has been renewed many times, but much of it is still laid on the original, hand-prepared grade. In several places, where later routes have bypassed the initial line, it is still possible to see the original track, abandoned in the wilderness.

During the tumultuous years between 1865 and World War I, the nation's population nearly tripled, from 35,700,000 to 103,400,000. At the same time, the American rail network increased by a factor of seven — to 253,626 miles — while gross operating revenues rose spectacularly — thirteen-fold, from $300 million in 1865 to approximately $4,000 million in 1917. In fact, the length of the American railroad track exceeded that of Europe even before the turn of the twentieth century.

In 1865, America's railroads were neither efficient nor integrated. But integration was to follow. An early example of how this occurred was provided by the Vanderbilt family, who skillfully brought together a collection of railroad roads to form the New York Central System.

This network reached from New York City to Chicago and eventually served Boston, Cincinnati, and St. Louis with main routes. The Vanderbilts plowed capital back into the railroad and demanded a high degree of efficiency from their operations. Not surprisingly, the managers of the Vanderbilt roads dutifully guarded their respective service areas against encroachment by competitors. In this so-called "trunk line region," only the Pennsylvania Railroad afforded any significant competition, although the Erie and a few other services plied the same region.

In the West, a similar pattern emerged as the Southern Pacific forged a crescent-shaped route structure stretching from Portland through San Francisco, Los Angeles, Tucson, San Antonio, and Houston, through to New Orleans. Led by the irrepressible Collis P. Huntington, the Southern Pacific survived the Panics of 1873 and 1893 and, by 1900, controlled a network

Above: New York Central and Hudson River Railroad No. 999 was built for the New York Central and Hudson River Railroad in 1893, which was intended to haul the railroad's Empire State Express train service between New York and Chicago. This locomotive is claimed to have been the first in the world to travel over 100 mph.

Left: Introduced in 1895, the D16 class followed the reputation established by the Pennsylvania Railroad for big locomotives that soon established reputation for high speed operations on the line between Camden and Atlantic City.

Above: Constructed in just 20 days by Baldwin Locomotive Works, B&O No4500 was the first USRA locomotive produced under federal management. No4500 was equipped with the latest technology of its time, including a superheater and stoker.

of over 14,000 miles. The network was particularly attractive to Edward H. Harriman, who gained control over the Union Pacific railroad when Huntington died. The railroad was connected the historic Overland Route at Ogden. "We have bought not a railroad, but an empire," Harriman exulted in 1901.

At first, the country's railroads built their primary arteries to link established cities, and the railroad companies often bought undeveloped areas that they could develop later. Competition on the rail routes was cutthroat. A fleshing out of the basic network followed later, with the opening of secondary lines and branches. These were built to open up farm land, to access stands of timber, to serve mines and quarries, to outflank rival companies, to make territorial claims, or to achieve a combination of all of these aims.

The railroad industry helped lay the foundations for the modern American economy, and pioneered many of the systems by which it is organized. These include advanced methods of raising finance, management, labor regulation, and competition. The industry opened up new fields of operation for financiers, bankers, and speculators and the expanding network connected manufacturer and consumer. As the railroad service became more dependable it had a huge impact on business practices. The system of inventory control was developed and the factory system was streamlined.

In short, the railroad (and its ally, the telegraph) were fundamental to America's development in both industry and agriculture. The West underwent an extraordinary transformation from an open plain to a vibrant and economically active area.

During World War I, the federal government took control of the nation's railroads and formed the United States Railroad Administration (USRA). This was to facilitate the efficient mobilization of troops and supplies. The USRA oversaw the mass production of standardized locomotives and the operation of all privately owned railroads. The USRA Locomotive Committee consisted of representatives from ALCO, Baldwin Locomotive Works, and the Lima Locomotive Works. It organized the manufacture of over 1,800 locomotives using cutting-edge technology. Although many railroads resented the USRA's control, the organization streamlined the railroad industry and made advances for railroad labor by increasing wages and decreasing the workday to eight hours. USRA control ended on March 1, 1920 but its durable locomotives continued to have a lasting influence on the railroad industry.

Diesel Power

1918 marked the precursor to the diesel locomotives that are still in service today. The steam-powered locomotives

Above: Baltimore & Ohio No50 is one of five experimental 1,800 hp passenger diesel locomotives built by EMC(later to become EMD) in 1935. They were the first non-articulated diesels to work on U.S. main line railroads.

Below: The GE 45-Ton Switcher is a four-axle diesel locomotive that was built by GE between 1940 and 1956.

engines that were the product of the Industrial Age became a necessity for some, and a luxury for others as people traveled across the nation. The inevitable march of progress meant that steam power morphed into the internal combustion engine. The American Locomotive Company (Alco) partnered with two major players in the industry, Ingersoll-Rand and General Electric, to design a diesel-powered motor car. This train was called the GM-50 and ran on the Jay Street Connecting Railroad No. 4 in New York City. It was the first diesel-electric powered vehicle to find its way onto America's railroad tracks. By 1924 the trio of companies had designed a more advanced diesel motor that powered a 60-ton boxcar. The Central Railroad of New Jersey purchased the engine (which produced 300 horsepower) and the Baltimore and Ohio Railroad followed suit. Working with B&O, the Electro-Motive Corporation (which was to become General Motor's Electro-Motive Division) fine-tuned the diesel-electric locomotive design. In the 1930s, B&O began running the resultant engines on North American railroads.

The diesel locomotive became popular because of its simple use of a diesel engine to generate power. This turns the traction motors which power the locomotive's wheels.

Above: F-Series Diesel Electric locomotives were introduced in 1939 by GM's Electromotive Division and manufactured at the La Grange plant in Illinois and at the GMDD plant in London, Ontario, Canada.

Below: The EMD GP15, technically listed by the builder as the GP15-1, was a late model of the General Purpose series intended by Electro-Motive Built for light duty use.

KCS 2
Kansas
City
Southern
Lines
DOG
TAGS
LICENSE
PLATES

Left: This Kansas City Southern F-series engine was built in 1954. The type was built between 1939 and 1960.

Each part of the diesel-electric motor serves its own purpose, and diesel-electric locomotives generate and utilize their own power to move the train. This power source meant easy maintenance as engine units could be easily replaced. This was a great step forward from the complete boiler strip downs that were involved with steam rebuilds. The other big advantage of diesels was the self-contained power source that meant they could run on non-electrified track.

Originally the diesel locomotive was used in boxcar or switcher configurations. But as diesel technology advanced, railroads used diesels to haul heavy passenger and freight trains across the country. They proved far more reliable and required less fueling than steam locomotives. This meant that the trains could keep moving instead of making frequent stops to refuel with water, coal and/or oil. Diesel-electric locomotives also required much less maintenance than steam-powered engines. This meant that they continued to earn revenue when steam locomotives were laid up in the workshop. This combination of efficiencies endeared diesels to America's railroad operators. Despite the protests of hardcore steam enthusiasts, diesels had mostly replaced steam locos by the 1950s.

Above left: A prototype GG-1 locomotive was built for the PRR with a streamlined casing by the famous industrial designer Raymond Loewy. Between 1935 and 1943, 139 of these GG-ls were built and remained in service until 1982.

Above: A Metro North electric unit typifies the East Coast Commuter line traffic.

Electric Power

At the same time that diesel-electric engines were being developed, true electric-powered engines were also being developed. These engines ran on current supplied by a

Right: A northbound Amtrak Acela Express passing through Old Saybrook, Connecticut in 2011

collection source along the track. This power source was being developed before the nineteenth century was out. In 1895 the Baltimore and Ohio Railroad electrified their mainline as it ran through the city of Baltimore and in particular the Howard Street tunnel which ran 1 1/4 miles under the city. The tunnel has a steep gradient and the powerful electric locos were used to assist steam-powered engines trough the tunnel to avoid excessive pollution.

Anti pollution laws in New York made it obligatory for inbound expresses to be electrified as they came into Penn and Grand Central stations. On January 28, 1935 the electric line was completed between Washington, D.C and New York City. PRR ran a special electric train that was pulled by a GG-1 electric locomotive. The line was opened for revenue service on February 10, 1935. The train made a round trip from D.C. to Philadelphia. On the return trip, the train set a speed record, arriving in D.C. just one hour and fifty minutes after leaving Philadelphia.

Electric locomotives were mainly geared to the high profit East Coast Commuter lines such as the PRR and those serving populous metropolitan areas such as Chicago. Elsewhere the daunting task of electrifying the track by providing a third rail or (later) overhead pantographs was not cost effective. This was particularly true for the transcontinental routes where the cost would be enormous. The third rail system proved unreliable. It was vulnerable to floods and adverse weather conditions. Overhead pantographs were even more costly, so this ruled out vast lengths of electrified prairie trackage.

Electric power sustained its green image through the fuel crisis of the 1970s and remains a major source of power to the nation's railroads.

Today, American rail transportation is dominated by freight. Passenger services, once a large and vital part of the nation's passenger transportation network, now plays a limited role compared to that in other countries. The U.S. rail industry has experienced repeated re-inventions due to the changing economic needs and the rise of automobile, bus, and air transport. Freight railroads continue to play an important role in the U.S. economy, especially for moving containerized imports and exports, including shipments of coal and oil.

Above and below: The Acela Express offers hourly service downtown to downtown during peak morning and afternoon rush hours between New York, Washington, DC, Baltimore, Philadelphia and other intermediate cities, as well as many convenient round-trips between New York and Boston.

The sole remaining intercity passenger railroad in the continental U.S. is Amtrak. Commuter rail systems still exist in more than a dozen metropolitan areas, but these systems are not extensively interconnected, so commuter rail cannot

be used alone to traverse the country. New commuter systems have been proposed for approximately two dozen other cities, but bottlenecks in local government administration and the ripple effects from the 2007–2012 global financial crisis have relegated these projects to the distant future.

The most notable exception to the general lack of passenger rail transport in the U.S. is the Northeast Corridor between Washington, Baltimore, Philadelphia, New York City and Boston. This network also has significant branches in Connecticut and Massachusetts. The corridor handles a frequent passenger service that is both Amtrak and commuter. New York City itself is noteworthy for its high usage of passenger rail transport, on both the subway and commuter rail network. This includes the Long Island Rail Road, Metro-North Railroad, and New Jersey Transit. The New York subway system is used by one third of all U.S. mass transit users.

Other major cities with substantial rail infrastructure include Boston, with it's MBTA, Philadelphia, with its SEPTA, and Chicago with its Metra (the elevated and commuter rail system). The commuter rail systems of San Diego and Los Angeles (Coaster and Metrolink) also connect at Oceanside, California.

In 1900, there were 132 Class I freight railroads in America. Today, as the result of mergers, bankruptcies, and major changes in the regulatory definition of Class I, there are only twelve Class 1 railroads in North America.

These include Amtrak and seven freight railroads are designated Class I based on 2011 measurements released in 2013. As of 2011, U.S. freight railroads operated 139,679 route-miles of standard gauge track in the U.S. Although Amtrak qualifies for Class I status under the revenue criteria, it is not considered a Class I railroad because it is not a freight railroad.

Remaining Class I Railroads in North America

Via Rail (has only Canadian trackage, with none in the U.S.)

Trackage in both the United States and Canada:
Amtrak
BNSF Railway
Canadian National Railway
Canadian Pacific Railway
CSX Transportation
Norfolk Southern Railway

Trackage in the United States (no tracks in Canada or Mexico):
Union Pacific Railroad

Right: Amtrak's Empire Builder coming into Browning, MT eastbound in the winter.

150
AMTRAK

Above: No410 was built in 1911 by the American Locomotive Company Manchester, NH for the Boston & Maine Railroad. It is one of only two remaining B&M 6 wheel switcher locomotives.

Below: The Pennsylvania Railroad S-1 class steam locomotive (nicknamed "The Big Engine") was a single experimental locomotive, the longest and heaviest rigid frame reciprocating steam locomotive ever built.

Trackage in both the United States and Mexico:
Kansas City Southern Railway (in Mexico via wholly owned and jointly operated subsidiary Kansas City Southern de México)

Trackage in Mexico (no tracks in the United States):
Ferromex (Forty percent owned by Union Pacific Railroad)
Kansas City Southern de México (owned by the Kansas City Southern Railway)

Conclusion

The railroad has served America very well over the last 190 years, supporting the growth of industry and leisure. The railroad has also consolidated the nation from coast to coast. The only question is if America will continue to support its railroads.

The Illustrated Directory of North American Locomotives catalogs the locomotives that have generated this tremendous success. These engines have been both audacious and prosaic. In the Steam chapters we have examined many examples from the humble steam switcher to the streamlined intercity expresses of the 1930s, from the American Type 4-4-0 which dominated the Prairie landscape in the 1880s to the general purpose freight and passenger K-4s of the PRR, from massive coal haulers such as the H-8 Allegheny class of the Chesapeake & Ohio, to the last attempts of steam power to stay current with turbine engines such as the Class M-1 of 1947.

The onslaught of Diesel power that began in the 1920s brought many new innovations. The manageability, economy, and instant power of the diesel engines soon won the battle over steam.

Diesels even became popular with hardcore steam fans. Now many of the early diesel examples that lie rusting in steel company yard lines are just as eagerly photographed

Below: Sixty H-8 "Allegheny" class locomotives were built for the Chesapeake and Ohio Railway (C&O) between 1941 and 1948 by the Lima Locomotive Works. The "Allegheny" name refers to the C&O locomotives' job of hauling coal trains over the Allegheny Mountains.

as Norfolk and Western's steamer J-1.

Electric-powered engines were gradually introduced in the early twentieth and remain with us to this day. They have become even more relevant with their green image.

The Illustrated Directory of North American Locomotives also shows the locomotives of the past that have been preserved. The book also highlights the preserved railroads, private supporters groups, and museums who keep these national treasures accessible to the public. The book aims to capture a slice of the fascinating American railroad. The book is not an official publication and can't be totally comprehensive, but it packs as much railroad lore as possible into its 432 pages to interest and entertain everyone who loves the American railroads.

Above: The mighty M-1 was the final attempt to keep steam viable by using a steam driven turbine to power electric motors. It proved expensive and overcomplicated and was soon overtaken by diesel power.

Below left: The Alco RS-3 is a 1,600 hp B-B road switcher diesel-electric locomotive. manufactured by American Locomotive Company and Montreal Locomotive Works (MLW) from May 1950 to August 1956, and 1,418 were produced

Below: F-Series Diesel Electric locomotives were introduced in 1939 by GM's Electromotive Division and manufactured at the La Grange plant in Illinois and at the GMDD plant in London, Ontario, Canada. This is an FL-9 variant.

Above: No611 is the sole survivor of Norfolk & Western's fourteen class "J" steam locomotives designed by their own mechanical engineers in 1940 and built at the railroad's Railway's East End Shops in Roanoke, Virginia between 1941 and 1950.

The Early Days: 1829-1899

Stourbridge Lion 0-4-0 (1829)

Some of the earliest locomotives in use on America's fledgling railroads were imported from England where the Industrial revolution was already well under way. The Stourbridge Lion was named for its home town of Stourbridge near Birmingham in England's "Black Country" so called for the soot and grime from the coal burning chimneys of its industry. The 0-4-0 was built by Foster & Rastrick who built an identical twin locomotive, the Agenona, at the same time.

On August 9th, 1829, the locomotive took to the tracks of the Delaware & Hudson Canal Company at Honesdale, Pennsylvania.

Horatio Allen , the company's master mechanic, who had previously visited England to order the locomotive, later wrote describing his feelings: "I took my position on the platform of the locomotive and with my hand on the throttle

Below: The Stourbridge Lion in its heyday, as it would have looked in 1829. The loco's excessive weight for the track of the day meant that it was consigned to a stationary engine role and never took to the track again.

SPECIFICATIONS
Gauge: 4ft 8 1/2 in
Tractive Effort: 23,900 lbs
Cylinders: (2) 20 x 26 in
Driving Wheel Diameter: 68 in
Total Weight: 281,000 lbs
Overall Length: 67 ft

said, 'If there is any danger in this ride it is not necessary that the life and limb of more than one be subjected to danger.' The locomotive, having no train behind it answered at once to the movement of the hand . . . soon I was out of sight in the three miles ride alone in the woods of Pennsylvania. I have never run one since."

Unfortunately the Lion weighed more than twice as much as had originally been specified and Allen feared that running on the light track of the 4ft 5in gauge horse-operated tramway of the Delaware & Hudson Canal Company which connected the canal with mines at Carbondale would lead to disaster.

Instead the proud loco suffered the ignominious fate of being consigned to use as a stationary engine. However we know that the Stourbridge Lion would have succeeded perfectly well as a locomotive, if it had been allowed to operate on the right track because its sister locomotive, the Ageonona, performed reliably on the Shutt End Railway in Stourbridge for over a quarter of a century. Happily, Agenona has survived and can be seen in York's railway museum. The most conspicuous difference between the Lion and Agenona was that the latter had a much longer smokestack.

The driving mechanism of the Stourbridge twins consisted of what amounted to a pair of single-cylinder beam engines. The beams and linkage effectively reduced the stroke of the cylinders from 36in to 27in. Loose eccentrics engaging with stops fixed to the rear axle worked the valves when running and there was provision to move them by hand for starting. The design in fact owed much more to William Hedley of Puffing Billy fame than to the Stephensons. In 1829 Puffing Billy had been running for 16 years and so the Lion was built to a well-matured design, the result of a good deal of experience. But it was not to prevail over the direct-drive system which was to be the main feature of some 99 percent of steam locomotives ever built. Some 120 years were to pass before locomotives with near-vertical cylinders and complicated transmission systems would supersede the Stephenson concept. An excellent replica of the Stourbridge Lion is kept at the Smithsonian Museum of Science & Technology at Washington, D.C.

Left: The Agenona survived and now stands in the National Railway Museum in York, England. The smokestack is much taller than the Stourbridge Lion.

Best Friend of Charleston 0-4-0 (1830)

SPECIFICATIONS
Gauge: 4 ft 8 1/2 in
Tractive effort: 453 lb
Axle load: 4,500 lb
Cylinders: (2) 6 x 16 in
Driving wheels: 54 in
Steam pressure: 50 psi
Grate area: 2.2 sq ft
Water: 165 US galls
Adhesive weight: 9,000 lb
Total weight: 9,000 lb
Length: 14 ft 9 in

Best Friend of Charleston, which entered service with the South Carolina Canal and Rail Road Company at the end of 1830, was the first American-built locomotive to run in regular service on a public railroad. It was a 0-4-0 tank engine with a vertical water-tube boiler at the rear and inclined cylinders at the front that drove a rear crank-axle. *Best Friend of Charleston* was used for regular passenger service around Charleston, but its career was very short; on June 17, 1831, its boiler exploded, killing the fireman and wounding the rest of the engine crew. The fireman had fastened the safety valve down to stop it hissing while the engineer was away from the engine. The remains of the locomotive were rebuilt with a conventional horizontal boiler and renamed *Phoenix*; in this form it ran until the Civil War.

A working replica, built in 1928, is currently on display at the Atlanta regional headquarters of Norfolk Southern Railway.

Peter Cooper's "Tom Thumb" (1830)

SPECIFICATIONS
Gauge: 4 ft 8 1/2 in
Tractive effort: 820 lb
Axleload: 5,800 lb
Cylinders: (1) 5 x 27 in
Driving wheels: 30 in
Heating surface: 40 sq ft
Steam pressure: 90 psi
Grate area: 2.7 sq ft
Fuel: 800 lb
Water: 52 US gall
Adhesive weight: 5,800 lb
Total weight: 10,800 lb
Length: 13 ft 2 in

In 1830 Peter Cooper built the locomotive "Tom Thumb" to demonstrate the effectiveness of steam locomotives to the owners of the Baltimore and Ohio Railroad. It was the first American-built locomotive to run on a public railroad. (It actually never had a name; the nickname "Tom Thumb" was coined by an 1875 retrospective and it stuck.) It had a vertical boiler using a belt-driven mechanical blower for draft, with a single vertical cylinder driving the front axle via gearing. No tender was used; coal was carried on board, while the boiler was refilled at intervals along the line. "Tom Thumb" was strictly an experimental machine, used for testing and demonstration runs until 1831; it never went into regular service. However, it did convince the B&O directors to purchase steam locomotives for the railroad, and their earliest machines were all vertical-boilered.

Cooper did not continue as a locomotive maker; his goal was to stimulate B&O's growth, which would benefit his own local business interests. He went on to become a major American industrialist, philanthropist, and political activist, even running for President of the United States in 1876.

"Tom Thumb" did not survive, but a replica was built in 1927 for the B&O's centennial and now resides at the B&O Railroad Museum. Most depictions of "Tom Thumb" are based on this replica, but it has major differences from the original.

Left: The "Tom Thumb" replica on display at the B&O roundhouse.

John Bull 0-4-0 (later 4-2-0) (1831)

SPECIFICATIONS
Gauge: 5ft 0 in
Tractive Effort: 765 lbs
Cylinders: (2) 9 x 20 in
Driving Wheel Diameter: 54 in
Total Weight: 10 ton
Overall Length: 37 ft 0 in

John Bull is a British-built railroad steam locomotive which was one of the first to operate in the United States. It was operated for the first time on September 15, 1831, and it became the oldest operable steam locomotive in the world when it was run by its current owners the Smithsonian Institution in 1981. Built by Robert Stephenson and Company, the John Bull was initially purchased and operated by the Camden and Amboy Railroad, the first railroad in New Jersey. The locomotive was assigned the number 1 and its first name, "Stevens." Due to poorer quality track than was the norm in its native England, the locomotive had much trouble with derailment. To cure this the C&A's engineers added a leading truck to help guide the engine into curves. The leading truck's mechanism necessitated the removal of the coupling rod between the two main axles, leaving only the rear axle powered. Effectively, the John Bull became a 4-2-0 (a locomotive with two unpowered axles, one powered main axle, and no trailing axles). Later, the C&A also added a cowcatcher to the lead truck. The cowcatcher is an angled assembly designed to deflect animals and debris off of the railroad track in front of the locomotive. To protect the locomotive's crew from the weather, the C&A also added a cab to the rear of the locomotive where the controls were located. C&A workshop crews also added safety features such as a bell and headlight. The C&A used the locomotive consistently from 1833 until 1866, when it was removed from active service and placed in storage. The Pennsylvania Railroad (PRR) acquired the assets of the Camden & Amboy in 1871, and subsequently refurbished and operated the

Above: The John Bull was brought out of retirement for the Chicago World's Fair in 1893.

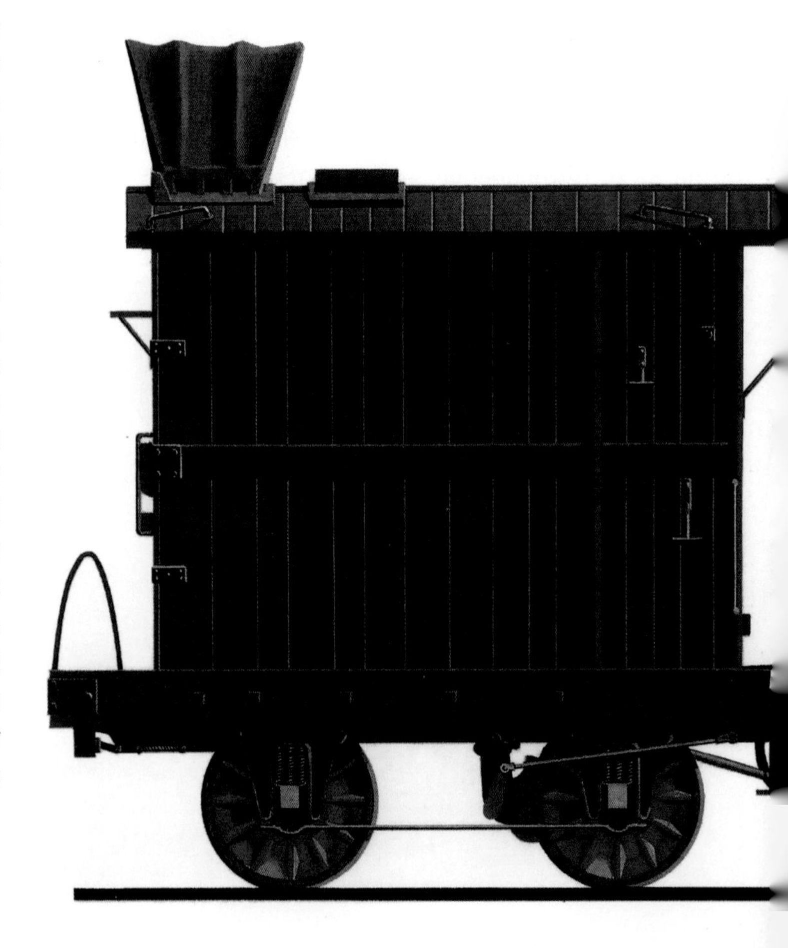

locomotive a few times for public displays: it was fired up for the Centennial Exposition in 1876 and again for the National Railway Appliance Exhibition in 1883. In 1884 the locomotive was purchased by the Smithsonian Institution as the museum's first major industrial exhibit.

In 1939 the employees at the Pennsylvania RR's Altoona workshops built an operable replica of the locomotive for further exhibition duties, as the Smithsonian desired to keep the original locomotive in a more controlled environment. The replica can be seen at the Railroad Museum of Pennsylvania. The original was on static display for the next 42 years, until in 1981 the Smithsonian commemorated the locomotive's 150th birthday by firing it up, making it the world's oldest surviving operable steam locomotive. Today, the original *John Bull* is on static display once more in the Smithsonian's National Museum of American History in Washington, D.C.

Above: On September 15, 1981, the original John Bull operated under steam on a few miles of branch line, making it the oldest operable steam locomotive in the world.

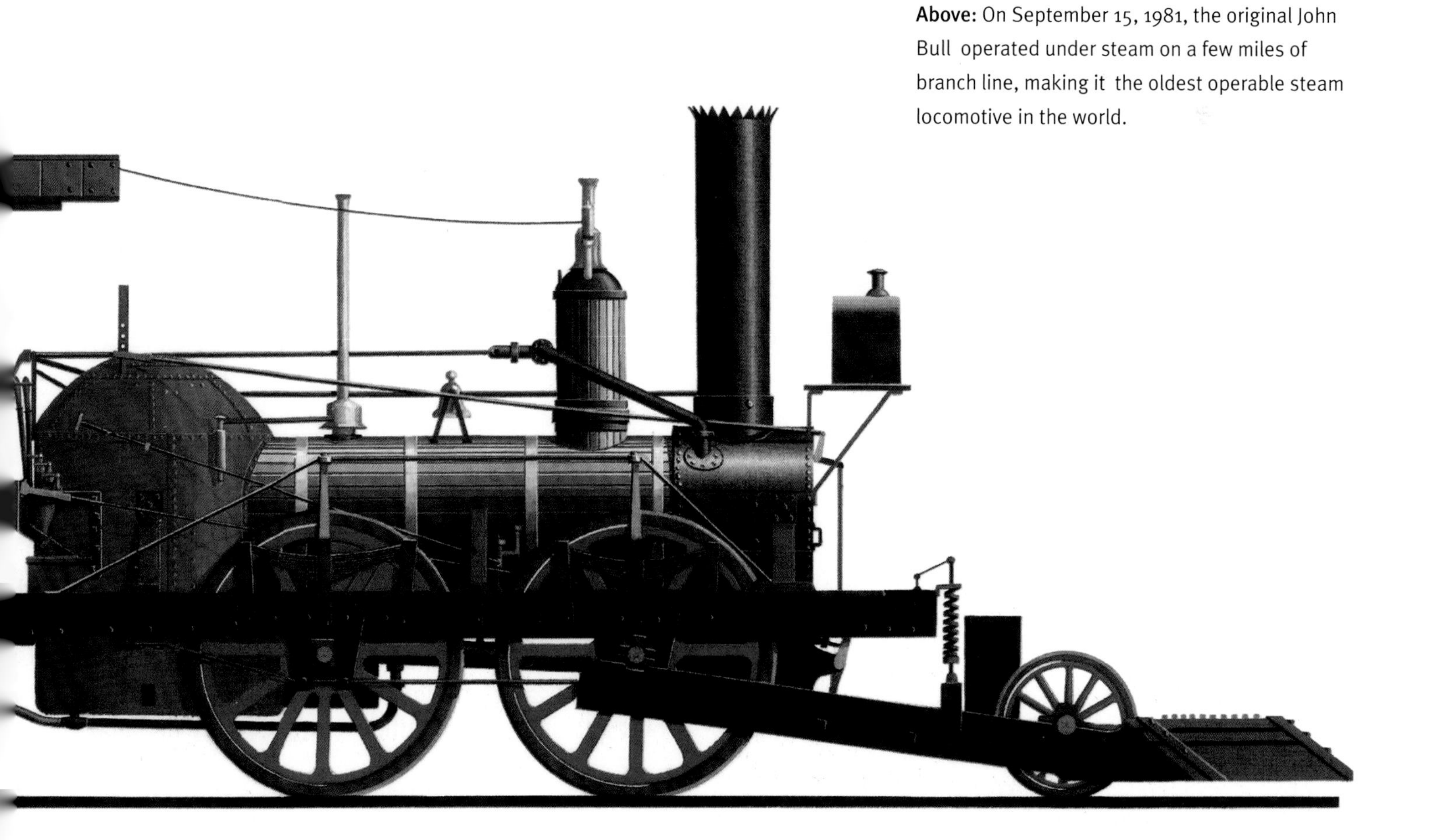

De Witt Clinton 0-4-0 (1831)

The De Witt Clinton was built in 1831 for the Mohawk and Hudson Railroad, a 14 mile stretch of track which ran between Albany and Schenectady. This was the first section of what was one day destined to become the giant New York Central Railroad. De Witt Clinton was the fourth locomotive built in the U.S. by the New York Foundry and the first to run in the state of New York. The previous operation of the road was horsedrawn but on June 25th, 1831, the De Witt Clinton was delivered by water along the Hudson River from New York City.

It had a similar layout to the New York Foundry's previous locomotive West Point. Like West Point, De Witt Clinton was an 0-4-0 with inclined cylinders at the rear next to the firebox, driving a crank axle at the front. De Witt Clinton was designed by John B. Jervis, Chief Engineer of the M&HRR, and like other first attempts it needed some fine tuning. The boiler was without a dome and hence there was a tendency for water to boil over and enter the cylinders; this was corrected by adding a dome. The exhaust pipes also needed some adjustment before they would draw the fire properly.

Above: A replica of the De Witt Clinton reenacting its inaugural trip.

By August 9, all was ready for the first scheduled run. Five stage coach bodies mounted on railroad wheels were provided coupled together with chains to form a train. It is said that hundreds wanting to travel were turned away; perhaps it was as well, for the journey was more exciting than comfortable. The slack action of the chains allowed the coaches—particularly the rear ones— to jerk violently into motion and upset the passengers a moment after the locomotive had begun its journey or when it slowed down. Because of the problems with drawing the fire sufficiently to burn coal, wood was being burnt as a temporary expedient. As a result red hot fragments poured out of the chimney, burning holes in the clothes of the passengers. Even so, it must have been fun doing the journey at a mind-bending speed of 25mph or so (the average speed was reportedly 18mph) watching spectators' horses bolting and, perhaps, being aware that history was being made that day. The performance was excellent and no damage was caused to the wooden rails strapped with iron which formed the track, but the relatively light construction of the locomotive meant that it was not destined for a long career and the locomotive served until 1833, when it was scrapped. A working replica was built in 1893 and today resides at the Henry Ford Museum.

Above and below: The replica, soon after it was built in 1893, and today at the Henry Ford Museum.

SPECIFICATIONS
Gauge: 4ft 8 1/2 in
Tractive Effort: 3,526 lbs
Cylinders: 5 1/2 x 16 in
Adhesive Weight: 6,750 lbs
Total Weight: 6,750 lbs
Overall Length: 19 ft 8 in (without tender)

West Point 0-4-0 (1831)

SPECIFICATIONS
Gauge: 4ft 8 1/2 in
Cylinders: (2) 6 x 16 in
Driving Wheel Diameter: 4 ft 6 in

West Point, which was built in 1830 and began service with the South Carolina Canal and Rail Road Company at the start of 1831, was the second American-built locomotive to serve on a public railroad, immediately after Best Friend of Charleston. In fact, both locomotives had the same builder, the West Point Foundry, and West Point shared the same frame and running gear as its sister locomotive. Unlike Best Friend of Charleston, West Point had a conventional horizontal fire-tube boiler and carried a tender, and its cylinders were located at the rear, next to the firebox. In general layout, West Point resembled the Stephenson 0-2-2 locomotives, but its cylinders drove a front crank axle and all its wheels were coupled, making it a 0-4-0.

Brother Jonathan 4-2-0 (1832)

SPECIFICATIONS
Gauge: 4ft 8 1/2 in
Tractive Effort: 1,023 lb
Axle load: 7,000lb
Cylinders: (2) 9 1/2 x 16 in
Driving Wheel Diameter: 60 in
Boiler Pressure: 50 psi
Total Weight: 14,000 lbs
Overall Length: 16 ft 5 1/2 in

John B Jarvis is credited with the innovative design that gave us the four wheel bogie at the front of passenger locomotives. The idea was for the bogie to provide guidance to the two following driving wheels allowing them full contact with the outer rail on curves as near as possible in a tangential attitude. For any particular radius, or even at a kink in the track, the bogie would take up an angle so that the three contact points between wheel and rail on each side were correctly lined up with the curve. This was particularly important on the light rough tracks of 1830s. This design was first tried out in 1832 on the locomotive originally known as the "Experiment" built at the West Point Foundry in New York and delivered to the Mohawk & Hudson River Railroad. On delivery the Experiment was renamed Brother Jonathan—then an impolite way of referring to the English; no doubt the name was a gesture of triumph at having thrown off any possible continued dependence on English locomotive design. The locomotive had a small boiler copied from Stephenson's Planet type, with the connecting rods passing between the sides of the firebox and the main frame. The single driving wheels were mounted behind the firebox giving a fairly stable layout. Although many of the Brother Jonathan's quirky features did not continue in the design of American locomotives in the following years, the leading bogie idea certainly did and there are countless examples in this book. . Brother Jonathan itself was successful in other ways; converted later to a 4-4-0 it had a long and useful life.

Atlantic 0-4-0 (1832)

SPECIFICATIONS
Gauge: 4 ft 8 1/2 in
Tractive Effort: 4245 lb
Axleload: 13.580 lb
Cylinders: (2) 12 x 22 in
Driving Wheel Diameter: 36 in
Steam Pressure: 75 psi
Adhesive weight: 27,160 lb
Total Weight: 27,160 lb
Overall Length: 14 ft 7 in

Atlantic was built by inventor and foundry owner Phineas Davis for the Baltimore and Ohio Railroad (B&O) in 1832. It is the first commercially successful and practical American-built locomotive and class prototype, and Davis' second constructed for the B&O, his first having won a design contest announced by the B&O in 1830.

Built at a cost of $4,500, the Atlantic weighed 6.5 tons and had two vertical cylinders. Despite Davis winning the competition the contract had been awarded to the inventor of the "Tom Thumb," Thomas Cooper. But when Cooper failed to deliver the first five locomotives commissioned B&O bought out the patents. A few of these ideas were incorporated in the Atlantic by Davis, whether by specification or because Davis wanted them is unclear. The locomotives he delivered before his death in 1835 were the first commercially viable, sufficiently efficient, coal burning steam locomotives produced domestically in the United States.

Ox teams were used to convey Atlantic to Baltimore, where it made a successful inaugural trip to Ellicott's Mills, Maryland, a distance of 13 miles. Nicknamed the 'Grasshopper' for its distinctive horizontal beam and long connecting rods, the locomotive carried 50 pounds of steam and burned a ton of anthracite coal on the 40-mile trip from Baltimore. Satisfied with this locomotive's operations, the B&O built 20 more locomotives of a similar design at its Mt. Clare workshops in Baltimore. Despite this success, the Atlantic prototype engine was scrapped in 1835 after the death of Phineas Davis. Why is unclear.

Below: The former Andrew Jackson, rebuilt to resemble the Atlantic, preserved on static display at the B&O Railroad Museum.

Old Ironsides 2-2-0 (1832)

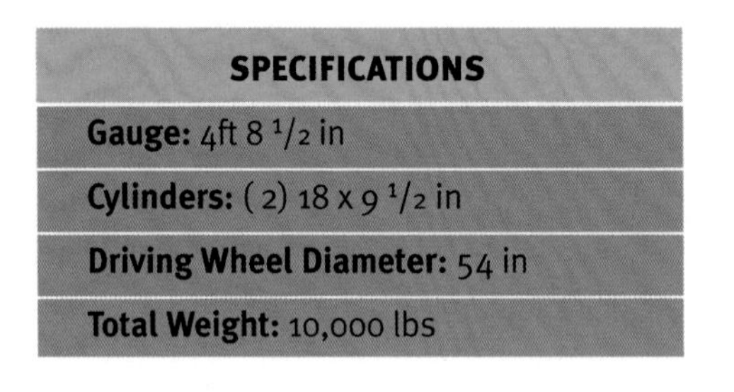

SPECIFICATIONS
Gauge: 4ft 8 1/2 in
Cylinders: (2) 18 x 9 1/2 in
Driving Wheel Diameter: 54 in
Total Weight: 10,000 lbs

Matthias W. Baldwin, the founder of the Baldwin Locomotive Works, originally learned the trade of a jeweler, but as that trade declined he entered a business supplying tools for bookbinding and printing in Philadelphia. This business was so successful that steam power became necessary to aid in manufacturing, and an engine was bought for the purpose. This proving unsatisfactory, Mr. Baldwin decided to design and construct one which should be specially adapted to the requirements of his shop. The design of the machine was compact and efficient and its workmanship was so excellent that Mr. Baldwin soon began receiving orders for similar engines. Thus, with his attention turned to steam engineering, the way was prepared for his grappling with the problem of the locomotive. In 1831 he constructed his first experimental steam locomotive. Based on designs first shown at the Rainhill Trials in England, Baldwin's prototype was a small demonstration engine that was displayed at Peale's Philadelphia City Museum. The engine was strong enough to pull a few cars that carried four passengers each. This locomotive was unusual for the time in that it burned coal, which was available locally, instead of wood.

The next year Baldwin built his first commissioned steam locomotive for the fledgling Philadelphia, Germantown & Norristown Railroad. This engine, nicknamed Old Ironsides, initially traveled at the rate of only 1 mile per hour in trials made on November 23, 1832, but the machine was later refined and improved so that a peak speed of 28 mph was attained.

Old Ironsides was a four-wheeled engine, modeled on the Stephenson design as shown in the "Planet" class, and weighed, in running order, something over five tons. The rear, or driving wheels, were 54 inches in diameter on a crank axle placed in front of the firebox. The cranks were 39 inches from center to center. The front wheels, which were simply carrying wheels, were 45 inches in diameter, on an axle placed just back of the cylinders. The cylinders were 9-1/2 inches in diameter by 18 inches stroke, and were attached horizontally to the outside of the smokebox, which was D-shaped, with the sides receding inward, so as to bring the center line of each cylinder in line with the center of the crank. The wheels were made with heavy cast-iron hubs, wooden spokes and rims, and wrought-iron tires. The frame was of wood, placed outside the wheels. The boiler was 30 inches in diameter, and contained 72 copper tubes, 1-1/2 inches in diameter and seven feet long. The tender was a four-wheeled platform, with wooden sides and back, carrying an iron box for a water tank, enclosed in a wooden casing, and with a space for fuel in front. The engine had no cab. A peculiarity in the exhaust of the "Ironsides" was that there was only a single straight pipe running across from one cylinder to the other, with an opening in the upper side of the pipe, midway between the cylinders, to which was attached at right angles the perpendicular pipe into the chimney. The cylinders, therefore, exhausted against each other; and it was found, after the engine had been put in use, that this was a problem. This defect was afterward remedied by turning each exhaust pipe upward into the chimney as is now done.

The price of the engine was to have been $4,000, but some difficulty was found in procuring a settlement. The PG&NRR claimed that the engine did not perform according to contract; and objection was also made to some of the defects alluded to. After these had been corrected as far as possible, however, Mr. Baldwin finally succeeded in effecting a compromise settlement, and received from the Company $3,500 for the machine.

On subsequent trials, however the " Ironsides" attained a speed of 30 miles per hour, with its usual train attached. So great were the wonder and curiosity which attached to such a prodigy, that people flocked to see the marvel, and eagerly bought the privilege of riding after the strange monster.

Campbell 4-4-0 (1837)

One of the most popular wheel arrangements in America started in Philadelphia when Henry Campbell, an engineer of the Philadelphia, Germanstown & Norriston Railroad had the idea of combining coupled driving wheels, as fitted to Best Friend of Charleston, with the leading bogie of Brother Jonathan. Although the locomotive was intended for carrying coal, it has its place here as the prototype of perhaps the most numerous and successful of all passenger-hauling wheel arrangements. Campbell followed the layout of Brother Jonathan but by adding an additional driving wheel coupled to the first by cranks outside the frames he created the first 4-4-0 in May 1837. After patenting the design, Campbell commissioned local Philly engineer James Brooks to produce it.

The 4-4-0 configuration gave twice the adhesive weight, while at the same time made a locomotive that could ride satisfactorily round sharp or irregular curves. The locomotive's high boiler pressure of 90 psi was also notable for the time. Whilst this remarkable locomotive demonstrated great potential, the flexibility provided in order to cope with poorly lined tracks was not accompanied with flexibility in a vertical plane to help with the humps and hollows in them. In consequence, Campbell's 4-4-0 was not in itself successful.

SPECIFICATIONS
Gauge: 4ft 8 1/2 in
Tractive Effort: 4,373 lbs
Axle Load: 8,000 lbs
Cylinders: (2) 14 x 15 3/4 in
Driving Wheel Diameter: 54 in
Steam Pressure: 90psi
Adhesive Weight: 16,000 lbs
Length: 16 ft 5 1/2 in (engine only)

Hercules 4-4-0 (1837)

SPECIFICATIONS
Gauge: 4 ft 8 1/2in
Tractive effort: 4,507 lb
Axle load: circa 10,000 lb
Cylinders: (2) 12 x 18 in
Driving wheels: 44 in
Steam pressure: 90 lb/sq in
Adhesive weight: circa 20,000 lb
Total weight: *30,000 lb (14t).
Length overall: * 18 ft ll in
* Without tender—boiler and tender details not recorded.

The Brother Jonathan—the first locomotive to have a four-wheel leading bogie—showed that it was possible to sit on the uneven tracks in the same way that a three legged stool sat on the floor. As locomotives progressed, the desire for more driving wheels became evident and Campbell's move toward a 4-4-0 configuration created problems in keeping all eight of his wheels on the uneven track at the same time. For that reason the Campbell loco was not a success. When, in 1836, the Beaver Meadows Railroad ordered a 4-4-0 from Garrett & Eastwick of Philadelphia, their workshop foreman, Joseph Harrison, had the idea of making his two pairs of driving wheels into a kind of non-swivelling bogie by connecting the axle bearings on each side by a large cast iron beam, pivoted at its centre. The pivots were connected to the mainframe of the locomotive by a large leaf spring on either side.

In this way eight wheels were made to support the body of the locomotive at three points. It was a clever idea which solved the problem of running a 4-4-0 on rough tracks and was the basis of the three-point compensated springing system which was applied to most of the world's locomotives from simple ones up to 4-12-2s.

Hercules was well named and many similar locomotives were supplied. As a result of his idea Joseph Harrison was made a partner in the firm which on Garrett's retiring became known as Eastwick & Harrison. The famous "American Standard" 4-4-0, of which 25,000 were built for the U.S. alone, was directly derived from this innovative engine.

Below: Hercules, built by Garrett & Eastwick of Philadelphia in 1836, marked an important step forward in locomotive development.

Lafayette 4-2-0 (1837)

William Norris advanced the design of locomotives significantly in 1836 when he built Washington County Farmer for the Philadelphia and Columbia Railroad. This 4-2-0 loco bore some resemblance to Brother Jonathan with a horizontal boiler, leading bogie and single driving wheel but the two cylinders were now outside of the frame and the wheels and the valves were on top of the cylinders. The driving wheels were now positioned in front of the firebox distributing the loco's weight directly above them, thus improving adhesion This began to look like a steam locomotive as we know it.

Norris's background was in drapery but he started building locomotives in 1831 in partnership with Colonel Stephen Long. Soon he set up on his own and by the beginning of 1836 had seven locomotives under his belt.

In 1827 the Baltimore & Ohio Railroad was the first public railroad for passengers and freight transport to receive a charter. It was opened for twelve miles out of Baltimore in 1830, but for a number of years horses provided haulage power—although there were trials with steam locomotives. Steam took over in 1834 in the form of vertical-boiler locomotives, known as the "Grasshopper" type. Directors of the Baltimore & Ohio Railroad were seeking more powerful locomotives to perform satisfactorily on the rugged terrain of their road and they were impressed by Norris's work and asked him to build a series of eight similar engines. The first was Lafayette, delivered in 1837; it was the first B&O locomotive to have a horizontal boiler. Edward Bury's circular domed firebox and bar frames were also features and the engine is said to have had cam-operated valves of a pattern devised by Ross Winans of the B&O.

The locomotives were a great success, giving much better performance at reduced fuel consumption. They were also relatively reliable and needed few repairs. The same year Norris built a similar locomotive for the Champlain & St. Lawrence Railway in Canada. This was the first proper locomotive exported from America, and the hill-climbing ability of these remarkable locomotives led to many further sales abroad.

SPECIFICATIONS
Axle Load: 13,000 lbs
Cylinders: (2) 10 1/2 x 18 in
Driving Wheel Diameter: 48 in
Heating Surface: 394 sq ft
Steam Pressure: 60 psi
Grate Area: 8.6 sq ft
Water: 540 gall
Adhesive Weight: 30,000 lbs
Total Weight: 44,000 lbs
Overall Length: 30 ft 4 1/2 in

Pioneer 4-2-0 (1837)

SPECIFICATIONS
Gauge: 4ft 8 1/2 in
Tractive Effort: 23,900 lbs
Cylinders: (2) 20 x 26 in
Driving Wheel Diameter: 68 in
Total Weight: 281,000 lbs
Overall Length: 67 ft

Left: Pioneer in 1898.

Pioneer is the name of the first railroad locomotive to operate in Chicago, Illinois. It was built in 1837 by Baldwin Locomotive Works for the Utica and Schenectady Railroad (U&S) in New York, The Utica and Schenectady Railroad originally named it Alert. It worked on the U&S for nine years before it was sold to the Michigan Central Railroad. Michigan Central added a cab and tender to the locomotive and used it for two years before selling it again in 1848 to William B. Ogden of the Galena and Chicago Union Railroad.(G&CU, the oldest predecessor of Chicago and North Western Railway). The locomotive arrived in Chicago by schooner on October 10, 1848, and it pulled the first train westbound out of the city on October 25, 1848.

The G&CU renamed the locomotive Pioneer and used it in the construction of the G&CU until 1850, at which time the locomotive was loaned to the Chicago, Burlington and Quincy Railroad for work in and around Chicago. The locomotive is now preserved and on display at the Chicago History Museum.

Mud-digger 0-8-0 (1844)

SPECIFICATIONS
Gauge: 4 ft 8 1/2 in
Axle Load: 12,925 lbs
Cylinders: (2) 17 x 24 in
Driving Wheel Diameter: 33 in
Adhesive Weight: 47,000 lbs
Total Weight: 47,000 lbs*
Overall Length: 49 ft 10 in*
* without tender

In the days when the Baltimore & Ohio Railroad relied on horse power a young horse dealer named Ross Winans called on them to ply his wares. With the advent of steam power Winans turned his hand to engineering and he was appointed assistant engineer of machinery in 1831 and four years later, as Gillingham & Winans, took over locomotive building in the B&O Mount Clare shops from Phineas Davis (Davis & Gartner). Here he completed building the vertical-boilered Grasshopper locos but by 1837 he was ready to produce a loco based on his own design—"Winans Crab." Because of the adoption of horizontal cylinders, the Crabs came one short step closer to conventionality when compared with the Grasshoppers. The separate crankshaft gearing and crankshaft remained however, as well as a vertical boiler. Because of the gearing the cranks turned the opposite way to the wheels, hence the supposedly crab-like gait. Two 0-4-0 crabs, McKim and Mazeppa, were built by Gillingham & Winans for the B&O and were delivered in 1838. They gave good service, lasting until 1863 and 1868 respectively.

Following the Crabs came the more ambitious 0-8-0 "Mud-diggers." This design featured a proper horizontal locomotive boiler, but the layout of the machinery was unchanged, apart from the two extra axles. The first one was appropriately named Hercules and there were 13 more, all built between 1844 and 1847.The name Mud-digger came from the design of the low slung outside cranks that moved so close to the ground that they were prone to churn up mud and stones as the locomotive moved along the tracks.

Seven of the Mud-diggers were rebuilt completely on more conventional lines between 1853 and 1856, losing their names in the process. The remainder soldiered on for many years, the last (No.41 *Elk)* being retired in 1880.

The last of these locomotives to be built (No.49 *Mount Clare)* was rather different from the others. It was built by the B&O itself rather than Winans, had no geared drive and was inside-connected. The cylinders were driven by a jackshaft mounted between the second and third axles. The jackshaft was connected to the wheels by outside cranks and an additional pair of coupling rods.

Below:The Mud-diggers remained in service right up until the early 1880s. This one has been rebuilt with a camelback-style cab astride the boiler.

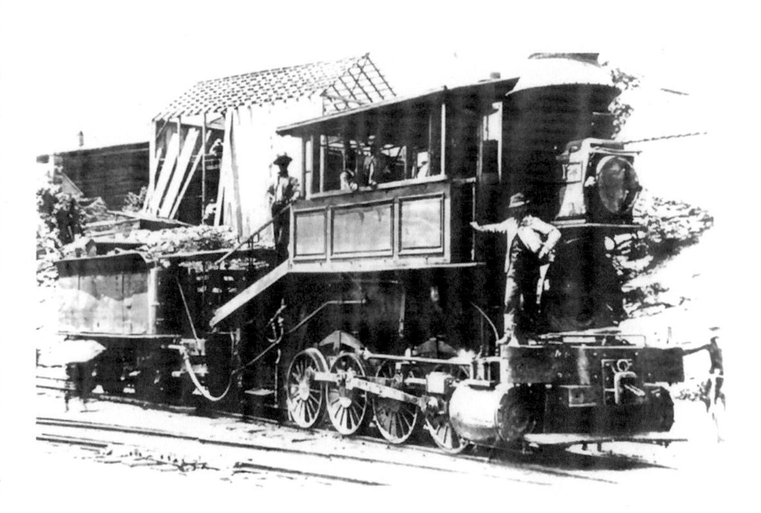

Crampton (1846)

SPECIFICATIONS
Gauge: 4 ft 8 1/2 in
Cylinders: (2) 13 x 34 in
Driving Wheel Diameter: 96 in
Total Weight: 50,000 lbs
Overall Length: 29 ft 6 in (without tender)

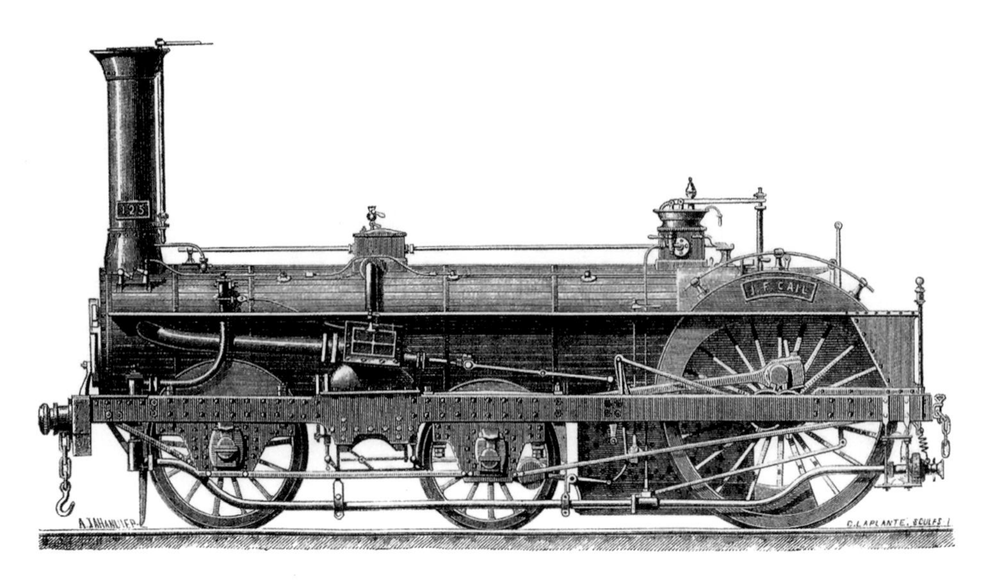

The Crampton locomotive was designed by Thomas Russell Crampton and built by various firms from 1846. The U.S. builders were Norris & Co of Philadelphia.

Notable features were a low boiler and large driving wheels. The crux of the Crampton patent was that the single driving axle was placed behind the firebox, so that the driving wheels could be very large and therefore keep rotational speed low. This helped to give this design a low center of gravity, so that it did not require a very broad-gauge track to travel safely at high speeds. Its wheel arrangement was usually 4-2-0 or 6-2-0.

Because the single driving axle was behind the firebox, most Crampton locomotives had outside cylinders. The

Crampton Loco Camden & Amboy RR

lightly constructed connecting rods required wire bracing. Another peculiarity on some Crampton locomotives was the use of a boiler of oval cross-section, to lower the center of gravity. It would nowadays be regarded as bad engineering practice because the internal pressure would tend to push the boiler into a circular cross-section and increase the risk of metal fatigue.

Thomas Crampton impressed many railroad managers particularly in France and Germany where over 300 locomotives were made to his design. For many years in France the phrase "prendre le Crampton," literally "take the train," showed how familiar his designs became.

Robert Stevens, president of the Camden & Amboy Railroad, visited Europe in 1845 and came back enthused with Crampton's design. He soon arranged for several similar, but even more amazing-looking, contrivances to be built by Morris & Co. The first was completed in 1849 and was named after John Stevens, the railroad pioneer who, in 1825, built a small model demonstration locomotive which he ran on a short circular railroad on his estate at Hoboken, New Jersey. Seven 6-2-Os were built in all over the next three years.

The Crampton engines were very fast but lacked adhesive weight so they were only suitable for lightly traveled trains; ironically, because the trains were fast they were popular and so ceased to be light. Even so, these particular examples lasted until the early-1860s, although by then some, if not all, had been rebuilt into more conventional 4-4-Os.

The John Stevens was intended to burn anthracite but both the grate area and the firebox would have been on the small size, even for bituminous coal. So there were problems of steam generation as well as a lack of adhesion. The oval boiler barrel was no larger in cross section than the diameter of the wheels of the leading truck. and smaller than the amazing smokestack. At the same time, a notable feature was the valve gear operated by a return crank just as in a modern steam locomotive. There were also double valves to control admission and cut-off separately.

Despite its peculiarities the Crampton loco John Stevens remained in service on the Camden & Amboy Railroad for more than 20 years.

Pioneer 2-2-2 (1844)

SPECIFICATIONS
Gauge: 4 ft 8 1/2 in
Tractive effort: 1,592 lb
Driving wheel: 54 in
Cylinders: (2) 8 1/2 x 14 in
Boiler pressure: 100 psi
Weight: 25,000 lb
Wheelbase: 13 ft 7 in

The Pioneer is a steam locomotive made in 1851 by Seth Wilmarth, the owner of a large machine shop in Boston who made a few locomotives. Pioneer was built just two decades after America's first domestically made locomotive and its design owed much to an earlier time.

It is an unusual locomotive and on first inspection would seem to be imperfect for service on an American railroad of the 1850s. In fact its general type was obsolete on almost all railroads in the U.S. by 1850.

In the standard type nomenclature for steam locomotives, Pioneer is a "2-2-2T" type, meaning that it has an unpowered leading pair of wheels; a single powered axle (the larger-diameter wheels, driven by the steam cylinders via connecting, (or "main"), rods); and another unpowered pair of wheels at the rear. Therefore it has no truck, and only one set of driving wheels, an arrangement which makes it very different from the highly successful standard 8-wheel engine of the 1850s. All six wheels of the Pioneer are rigidly attached to the frame. It is only half the size of an 8-wheel engine of 1851 and about the same size of the 4-2-0 so common in this country some 20 years earlier. Its general arrangement is that of the rigid English locomotive which had, years earlier, proven unsuitable for use on U.S. railroads. The "T" stands for "tank engine," meaning one that has no separate tender for carrying its fuel (wood) and water for the boiler; fuel and water is carried on the same single chassis as the boiler, cab, and running gear.

Despite this Pioneer served the Cumberland Valley RR successfully for a number of years, connecting Harrisburg, Pennsylvania with Hagerstown, Maryland and Winchester, Virginia. The locomotive was designed specifically to pull two-car passenger trains. Pioneer was one of several locomotives badly damaged by fire during the Civil War, during a Confederate raid on the CVRR roundhouse at Chambersburg, Pennsylvania. The CVRR rebuilt the engine, operating it on light one- and two-car passenger trains until the mid 1880s, and then saved and exhibited it as an historic relic. The Pennsylvania RR (then one of the nation's largest) absorbed the CVRR soon after. The PRR entirely repainted Pioneer in 1947 for the 1947-48 Chicago Railroad Fair. The lettering on the fenders, "PIONEER," is inauthentic. A replica headlight was added by NMAH (then NMHT) in December 1965.

Camel 4-6-0 (1848)

SPECIFICATIONS
Gauge: 4 ft 8 1/2 in
Axle Load: 27,716 lbs
Cylinders: (2) 19 x 22 in
Driving Wheel Diameter: 50 in
Tractive Effort: 6,775 lbs
Steam Pressure: 65 psi
Grate Area: 17.2 sq ft
Adhesive Weight: 56,500 lbs
Total Weight: 77,100 lbs
Overall Length: 52 ft 5 in

It was 25 years after the Baltimore & Ohio Railroad received its charter that the line finally reached the Ohio River at Wheeling, Virginia. To handle the expected increase of traffic over sections which included grades up to 2.2 percent (l-in-45), Samuel Hayes, who was "Master of Machinery", devised these ten-wheeler 4-6-0s. They were a development of Ross Winans 0-8-0 *Camel* design, of which some 20 had been built between 1848 and 1852. The 0-8-0s were prone to derailing when required to run at above coal-train speeds and so a leading truck was necessary, yet something with more tractive effort and power than a 4-4-0 was also needed.

As on the older 0-8-0s, the firebox of the new engines was a long one situated behind the driving wheels. The fireman's platform was located low down and close to the wheels. He himself was separated from the engineer, who drove the loco from a cab mounted on top of and half-way along the boiler. The big firebox was designed to be specially suitable for burning the anthracite coal then mined in huge quantities in the eastern states.

The result was this unusual and impressive design, of which 17 were built in 1853 and 1854. For the first time a whole locomotive class on the B&O failed to receive names. Even so, they were clearly excellent, several lasting in normal service for over 40 years. Another pointer to their qualities was that Hayes' successor, Henry Tyson, had another batch of nine built in 1857, similar in appearance but different in design, and these also had long lives. A further group of 109 was built between 1869 and 1875, under the regime of Master Mechanic J. C. Davies.

American Type 4-4-0

The 4-4-0, variously called the American type or the standard eight-wheel engine, was the most popular wheel arrangement in nineteenth-century America. Originated by Henry Campbell engineer to the Philadelphia, Germantown & Norristown Railroad the 4-4-0 succeeded because it met every requirement of early United States railroads. It was well suited to all service, including passenger, freight, and switching. It was flexible; having three-point suspension and a leading bogie, and it operated well on uneven tracks. It was simple, having relatively few parts, which made it easy to repair. It was low in cost, and it was relatively powerful because of its four connected driving wheels. Thomas Rogers of Paterson, New Jersey perfected the design and was responsible for introducing most of the features which made it a true success. The most significant development, so far as the US was concerned was the general introduction of Stephenson's link motion, which permitted the expansive use of steam. This was in place of the "hook" reversing gears used until then, which permitted only "full forward" and "full backward" positions. In other aspects of design, Rogers gained his success by good proportions and good detail rather than innovation. An example was the provision of adequate space between the cylinders and the driving wheels, which reduced the maximum angularity of the connecting rods and hence the up-and-down forces on the slide bars. A long wheelbase leading bogie allowed the cylinders to be horizontal and still clear the wheels. This permitted direct attachment to the bar frames, which raised inclined cylinders did not. To allow flexibility on curves, early examples of the breed inherited flangeless leading driving wheels from their progenitors, but by the late 1850s the leading bogies were being given side movement to produce the same effect. Naturally the compensated spring suspension system giving three-point support to the locomotive was continued. Wood-burning was also nearly universal in these early years of the type, and the need to catch the sparks led to the characteristic Western style

spark-arresting smokestacks.

Within two or three years other makers such as Baldwin, Grant, Brooks, Mason, Danforth and Hinkley began offering similar locomotives. To buy one of these locomotives must have been rather like ordering a car today. A form was filled in specifying your options and very soon an adequate and reliable machine was delivered.

There was another revolution taking place too. The earlier years of the type were characterised by romantic names and wonderful brass, copper and paint work, but after the Civil War there was a time of cut-throat competition, with weaker railroads going to the wall. So austerity took over and locomotives reflected this. Some were not even named.

For most of the second half of the nineteenth century, this one type of locomotive dominated railroad operations in the U.S. It was therefore known as the "American Standard" and about 25,000 of them were built, differing only marginally in design. The main things that varied were the decor and the details. They were simple, ruggedly constructed machines appropriate for what was then a developing country; at the same time a leading bogie and compensated springing made them suitable for the rough tracks of a frontier land. Speeds on the rough light tracks of the frontier were not high—average speeds of 25mph start-to-stop, implying a maximum of 40mph, were typical of the best expresses. Although the 4-4-0s were completely stable at high speeds, the need for more power meant that by the 1880s a bigger breed of 4-4-0 as well as "Ten-wheelers" (4-6-0s) were taking over from the "American."

The General was built by Thomas Rogers of Paterson, New Jersey in 1855 and it is perhaps the most famous of the 4-4-0s. It came to fame when hijacked by a group of Union soldiers who had penetrated behind Confederate lines during the Civil War. The plan was to disrupt communications, in particular the 135 mile long (5ft gauge) track connecting Atlanta with Chattanooga. The Union forces were closing in on Chattanooga after their victory at Shiloh and the

Below: 4-4-0 "American" locomotive "Inyo" located at the Nevada State Railroad Museum in Carson City, Nevada. The Inyo was built by the Baldwin Locomotive Works in 1875, and pulled both passenger and freight trains. The Inyo weighs 68,000 lbs. Its 57 in driving wheels deliver 11,920 lbs of tractive force. In 1877 it was fitted with air brakes, and in 1910 it was converted to burn oil rather than wood.

SPECIFICATIONS
Gauge: Various (see text)
Tractive Effort: 6,885 lbs
Axle Load: 21,000 lbs
Cylinders: (2) 15 x 24 in
Driving Wheel Diameter: 60 in
Heating Surface: 98.0 sq ft
Superheater: None
Steam Pressure: 90 psi
Grate Area: 14.5 sq ft
Fuel: wood (2 cords)
Water: 2,000 US
Adhesive Weight: 43,000 lbs
Total Weight: 90,000 lbs
Overall Length: 52 ft 3 in

Confederates were expected to bring up reinforcements by rail. There was a trestle bridge at a place called Oostenaula and the intention was to steal a train, take it to the site and burn the bridge. A replacement would take weeks to build.

The Union raiders, twenty in number under the command of a Captain Andrews, stayed overnight at Marietta and bought tickets to travel on the train, stealing the locomotive at a place called Big Shanty, some 30 miles north of Atlanta, while the passengers and crew were having breakfast in the depot's eating house. The conductor of the train, whose name was Fuller, gave chase, first on a handcart and then on a small private ironworks loco, the Yonah.

The raiders' intention was to cut telegraph wires behind them, remove the occasional rail and demand immediate passage at stations they came to in the name of Confederate General Beauregard. However they faced a 1 1/2 hour delay at Kingston due to the presence of trains coming the other way on the single track.

In the end the Yonah arrived at Kingston only four minutes after Andrews and the General had left. Here Fuller took over another "American" 4-4-0, the Texas, which he drove in reverse in hot pursuit, and after this Andrews never had enough time to block the track before his Confederate pursuers caught up. In the end, after eight hours and 87 miles the General was abandoned when it ran out of fuel; the Union group then scattered into the woods. All were later captured and seven of the senior men shot.

Two factors which displayed the characteristics of the "American" type emerged from this debacle. First, in spite of the rough track, high maximum speeds of around 60mph were reached during the chase and both locomotives stayed on the rails. The second thing was that the range between fuel stops was very short. A full load of two cords of wood fuel (a cord is 128cu ft) would last for just 50 miles.

Both the General and the Texas (or what purports to be them) have survived. The former is in the Southern Museum of Civil War and Locomotive History at Kennesaw, Georgia. The Texas, as befits a Confederate conqueror, has an honoured place in Grant Park at Atlanta. Both were converted from the 5ft gauge of the Western & Atlantic Railroad after the war was over.

The Civil War was one of the first wars to be fought using railway transportation, on both sides represented mainly by the American type. The earliest transcontinental railroads were first built and then operated by them; the well-known picture of the last spike ceremony at Promontory, Utah, has placed the Central Pacific's Jupiter and the Union Pacific No. 119 second only to the General on the scale of locomotive fame. It is said that "America built the railroads and the railroads built America;" substitute "American 4-4-0" for "railroad" and the saying is equally true.

The American type was a universal loco; the only difference between those built for passenger traffic and those for freight was between 66in diameter driving wheels and 60in. It also served all the thousands of railroad companies who then operated America's 100,000 miles of line, from roads thousands of miles long to those a mere ten.

The last "American" class did not retire from normal line service for more than a century after Campbell put the first on the rails in 1837.

Left: The General now resides at the Southern Museum of the Civil War and Locomotive History.

William Mason No25 B&O (1856)

This wood burning locomotive is one of two built by the Mason Machine Works in Taunton, Massachusetts, in 1856 for the as Baltimore & Ohio Railroad and was road numbered No 25.It is an example of the most popular locomotive design on U.S. railroads, the American type 4-4-0. No25 was the first B&O locomotive to have Stephenson's link motion valve gear and a round smokebox set on a cylinder saddle.

William Mason of the Mason Machine Works wanted to improve the symmetry of the American locomotive, and his designs produced from 1853 until his death in 1883 had far less of the ornamentation that was typical of the day. Renumbered No55 in 1882, it returned to No25 in 1892 and was named "William Mason" in 1927 in honor of its builder when it appeared at the Fair of the Iron Horse. The locomotive has also appeared in many movies, featuring in *The Swan* (1956) with Grace Kelley and Alec Guinness, as The General in Disney's *The Great Locomotive Chase* (1956) and in the Civil War drama *Raintree County* (1957).

More recently, it was rebuilt as The Wanderer at the Strasburg Railroad in Strasburg, PA, for the 1999 Warner Brothers remake of the movie *The Wild, Wild West* starring Will Smith and Kevin Kline and also featured in *Tuck Everlasting* (2002) and *Gods and Generals* (2003). It is one of the oldest operating steam locomotives in the U.S. and has had a long career. It also appeared at the 1939 New York World's Fair and the 1948-49 Chicago Railroad Fair. The Mason Machine Works turned out over 750 steam locomotives until it ceased production in 1889. Producing textile machinery then became the company's core business until its decline in the 1920s. It finally went out of business in 1944.

SPECIFICATIONS
Gauge: 4 ft 8 1/2 in
Driving wheels: 60 in
Cylinders: (2) 16 x 22in
Grate: 15 sq ft
Firebox: 86 1/2 sq ft
Total heating surface: 784 sq ft
Steam Pressure: 100 psi
Tractive effort: 6,225 lb

Governor Stanford 4-4-0 (1862)

Governor Stanford is a 4-4-0 steam locomotive originally built in 1862 by Norris Locomotive Works. It entered service on November 9, 1863, and it was used in the construction of the First Transcontinental Railroad in North America by Central Pacific Railroad bearing road number 1. It was Central Pacific's first locomotive and it is named in honor of the road's first president and ex-California governor, Leland Stanford.

The locomotive was rebuilt in 1878 with larger cylinders and an increased boiler pressure, which increased its tractive effort to 11,081 pounds force. In 1891 the locomotive was renumbered to 1174. The locomotive was retired from regular service on July 20, 1895, and then donated to Stanford University; however, it was not delivered to the university until 1899. The locomotive was disassembled and stored during World War II but was returned to display at the university after reassembly by retired Southern Pacific engineer Billy Jones. In the 1960s, the university needed the space occupied by the engine for other uses, so the engine was removed and loaned to the Railway & Locomotive Historical Society, which had been in the process of collecting historic locomotives and rolling stock to be displayed in what would ultimately become the California State Railroad Museum in Sacramento. The locomotive is currently a centerpiece at the museum where it has been cosmetically restored to its 1899 appearance.

SPECIFICATIONS
Gauge: 4ft 8 1/2 in
Tractive Effort: 7,791 lbs
Cylinders: (2) 15 x 22 in
Adhesive Weight: 35,700 lbs
Total Weight: 50,000 lbs

Thatcher Perkins 4-6-0 (1863)

SPECIFICATIONS
Gauge: 4ft 81/2in
Weight: 90,700 lbs
Driving wheel: 60 inches (original) 58 inches (present)
Cylinders: (2) 18 x 26 in (original) 19 x 26 in (present)
Tractive Effort: 10,350 lbs (present)

In 1853, the Baltimore & Ohio Railroad built its first "Ten Wheeler" locomotives to tackle the tough mountain grades in what is now West Virginia. A decade later, this locomotive type was needed to meet the demand caused by the Civil War and increased freight and passenger traffic on the B&O. The No. 147 was part of the first series of "Ten Wheelers" designed by Master of Machinery, Thatcher Perkins, in 1863. After moving Union troops during the war, the versatile No. 147 continued to pull both passenger and freight trains.

Originally built as No. 147 and later renumbered as No. 282, the engine was preserved by the railroad in 1892 for public relations and exhibition purposes. At this time, the railroad renumbered the engine to represent another Perkins "Ten Wheeler" built in 1863, the No. 117. The railroad applied the name "Thatcher Perkins" to the engine during the B&O's 1927 Fair of the Iron Horse centennial celebration.

It was badly damaged during the museum's unfortunate 2003 Roundhouse roof collapse but was restored and renumbered to its original No. 147 in 2010.

Mogul 2-6-0 (1863)

SPECIFICATIONS
Gauge: 4 ft 8 1/2 in
Tractive Effort: 13,055 lbs
Axle Load: 20,900 lbs
Cylinders: (2) 16 x 24 in
Driving Wheel Diameter: 48 in
Heating Surface: 1,255 sq ft
Steam Pressure: 120 psi
Grate Area: 17 sq ft
Water: 2,880 US gal
Adhesive Weight: 57,000 lbs
Weight: 7,000 lbs (without tender)
Length: 55 ft (without tender)

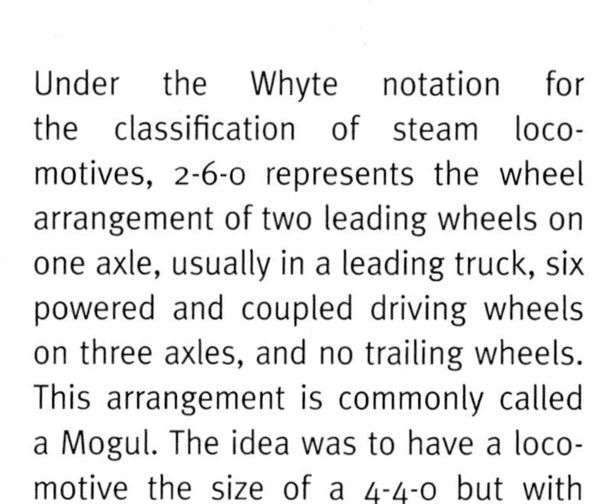

Under the Whyte notation for the classification of steam locomotives, 2-6-0 represents the wheel arrangement of two leading wheels on one axle, usually in a leading truck, six powered and coupled driving wheels on three axles, and no trailing wheels. This arrangement is commonly called a Mogul. The idea was to have a locomotive the size of a 4-4-0 but with one guiding and three driving axles instead of two and two respectively. In principle there would then be 50 percent greater adhesion and consequently 50 percent more pulling ability. This was the idea of a New Yorker named Levi Bissel, who in 1858 patented a new kind of leading truck which could move sideways as well as swivel. Both two- and four-wheel trucks were provided for. Inclined planes gave a self-centering action and in this way the two essential and previously incompatible elements of flexibility and stability could both be present.

The New Jersey Locomotive and Machine Company built their first 2-6-0 in 1861, as the "Passaic" for the Central Railroad of New Jersey. The Erie Railroad followed in 1862 with the first large order of this locomotive type.

In 1863, Rogers Locomotive and Machine Works built what some cite as the first true 2-6-0 built in the United States, for the New Jersey Railroad and Transportation Company. William Hudson, chief engineer of the Rogers Locomotive Works in Paterson, New Jersey, developed the principle by replacing the inclined planes with swing links and also connected the truck springs to the main springs by a big compensating lever. This meant that the weight distribution would remain sensibly constant on rough tracks.

The Baltimore and Ohio Railroad (B&O) No600, a 2-6-0 Mogul built at the B&O's Mount Clare Shops in 1875, won first prize the following year at the 1876 Centennial Exposition in Philadelphia. It is preserved at the B&O Railroad Museum, housed in the former Mount Clare shops in Baltimore.

Over the next 50 years 11,000 Moguls were built for U.S. railroads. They filled a general-purpose role between 4-4-0 passenger locomotives and 2-8-0 freight units.

Very few of these classic steam locomotives still exist, most of them having been scrapped as newer, faster, and more powerful steam engines were developed in the twentieth century. The USRA standard designs of 1914 did not even include a 2-6-0.

Out of the 11,000 produced only four 2-6-0 locomotives are still in operation in the United States.

Consolidation 2-8-0

The name "Consolidation," now universally applied to the type, came from the first of some heavy freight locomotives supplied to the Lehigh Valley Railroad in 1866, just as that company had been formed by means of a consolidation of a number of smaller railroads in the area. They were intended to work trains up the Mahoney Hill which had a grade of 2.5 percent (1-in-40).

To traverse curves easily, the two center pairs of driving wheels were flangeless. The connecting rods drove on to the third axle, the eccentric rods and links of the Stephenson link motion being shaped and positioned clear of the leading axles. Otherwise the successful U.S. 19th Century recipe for the steam locomotive was applied in its entirety.

Locomotives of the 2-8-0 wheel arrangement were built in the U.S. from 1866 onwards and by 1946 about 24,000 had been supplied to railroads all over the world. Tractive effort varied from 14,000lbs for narrow-gauge examples up to 94,000lbs for ones fitted with a booster tender and built for the Delaware & Hudson Railway in the 1920s. They were all intended for freight work, with occasional use hauling passenger trains on steeply-graded lines.

One important feature which became standard about this time was the casting of each cylinder integral with half the smokebox saddle. When bolted together, a very strong front end was produced. At this time also, Baldwins had begun making a great effort to standardize parts and fittings as between any particular locomotive and another similar one. It was a pet project of the founder, Mathias Baldwin—who sadly died at the age of 70 in the same year that Consolidation was built — and was certainly a major factor in the dominant position in locomotive manufacture that his firm was to reach.

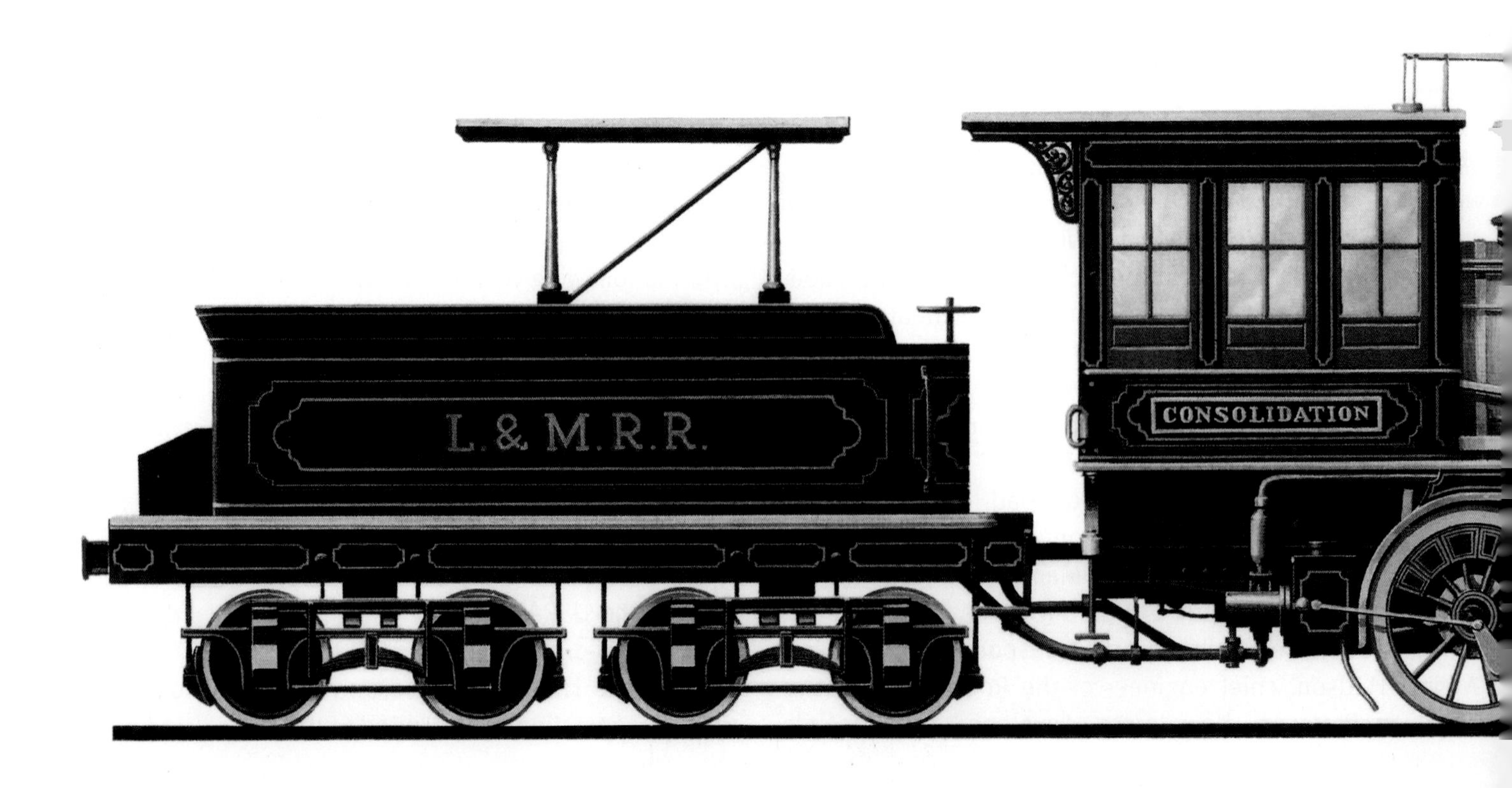

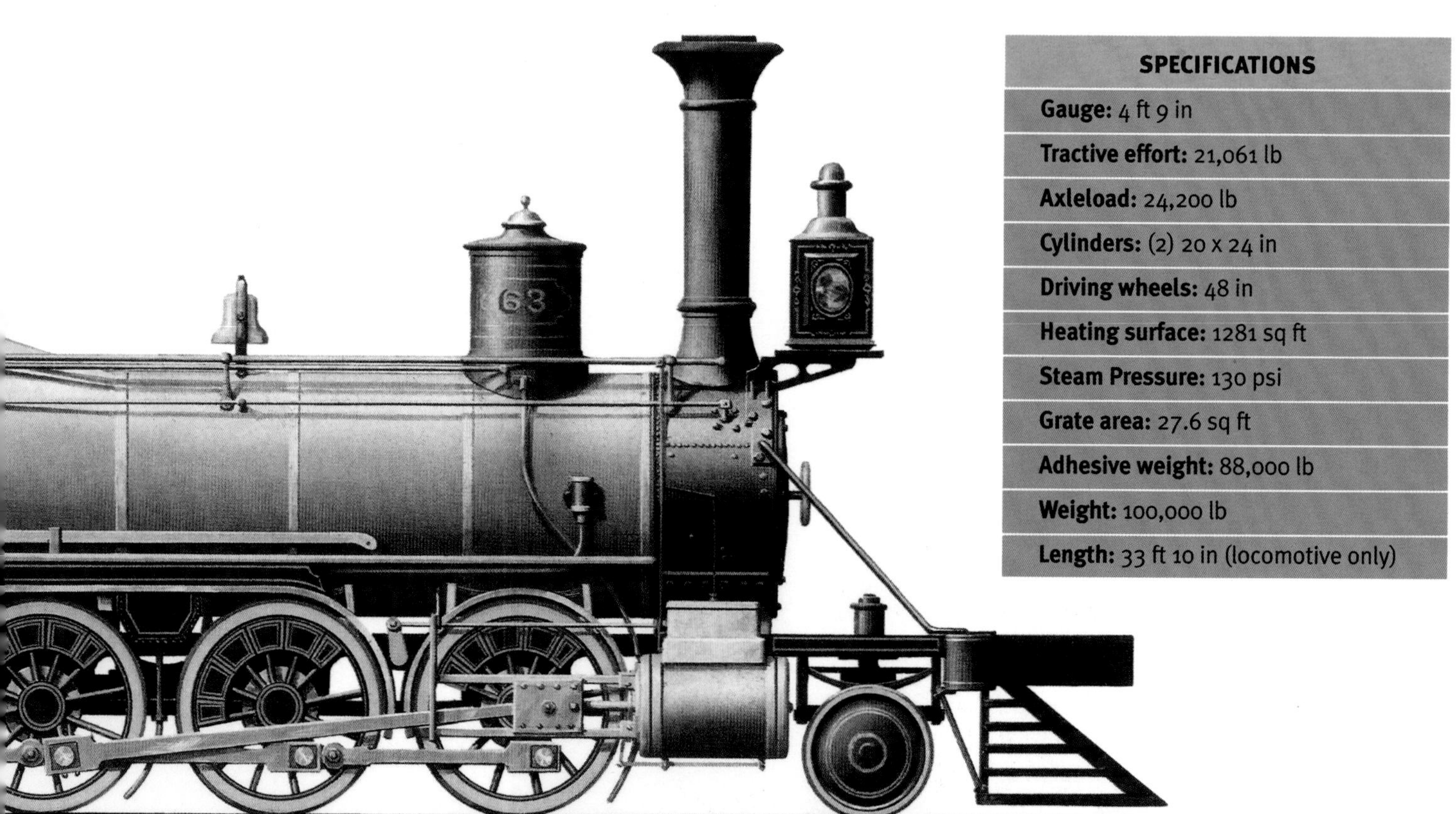

SPECIFICATIONS
Gauge: 4 ft 9 in
Tractive effort: 21,061 lb
Axleload: 24,200 lb
Cylinders: (2) 20 x 24 in
Driving wheels: 48 in
Heating surface: 1281 sq ft
Steam Pressure: 130 psi
Grate area: 27.6 sq ft
Adhesive weight: 88,000 lb
Weight: 100,000 lb
Length: 33 ft 10 in (locomotive only)

Reuben Wells 0-10-0 (1868)

Above: The Reuben Wells languishing at the Penn Central Railroad Company yards in the 1950s.

Named for its designer the Reuben Wells is a helper locomotive that was built in 1868. It was designed to push train cars up the 5.89 percent incline of Madison Hill in Madison, Indiana, the steepest segment of standard-gauge main-track in the United States. Weighing 10,000lbs it was the most powerful locomotive in the world at the time. It is 35 feet long.

The Jeffersonville, Madison and Indianapolis Railroad tried many different methods to get train cars up the hill. The problem was that the couplings used to connect the train cars were not strong enough to withstand being pulled up the hill, making it necessary for the cars to be pushed. A cogwheel system was in use for about twenty years but the Reuben Wells was the first steam engine to work the grade by adhesion alone, pushing the cars up the hill, as well as supporting them on the descent.

The Reuben Wells pushed train cars up Madison Hill for thirty years before it was retired in 1898. It stayed in reserve for another seven years before it was retired permanently and sent to Purdue University in 1905. In the years that followed the Reuben Wells was included in several exhibitions, including the Chicago World's Fair in 1933–34 and the Chicago Railroad Fair in 1948–49. Afterward, it remained in Pennsylvania at the Penn Central Railroad Company railroad yards. In 1968, the Reuben Wells was brought back to Indiana, where it was placed on permanent display at the Children's Museum of Indianapolis.

Left: Today, The Reuben Wells is preserved at the Children's Museum of Indianapolis.

SPECIFICATIONS
Gauge: 4ft 8 1/2 in
Total Weight: 10,000 lbs
Overall Length: 35 ft

Lovatt Eames 4-2-2 (1880)

SPECIFICATIONS
Gauge: 4 ft 8 1/2 in
Tractive effort: 11,000 lb
Axleload: 35,000 lb
Cylinders: (2) 18 x 24 in
Driving wheels: 78 in
Heating surface: 1,400 sq ft
Steam pressure: 150 psi
Grate area: 56 sq ft
Fuel: 14,000 lb
Water: 4,000 US gall
Adhesive weight: 45,000 lb
Total weight: 149,000 lb
Length overall: 55 ft 2 in

This locomotive was ordered by the Philadelphia & Reading Railroad—better known simply as the 'Reading'—as power for light high-speed trains, notably between Philadelphia and New Jersey. It had a 4-2-2 wheel configuration known as a "Bicycle" for its large single driving wheel. This was the Baldwin Locomotive Works 5,000th loco and whilst it was not considered a success by history it had some unusual features which had an influence on future locomotive specifications. The single-driver arrangement had the advantage of simplicity—and, in addition, the fewer rods there were to thrash around at high revolutions, the better. However the single driving wheel raised the objection of lack of adhesive weight that the engine designers catered for in a very creative manner. There was an interesting device fitted to the locomotive which, by means of a steam servo-cylinder mounted between the frames, could move the effective central pivot position of the equalizing levers connecting the driving wheels with the rear wheels. In this way the adhesive weight could be temporarily increased from 35,000 to 45,000lbs to assist with traction whilst starting. Provided that the engineer remembered to return the loco to two wheel drive before any speed was reached this worked well.

The locomotive also featured a wide firebox which was supported by a pair of idle wheels underneath. Desirable for any type of fuel, a big grate was essential when burning the anthracite coal which happened to be mined in the area served by the Reading Railroad.

In a twist of fate the Reading failed to pay Baldwin for the locomotive and it was sold instead to the Eames Vacuum Brake Company who used it for a demonstration vehicle for their products naming it the Lovatt Eames. At the time the well-known Westinghouse air brake was leading the market in the U.S. and rival firms like Eames were seeking new overseas markets for their products.

Both types of brake had the desirable automatic or "fail-safe" qualities but the vacuum brake was very much simpler. On the other hand it had problems associated with the fact that only some 12psi of differential pressure (much less at any altitude) was available, whereas the air brake had six times as much and more. In the USA, then, the air brake became standard; abroad, however, there was still scope for extending the use of vacuum. So the Lovatt Eames was sent across the sea to Britain and over a short trial period played a small part in influencing several companies in making the decision to adopt the vacuum brake.

The ultimate fate of the Lovett Eames is not recorded and it must be assumed that the engine was broken up after a few years.

Central Pacific 4-10-0 (1884)

The Central Pacific Railroad needed powerful locomotives to haul trains up the 1-in-36 grade from Sacramento (virtually at sea level) to 7,000ft at the Summit Tunnel through the Sierra Nevada mountains.

When the rails across the continent were joined on April 28, 1869, Central Pacific had nearly 200 locomotives. Most were standard American type 4-4-Os but ten-wheeler 4-6-os also featured on the roster.

From 1872 onwards CP began building locomotives in its own Sacramento shops to the design of Master Mechanic A.J. Stevens. In 1882, to meet the demands of "The Hill"— as the Sierra Nevada was known— Stevens introduced a twelve-wheeler or 4-8-0 designated No229.

In 1884, Stevens tried to go one step further with the superbly impressive 4-10-0 El Gobernador This magnificent machine was on paper and at the time the largest and most powerful loco in the world. It was built by Central Pacific Railroad at the railroad's Sacramento, California, shops. It was the last of Central Pacific's locomotives to receive an official name (its number was 237) and was also the only locomotive of this wheel arrangement to operate on United States rails. Its name is reminiscent of the railroad's first

Below: No. 229 was the first 4-8-0 ever built.

SPECIFICATIONS
Gauge: 4 ft 8 1/2 in
Tractive effort: 34,546 lb
Axleload: 26,750 lb
Cylinders: (2) 21 x 36 in
Driving wheels: 57 in
Steam pressure: 135 psi
Fuel: 10,000 lb (4.5t).
Water: 3,000 US gall
Adhesive weight: 121,600 lb
Weight: 239,650 lb
Length: 65 ft 5 in

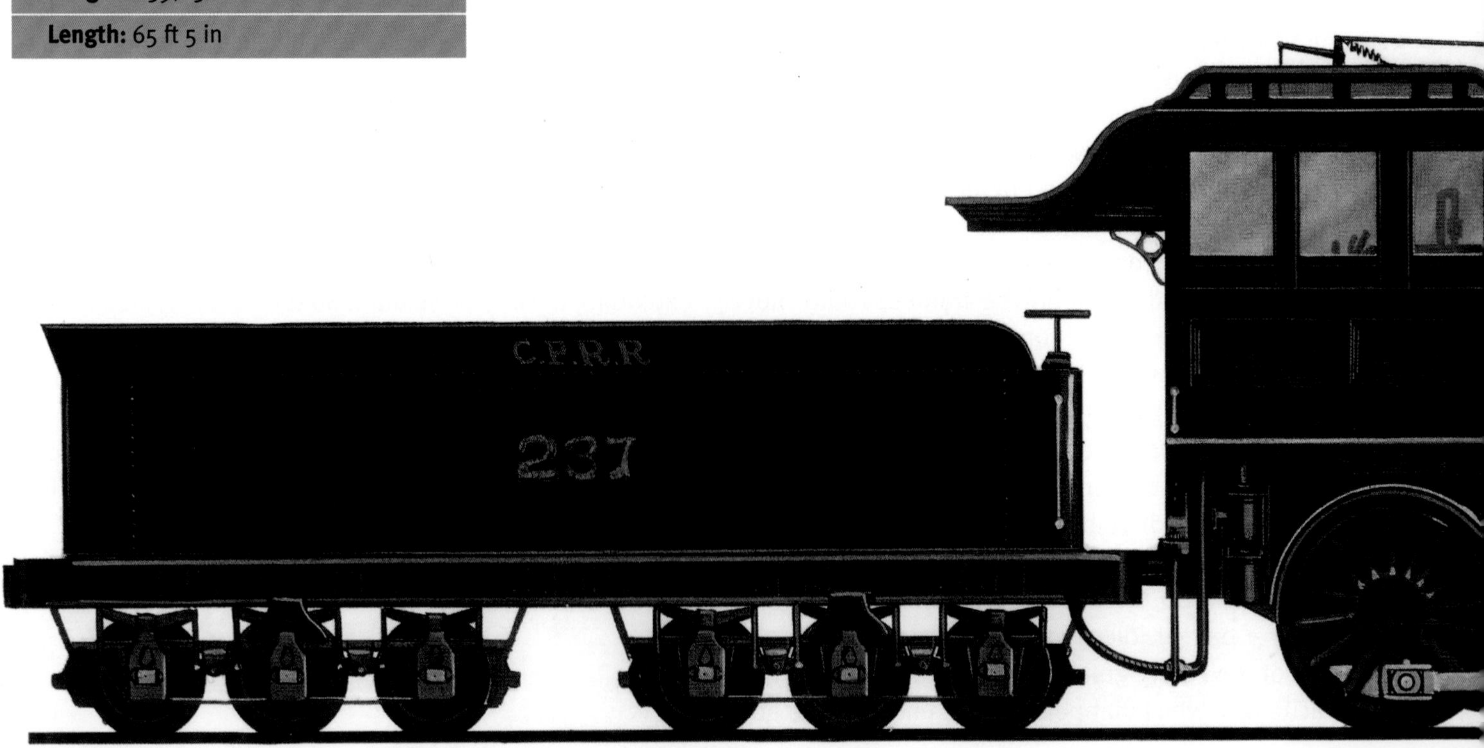

pair of driving wheels.

El Gobernador was completed in February 1883, amid much fanfare from the railroad, but it didn't enter service until March 1884, just over a year later. During this time, while still in Sacramento, the gigantic engine was used as an advertising tool by the railroad, to spectacular effect. The engine was kept under steam near the Central Pacific's passenger depot, where it would await the arrival of passenger trains coming in from the east. As a train arrived, El Gobernador would steam past the depot dragging a long line of empty freight cars behind it and causing quite a stir in the process. The engine would then be decoupled and placed on adjacent track where the passengers could get a good look at the monster up close.The locomotive was not a success and was reputedly scrapped in 1894.

locomotive, Governor Stanford, as El Gobernador is Spanish for The Governor. This locomotive is sometimes mistakenly referred to as a "Mastodon" type. However, this was the unofficial name for the earlier engine, No. 229, the first 4-8-0 ever built. Both engines looked nearly identical, except that El Gobernador was longer and had an additional

Above: No 237 at the roundhouse in Sacramento.

C-16/18/19 Class 2-8-0 (1882)

SPECIFICATIONS
Gauge: 3 ft
Tractive effort: 16,800 lb
Axleload: 13,818 lb
Cylinders: (2) 15 x 20 in
Driving wheels: 37 in
Heating surface: 834 sq ft
Steam pressure: 160 psi
Grate area: 14 sq ft
Fuel: 12,000 lb
Water: 2,500 US gall
Adhesive weight: 50,250 lb
Total weight: 111,600 lb
Length overall: 52 ft 2 3/4 in

Denver and Rio Grande Western 346 is one of a class of 28 2-8-0, Consolidation type, narrow gauge steam railway locomotives built for the D&RG Railroad by the Baldwin Locomotive Works in 1881–82.Originally it was named Cumbres and given the road number 406. Grant built a number of locomotives in the same class at the same time. In the 1870s General William Jackson Palmer of the D&RG set out to build a vast railroad system radiating out of Denver to serve the length and breadth of Colorado. It would have to cope with mountainous terrain so he decided on 3ft gauge following the example of the pioneering of the narrow-gauge Festiniog Railway in Wales. Jackson had soon taken his narrow tracks to the highest elevation then attained in the U.S. The altitude of the line at La Veta pass was 9,400ft, reached by a route which included 4 percent (1-in-25) gradients and 30° curves. Stronger locomotives than the 2-4-0s, 4-4-0s and 2-6-0s originally supplied were then required.

Number 346 was needed for the then new "San Juan Extension" of the D&RG, which ran from Alamosa, over Cumbres Pass, to Durango and Silverton. It was one of the Denver and Rio Grande's heavier Consolidations, a type C19—"C" for Consolidation, and "19" for 19,000 pounds of tractive effort. The loco was primarily used in helper service. In 1923 it was renumbered 346, and in 1924 it was fitted with a new boiler. The engine is now preserved at the Colorado Railroad Museum.

Below left: Denver & Rio Grande Western loco No 318 at Montrose,Colorado July, 1951.

Below: No 318, Class C-18, was built by Baldwin in March 1896, and is now preserved at the Colorado Railroad Museum.

No8 Maude 0-6-0 (1886)

SPECIFICATIONS
Gauge: 4ft 81/2in
Cylinders: (2) 18 x 24in
Driving wheels: 51 in
Steam Pressure: 145 psi
Tractive effort: 19,168 lb
Weight: 100,000 lb

0-6-0T No8 is the oldest Central of Georgia locomotive in existence. It was built by Baldwin in 1886 as a conventional steam locomotive (i.e. with tender) but was converted to a saddle tank by the CG shops in Macon, Georgia, in 1909.

It served as a switch engine in the Macon area until 1953, and was christened "Maude" after the stubborn, back-kicking comic strip mule in Frederick Opper's "And Her Name Was Maud." The locomotive is preserved at the Savannah Round House Museum in Georgia.

No999 4-4-0 (1893)

SPECIFICATIONS
Gauge: 4 ft 8 1/2 in
Tractive effort: 16,270 lb
Axle load: 42,000 lb (19t).
Cylinders: (2) 19 x 24 in
Driving wheels: 86 in
Heating surface: 1,927 sq ft
Superheater: None
Steam pressure: 190 psi
Grate area: 30.7 sq ft
Fuel: 15,400lb
Water: 3,500 galls
Adhesive weight: 84,000 lb
Total weight: 204,000lb
Overall length: 57 ft 10 in

New York Central and Hudson River Railroad No999 is a 4-4-0 steam locomotive built for the New York Central and Hudson River Railroad in 1893, which was intended to haul the railroad's Empire State Express train service between New York and Chicago. This locomotive is reputed to have been the first in the world to travel over 100 mph on May 10,1893.

The conductor timed the train travelling between two marks a mile apart. With four heavy Wagner cars weighing 50-55 tons each, about 2,000 cylinder horsepower would be needed and this would seem to be just a little too much to expect; not so much as regards steam production at a corresponding rate, but in getting that steam in and out of the cylinders in such quantities. A speed of 102.8mph over 5 miles, timed the previous night, is a little more credible, but both must, alas, be regarded as "not proven."

This combination of speed and luxury was shortly to result in one of the most famous trains of the world, the legendary year-round "Twentieth Century Limited," running daily from New York to Chicago.

No999 was specially built for the job and the train name was even painted on the tender. The NYC&HRR shops at West Albany turned out this single big-wheeled version of the railroad's standard 4-4-0s, themselves typical of the American class of their day, with slide-valves, Stephenson's valve gear and more normal 78in wheels. On account of the record exploit, No999's fame was worldwide; the locomotive even figured on a U.S. two-cent stamp in 1900. Today, much rebuilt and with its original 86in driving wheels replaced by modest 78in ones, No.999 is on display at the Chicago Museum of Science and Industry.

Below: No999 as seen today in the Chicago Museum of Science and Industry. Note how the replacement 78 in wheels alter the locomotive's profile compared to the original 86 in ones as seen in the earlier photo.

Cog Locomotive

SPECIFICATIONS
Gauge: 4ft 8 $^1/_2$ in
Tractive effort: 22,040 lb
Cylinders, HP: (2) 10 x 20 in
Cylinders, LP: (2) 15 x 22 in
Driving wheels: 22 in
Heating surface: 575 sq ft
Steam pressure: 200 psi
Grate area: 19 sq ft
Weight: 52,700 lb
Length: 22 ft 7 $^1/_2$ in

Cog-wheel locomotives are almost as old as steam traction, for in 1812 John Blenkinsop had one built by Fenton, Murray & Woods of Leeds, England, for the Middleton Colliery Railway. This occurred the year before William Hedley built *Puffing Billy,* the world's first really-practical steam locomotive. It was soon shown that for normally-graded railways the bite of an iron wheel on an iron rail was sufficient and the world then put aside cog railways for half-a-century when this form of traction was used again for mountainous gradients.

The Manitou and Pike's Peak Railway (also known as the Pikes Peak Cog Railway) is an Abt rack & pinion system cog railway with 4 ft 8 ½ in standard gauge track in Colorado, climbing the well-known mountain, Pikes Peak. The base station is in Manitou Springs, Colorado near Colorado Springs. It was built and is operated solely for the tourist trade.

Completed in 1890 the railroad ran from Manitou Springs, Colorado, to the 14,100ft summit of Pike's Peak. This altitude is the highest reached by a railway in North America. The 8.9 mile line is still open and runs daily (now using diesel traction) throughout the season from May to October. The maximum grade is 25 percent (1-in-4).

To work the line originally, the Baldwin Locomotive Works supplied three six-pinion rack locomotives. The boiler was mounted sloping steeply downwards towards the front so that it would be acceptably level on l-in-4 gradients. Steam was fed to two pairs of 17in bore by 20in stroke cylinders which drove three double rack pinions mounted on axles separate from the wheels which were not driven. There was a separate crankshaft driving the two rear rack pinions directly through gearing with a 1:1 ratio. The leading rack-pinion was driven from the middle one by a pair of short coupling rods.

In 1893 the locomotives were rebuilt to a new pattern with coupling rods that drove two double pinions from the outside.

The double-sided rack which was located between the tracks has two sets of teeth which are staggered to ensure that the two double pinions are always in mesh. The fleet was modified yet again in 1912 with changes to the drive train which increased the gearing of the locos positively.

Above: Pikes Peak cog railway locomotive and car around the turn of the 20th century.

Left: Preserved Baldwin Engine No. 1 on public display at the Colorado Railroad Museum.

Forney 0-4-4T

SPECIFICATIONS
Gauge: 4 ft 8 1/2in
Axleload: 20,000 lb
Cylinders, HP: (2) 9 x 16 in
Cylinders, LP: (2) 15 x 16 in
Driving wheels: 42 in
Heating surface: 555 sq ft
Steam pressure: 160 psi
Grate area: 19 sq ft
Water: 750 gall
Adhesive weight: 40,000 lb
Weight: 58,000 lb
Length: 27 ft 6in

The Forney is a type of tank locomotive patented by Matthias N. Forney between 1861 and 1864. The locomotives were set up to run cab first, effectively as a 4-4-0. The 4-4-0 wheel arrangement, with its leading truck giving three-point suspension, was noted for its good tracking ability, while the flangeless middle wheels allowed the locomotive to round tight curves. Placing the fuel and water over the truck rather than the driving wheels meant the locos had a constant adhesive weight, something other forms of tank locomotive did not.

Large numbers of Forney locos were built for the surface and elevated commuter railroads that were built in cities such as New York, Chicago and Boston. These railroads required a small, fast locomotive that tracked well and could deal with tight curves. Their short runs meant the limited fuel and water capacity was not a problem, making the Forney ideal. Forneys were used extensively on both the New York and Chicago elevated lines when these were operated by steam. One of the last urban routes to be equipped with them was the Chicago & South Side Rapid Transit Railroad, which was opened in June 1892. Baldwins supplied 45 exceedingly neat little Vauclain compound Forneys. However, as these

Above: A northbound train pauses at the Indiana Avenue station in this view looking east along an unpaved 40th Street. Note the surface track of the Chicago Junction Railway.

railroads began to electrify or were replaced by subways at the end of the 19th Century, Forneys began to disappear, and by the turn of the century all the Forneys had been sold or scrapped.

Forneys were also popular on the 2ft narrow gauge railroads of Maine. The use of these locomotives differed in that they were run smoke stack leading, like a conventional locomotive, and all driving wheels were flanged. The latter resulted in Maine narrow gauge railroads having comparatively broad radius curves. Further developments included the introduction of locomotives with a leading pony truck, giving a 2-4-4 wheel arrangement. This was done to improve tracking ability in these locomotives.

Below: Forneys were pre-eminent with elevated railways at the end of the 19th century.

The 0-4-4T type was a precursor of other designs which may have drawn on the Forney, such as the Boston & Albany and Central of New Jersey 4-6-4T, which have been only called "tank engines." The notable three-foot gauge 2-6-6T Forney was actually an articulated locomotive, in which the boiler and fuel and water tank were on one frame and the engine was articulated to pivot beneath the boiler, such that the valve connecting rods had to be raised in a frame up over the boiler and wide enough to accommodate the articulation of the engine under the boiler.

Today Forney locomotives can still be seen at the Maine Narrow Gauge Railroad Museum, at the Forney Transportation Museum, and on the Disneyland Railroad.

Left: The Forney Museum of Transportation has preserved this 1897 locomotive originally built by Porter in Pittsburgh.

No382 4-6-0 (1896)

SPECIFICATIONS
Gauge: 4 ft 8 1/2 in
Tractive effort: 21,930 lb
Axleload: 36,923 lb
Cylinders: (2) 19 x 26 in
Driving wheels: 69 in
Heating surface: 1,892 sq ft
Steam pressure: 160 psi
Grate area: 31.5 sq ft (2-9m²).
Fuel: 18,000 lb
Water: 5,000 gall
Adhesive weight: 100,700 lb
Weight: 205,550 lb
Length: 60 ft 6 in

Above left: Casey Jones in the cab of No382 shortly before his death.

On April 30, 1900, at 3:52 am a south bound passenger train of the Illinois & Central RR crashed into four cars of a freight train at Vaughan, Mississippi.

Jonathan Luther "Casey" Jones was born in Missouri in 1863 and in 1876 moved to Cayce, Kentucky. At age 15, he left home for Columbus, Kentucky, to work for the Mobile and Ohio Railroad as a telegrapher and later as a brakeman and fireman.

Jones moved to Jackson, Tennessee, still working for the Mobile and Ohio. When asked by a fellow railroad man where he was from, Jones said he was from Cayce, Kentucky, and the nickname "Casey" was born.

In 1888 he was hired by the Illinois Central Railroad. On February 23, 1891, Casey was promoted to engineer and was later assigned to passenger runs between Memphis, Tennessee, and Canton, Mississippi, a run of about five hours. This was one link of a four train run between Chicago and New Orleans.

On the morning of April 29, 1900, Jones pulled into Memphis from Canton where he was to lay over until the next day. The regular engineer who was to make the night run was ill so Casey agreed to take his place. Engine No. 382, with Jones in the cab, departed about an hour and a half late.

Casey Jones was known for his insistence that he "get there on the advertised" time and as he approached Vaughan, Mississippi he was only running a couple of minutes late. Two freight trains were on a siding but their combined length was longer than the siding. As they attempted to clear the main track, an air hose on No. 72 broke, locking the brakes and leaving four cars of No. 83 extending onto the track at the north end.

When Sim Webb, Casey's fireman, saw the caboose on the track he jumped from the cab, but Casey did not. Some say Casey Jones stayed with his engine because of his sense of duty and the value he put on human life. Jones died in the accident, but no other person was killed or seriously injured.

Wallace Saunders, who worked in the Canton roundhouse, wrote a tune remembering Casey that became a favorite of fellow workers. Bert and Frank Leighton, a couple of vaudeville performers, spread the "Ballad of Casey Jones" across the country. The song was copyrighted in 1909. In the 1930s, a book, a motion picture and a radio series added to the legend. In 1962, Johnny Cash released his version of "Casey Jones."

The official accident report said the "Engineer Jones was solely responsible having disregarded the signals given by Flagman Newberry." Until his death in 1957, Sim Webb, Casey's fireman, maintained that "we saw no flagman or flare, we heard no torpedoes."

Camelback Class 4-4-2 (1896)

SPECIFICATIONS
Gauge: 4 ft 8 1/2 in
Tractive effort: 22,906 lb
Axle load: 40,000 lb (18t).
Cylinders: (4) see text.
Driving wheels: 84 in
Heating surface: 1,835 sq ft
Superheater: none
Steam pressure: 200 psi
Grate area: 76 sq ft
Water: 4,000
Adhesive weight: 79,000 lb
Total weight: 218,000 lb

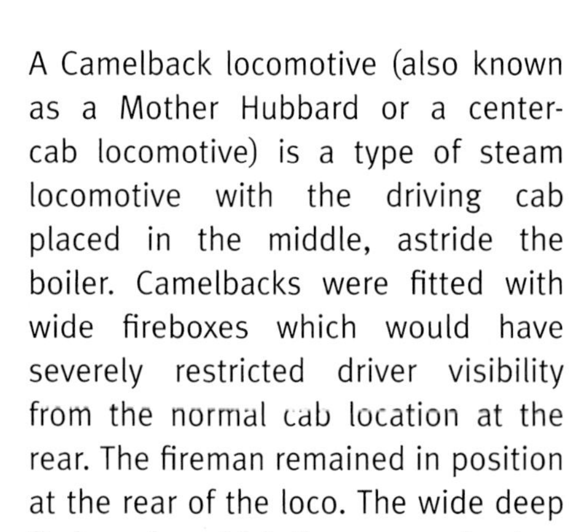

A Camelback locomotive (also known as a Mother Hubbard or a center-cab locomotive) is a type of steam locomotive with the driving cab placed in the middle, astride the boiler. Camelbacks were fitted with wide fireboxes which would have severely restricted driver visibility from the normal cab location at the rear. The fireman remained in position at the rear of the loco. The wide deep firebox, for which the 4-4-2 wheel arrangement is wholly appropriate, was adopted in order to allow anthracite coal to be burnt satisfactorily, but later it was realized that a large grate was also an advantage with bituminous coal and even with oil. The Camelbacks had pairs of compound cylinders on each side, driving through a common cross-head . The arrangement was named after Samuel Vauclain head of the Baldwin Locomotive Works, and his object was to attain the advantages of compounding without its complexities. In this case the high-pressure cylinders, 13in bore by 26in stroke, were mounted on top and the low-pressure ones 22in bore x 26in stroke below. A single set of valve gear and a single connecting rod served both cylinders of each compound pair. Problems were experienced with synchronizing the high and low pressure cylinders however.

The Atlantic City Railroad (ACR) ran them on fast trains which took people from the metropolis of Philadelphia to resorts on the New]ersey coast. It was a 55 1/2 mile run from Camden, New Jersey (across the river from Philadelphia) to Atlantic City and there was intense competition from the mighty Pennsylvania Railroad which had direct access into the big city. In July and August, for example, it was noted that the booked time of 50 minutes was kept or improved upon each day. On one day the run is reported to have been made in 46 1/2 minutes start-to-stop, an average speed of 71.6mph This certainly implies a steady running speed of 90mph or more, but reports of 100mph plus speeds with these trains should be regarded as pure conjecture. Despite this the "Atlantic City Flier" was certainly the fastest scheduled train in the world at that time.

D16ab 4-4-0

Introduced in 1895, the D16 class followed the reputation established by the Pennsylvania Railroad for big loco-motives mostly built in the company's Altoona workshops.The Belpaire firebox used by the Pennsy gave their locos a characteristically high boiler stance than was usual at the time.

Two types were built one with 80 inch driving wheels used on the more level sections of the road and a 68 inch variant to be used on the hillier parts. They soon established reputation for high speed operations on the line between Camden and Atlantic City where they were in direct competition with the Atlantic City Railroad. Speeds of over 100 mph were not uncommon and a Presidential Special train clocked an average speed of 72 mph on the 90 mile stretch between Philadelphia and Jersey City.

SPECIFICATIONS
Gauge: 4 ft 8 1/2 in
Tractive Effort: 23,900 lbs
Cylinders: (2) 20 x 26 in
Driving Wheel Diameter: 68 in
Total Weight: 281,000 lbs
Overall Length: 67 ft 0 in

Steam In Charge: 1900-1950

I-1 Class 4-6-0 (1900)

SPECIFICATIONS
Gauge: 4ft 8 1/2 in
Tractive effort: 23,800 lb
Axle load: 45,000 lb
Driving wheels: 80 in
Heating surface: 575 sq ft
Steam pressure: 200 psi
Weight: 300,000 lb
Length: 62 ft 3 in

The I-1 Class locomotive, with 80 inch drivers, was built by the Brooks Locomotive Works of Dunkirk, New York State in 1900 for the Lake Shore & Michigan Southern Railroad. These high wheeled locomotives were intended to haul the express trains of the Western part of the New York to Chicago main line belonging to what was soon to become the New York Central Railroad.

The American standard 4-4-0 configuration has been successful in hauling express trains from the 1850s until the 1880s. However, by the late 1880s loads began to out grow the capacity of a locomotive with only two driven axles.

The obvious solution was simply to add a third coupled axle, and this is what was done. Some of the best features of the 4-4-0 were retained in the 4-6-0 such as the stable leading truck that guided the locomotive, but in other areas problems arose. For example the ashpan was liable to foul the rearmost axle and the gap between the leading driving wheels and the cylinders became too tight for easy maintenance .Despite these shortcomings there was a period at the end of the nineteenth century when the 4-6-0 ruled the express passenger scene in the US. About 16,000 examples went into service between 1880 and 1910.

Remarkably, shortly after the arrival of the 4-6-0s the LS&MS, anxious for more power, ordered some 2-6-2s with wide fireboxes over the trailing pony trucks to overcome the problems experienced by the I-1s. However, they were not a success, as the 2 wheeled leading truck didn't have the stability of the I-1s 4 wheel truck and tended to ride up over the head of the rails at high speeds thus putting the proven 4-6-0s back in charge of the legendary Twentieth Century Limited Pullman express. This service ran between New York and Chicago and was introduced on 15 June 1902.

The timing over the 960 miles between New York's Grand Central Terminal and La Salle Street station in Chicago was 20 hours, an average speed of 48mph This included several stops for servicing and changing locomotives and much slow running in such places as Syracuse, New York where the main line in those days ran along the main street.

The train originally consisted of a buffet-library car, dining car and three sleeping cars, the last of which had an observation saloon complete with brass-railed open platform. The comforts offered were the equivalent of the highest grade of hotel.

However providing this amount of comfort and luxury resulted in these 80ft Pullman cars being extremely heavy, thus limiting the train to maximum of only five cars. Increased popularity of the service meant that it was necessary to increase the amount of accommodation provided and accordingly these locomotives had to be replaced by larger more powerful locomotives but the I-1s had served their purpose in establishing the route.

Above: I-1 class 4-6-0 No.604 at the head of the "Twentieth Century Limited"

Class Q 4-6-2 Pacific (1901)

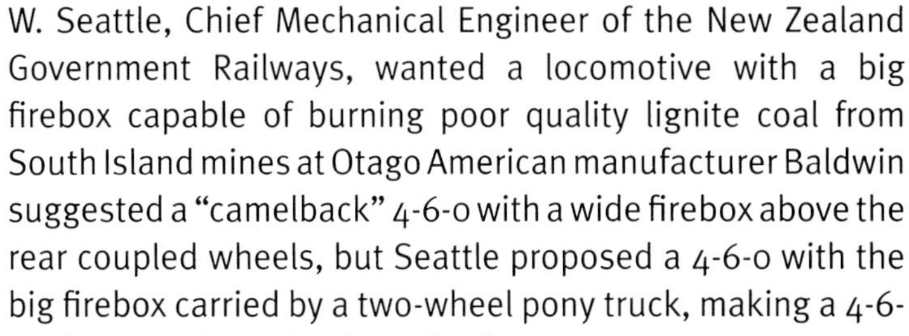

W. Seattle, Chief Mechanical Engineer of the New Zealand Government Railways, wanted a locomotive with a big firebox capable of burning poor quality lignite coal from South Island mines at Otago American manufacturer Baldwin suggested a "camelback" 4-6-0 with a wide firebox above the rear coupled wheels, but Seattle proposed a 4-6-0 with the big firebox carried by a two-wheel pony truck, making a 4-6-2. Thus was born the first of a famous type—arguably *the* most famous type — of steam express passenger locomotive, which was to go on being built until the end of the steam era.. The origin of the name Pacific, which classifies this type of locomotive, is given various explanations but one is due to this very early order for the type from New Zealand which is of course across the Pacific from America. Baldwin completed the 13 locomotives on the order and despatched them to New Zealand.

Also new on the Q Class was Walschaert's valve gear.

Named for Belgian engineer Egide Walschaert who had devised it back in 1844 but this application marked its entry into general use outside continental Europe. The gear gave good steam distribution, but the main advantage lay in its simplicity, as well as in the fact that it could conveniently be fitted outside the frames in the position most accessible for maintenance. In this case the gear was arranged to work outside-admission piston valves, which piston valves themselves were in the forefront of steam technology at the beginning of the 20th century

It should be said that this class of engine came closer than ever before to the ideal form of steam locomotive. Only two fundamental improvements were still to be applied generally—inside-admission piston valves instead of outside, and superheating.

After some minor modification the "Q" class gave long and faithful service, the last of them not ceasing work until 1957. During their prime, in addition to working the principal trains on the South Island main line, some came to the North Island for use on the Rotorua Express, running between Auckland and the famous hot springs of the same name.

SPECIFICATIONS
Gauge: 3 ft 6 in
Tractive effort: 19,540 lb
Axle load: 23,500 lb
Cylinders: (2) 16 x 22 in
Driving wheels: 49 in (1,245 mm)
Heating surface: 1,673 sq ft
Steam pressure: 200 psi
Grate area: 40 sq ft
Fuel:11,000 lb
Water: 2,000 gall
Adhesive weight: 69,500 lb
Total weight: 165,000 lb
Length overall: 55 ft 4 1/2 in

Above: The Baldwin works photo of the Class Q.

F-15 Class 4-6-2 (1902)

SPECIFICATIONS
Gauge: 4 ft 8 1/2 in
Tractive effort: 32,400 lb
Axle load: 52,500 lb
Cylinders: (2) 23 1/2 x 28 in
Driving wheels: 72 in
Heating surface: 2,938 sq ft
Superheater: None
Steam pressure: 150 psi
Grate area: 47sq ft
Fuel: 30,000 lb
Water: 9,000 galls
Adhesive weight: 157,000 lb
Weight: 408,000 lb
Length: 74 ft 0 in

Only a few weeks after the Missouri Pacific RR got the first of their 4-6-2s, the Chesapeake & Ohio Railroad took delivery from the American Locomotive Company of the prototype of their famous F-15 class Pacifics signifying that the standard North American express passenger locomotive of the twentieth century had finally arrived. This path-finding C&O No147 was also fitted with piston valves, but it still had Stephenson's link valve motion between the frames. Naturally no superheater, but her size and power set a new standard, A further 26 followed during the years 1903-11. Most survived until the C&O turned to diesels in the early 1950s and that said a great deal for the equalities of the F-15 class. Of course, as the years went by, top-line express work was passed on to their successors, yet there were routes whose weak bridges meant that these comparatively light yet powerful locomotives continued being used on prime routes almost to the end of steam. During the 1920s all the F-15 locomotives were modernised with Walschaert's valve gear, superheaters, larger tenders, different cabs, mechanical stokers, new cylinders and, in some cases, even new frames.

In addition to setting the style for nearly 7,000 U.S. 4-6-2s to follow, the F 15 founded a dynasty on their own road. The F164-6-2sof 1913 represented a 34 per cent increase in tractive effort and a 28 per cent increase of grate area,while for the F 17

of 1914 these increases were 45 per cent and 71 per cent respectively, in each case for a penalty of a 27 per cent increase in axle load. After World War I, classes F18 and F19 appeared, notable for 18,000 gallon 12-wheel tenders. These 61 4-6-2s handled all C&O's express passenger assignments until the coming of 4-6-4s in 1941.

Below: A later version of the F-15, the F-16 was built by Baldwin in 1937.

G-7 Class Chesapeake & Ohio 2-8-0 (1903)

The Chesapeake & Ohio Railroad began its love affair with 2-8-0 Consolidation type locomotives in 1881 when it acquired the G-1 class from the Cooke Locomotive Works.

It was the company's first venture away from 4-4-0 Americans and the 4-6-0 ten wheelers which were standard for both passenger and freight up until that time. The taxing 13 mile climb over the Allegheny mountains and the increasing coal traffic demanded a more powerful locomotive. C&O bought 42 2-8-0s between 1881 and 1883.

Above: No569 was a product of Alco's Richmond Works in 1906, seen here powering a passenger train at an undetermined location circa 1915. The crew posed for the photographer as laborers struggled to unload the train in the background. Note the big oil headlight, slanted cylinders, large pilot, the air pump slung rather openly on the boiler, and the large cab with its arched window. After the renumbering of 1924 this locomotive became 898.

SPECIFICATIONS
Gauge: 4ft 8 1/2 in
Tractive effort: 23,800 lb
Axle load: 45,000 lb
Driving wheels: 80 in
Heating surface: 575 sq ft
Steam pressure: 200 psi
Weight: 300,000 lb
Length: 62 ft 3 in

The type proved very satisfactory and as a result ,they continued buying them over the next three decades.Over this 30 year period the type was improved, enlarged and developed resulting in the G-7 in 1903.

In the era that they were conceived, the C&O was forwarding coal trains east from Hinton at a furious rate, delivering about 1,000 cars per day to Clifton Forge in trains of 45-60 cars using several Consolidations as power over the 0.56 Allegheny grade, including pushers from Ronceverte to Allegheny. That amounted to probably 15-20 loaded trains east and about that many west, per day, plus four fast freights, two local freights and eight sets of passenger trains, making this line congested, even though by the turn of the century much of it was double tracked. Some bottlenecks remained, including Lewis Tunnel near the summit, and Big Bend Tunnel at Talcott, eight miles east of Hinton, both of which constricted the line to a single track. The arrival of the G-7s allowed larger trains which improved the flow of traffic. The C&O seems to have been happy with the G-7s and continued ordering them through 1907, building up a roster of 205 of the engines.

A further development was the G-8 class, which appeared in 1907 just after the C&O received the last of the G-7 class. This class had 56-inch drivers, and 206,000 pounds engine weight, and developed 40,775 pounds tractive effort. These locomotives had a different boiler from the G-7. Whereas the G-7 had a noticeably tapered boiler, the G-8 has a straight boiler, eight inches higher. They also had Walschaerts valve gear and outside piston valves, therefore the cylinders were straight rather than the slanted ones seen on the G-7. Richmond Locomotive Works, the C&O's favorite "hometown" builder built only two locomotives for this class. Because only two locomotives were built, it would seem to indicate that G-8 was an experimental type. It was probably an effort to go beyond the G-7 in performance, but apparently the C&O's mechanical department wasn't that pleased with the G-8, and as a result no more were built. Two years later the C&O got its G-9 class 2-8-0s, the final evolution of the Consolidation type on the line. These 50 Richmond-built locomotives arrived in 1909, and were almost exact copies of the latest G-7s, except for their use of Walschaerts valve gear and outside piston valves, like those on the G-8s.

Then, in 1911, the motive power picture began to change, as the C&O started its heavy acquisition of 2-6-6-2 as well as adding the 2-8-2 wheel arrangement in the form of the K-l. The K-l bested the G-7 and G-9 by 53 percent in tractive effort, while the H-4 2-6-6-2s exerted a tractive force 80 percent greater than a G-7/9.

Despite this a further batch of six G-7s were ordered in 1916. They had the added improvement of a superheater. designated G-7S, many earlier G-7s were rebuilt with superheaters. The last G-7S types of 1916 were the last C&O 2-8-0s. Many of their class lasted up to the end of steam, as did almost all the G-9 class. In the later years, they were often used as work trains, local freights, and even some were used on mine run trains.

The use of pairs of 2-8-0s continued out of Thurmond on the Keeney's Creek mine runs up to dieselization. Some G-7S class locomotives even provided passenger service on branch lines. Others performed switching work.

B&O Railroad No2400 0-6-6-0 (1903)

SPECIFICATIONS
Gauge: 4 ft 8 1/2 in
Tractive effort: 96,600 lb
Axleload: 61,325 lb
Cylinders: (2) 22 x 32 in
Cylinders: LP: (2) 32 x 32 in
Driving wheels: 56 in
Heating surface: 5,586 sq ft
Steam pressure: 235 psi
Grate area: 72.2 sq ft
Fuel: 30,000 lb
Water: 7,000 gall
Adhesive weight: 335,104 lb
Total weight: 477,500 lb

The Baltimore & Ohio Railroad had a problem in taking heavy trains up the Sand Patch incline, 16 miles long, graded at 1 per cent (1-in-100). There were sharp curves and this limited the number of driving wheels which a straight unhinged locomotive could have. The idea was to replace the two 2-8-0 helpers necessary for a 2,000-ton train by a single locomotive.. The B&O management took their cue from Europe where Anatole Mallet ,a Frenchman, had taken out a patent in 1884 for a locomotive with the front part hinged. His idea was geared toward small tank engines intended for sharply-curved local and industrial railways. Ironically it ended up being used for the largest, heaviest and most powerful locomotives ever built.The B&O directors went to the American Locomotive Co and asked them to build first ever US Mallet compound. It was a turning point in US locomotive history.

No2400 launched another innovation as far *as* the US was concerned—one which was to affect locomotive practice even more than the Mallet arrangement, for this pathfinding machine was the first significant US locomotive to have outside Walschaert valve gear. One set worked the inside-admission piston valves of the rear high-pressure cylinders and another had the slightly different arrangement appropriate to slide valves which are inherently outside-admission.

This simple, accessible and robust mechanism, which produced excellent valve efficiency, was very soon to become the standard valve gear for most North American locomotives. Another feature to become standard later was the steam-powered reversing gear. With four valve gears, the effort required was too great for a manual arrangement. Later, the mechanism of straight two-cylinder engines became so massive that power-reverse became general on all except the smallest.

The Mallet principle could be described as building a normal locomotive with a powered leading truck, for the frame of the rearhigh-pressure engine has the boiler attached rigidly to it. The large low-pressure cylinders can if required be placed in front of the smokebox, and thereby they can be freed of any restrictions on their size. The pivoting arrangements for the front engine are relatively simple, with a hinge at the rear and a slide at the front.

No463 K-27 2-8-2 (1903)

No463 is a 3 foot narrow gauge, Mikado type, 2-8-2 steam railway locomotive built for the Denver and Rio Grande Railroad by the Baldwin Locomotive Works in 1903. The class eventually became known by the nickname "Mudhens." It is one of two remaining locomotives of D&RGW class K-27. Her whistle is similar to hooter whistles on the East Coast of the USA but has a slightly different sound.

Fifteen locomotives were built, originally class 125, then reclassified K-27 in 1924 when D&RG became the Denver and Rio Grande Western Railroad. The K-27s were built as Vauclain compounds, with two cylinders on each side, expanding the steam once in the smaller cylinder and then a second time in the larger one. The extra maintenance costs of the two cylinders were greater than the fuel saving, so they were converted to simple expansion in 1907–1909. They were Rio Grande's last purchase of compound locomotives. They pulled freight, passenger and mixed trains on the D&RGW in and over the Colorado Rocky Mountains, traversing the entire length of the railroad. They were built with their main structural frames outside the driving wheels, with the counterweights and rods attached outside the frames.

No463 was in Chama, New Mexico in 1923, in Alamosa, Colorado and Montrose, Colorado in the 1930s, and spent most of the 1940s and early fifties on Rio Grande Southern tracks and on Silverton. The locomotive operated on Farmington, New Mexico branch in 1947.

No463 was sold to cowboy actor and singer Gene Autry in May 1955 who never used the engine but donated it the City of Antonito, Colorado. It was restored by and entered into

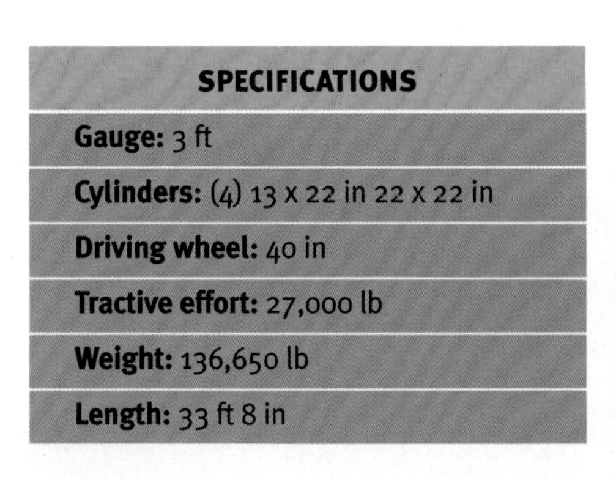

SPECIFICATIONS
Gauge: 3 ft
Cylinders: (4) 13 x 22 in 22 x 22 in
Driving wheel: 40 in
Tractive effort: 27,000 lb
Weight: 136,650 lb
Length: 33 ft 8 in

service on the Cumbres and Toltec Scenic Railroad in 1994. It was taken out of service with a broken side rod in 2002. In 2009, it was moved to the railroad's shop at Chama, New Mexico where a major rebuild was undertaken, completed in Spring 2013. On May 20, the locomotive was steamed up, running in the lead with D&RGW K-36 No487, to pull the railroad's annual deadhead, or move empty cars, on its first run.

No463 was added to the National Register of Historic Places in 1975 as Engine No463.

900 2-10-2 Santa Fe (1903)

SPECIFICATIONS
Gauge: 4 ft 8 1/2 in
Tractive effort: 62,560 lb
Axleload: 46,800 lb
Cylinders, HP: (2) 19 x 32 in
Cylinders, LP: (2) 32 x 32 in
Driving wheels: 57 in
Heating surface: 4,817 sqft
Steam pressure: 225 psi
Grate area: 59.5 sq ft
Fuel:30,000 lb
Water: 17,000 gall
Adhesive weight: 234,000 lb
Weight: 433,000 lb
Length: 81 ft 5 in

By the turn of the 20th century, locomotive engineering had moved on apace since the failure of such early ten-coupled locomotives as Southern Pacific's El Gobernador of 1883. The Atchison, Topeka & Santa Fe Railway had an equal if not harder task in moving freight over the mountain barriers on its way to and from California. The approach to Santa Fe's Raton Pass was graded at 3.5 percent and two large Decapod 2-10-Os were built as pushers for use on this horrendous incline; they were very successful. The only problem occurred when running back down after helping a train up the hill; it was found that guiding wheels were advisable to assist the limited tracking qualities of the long rigid wheelbase. As a result the company asked Baldwin to build the world's first 2-10-2, and so history was made.

The 2-10-2 type, named "Santa Fe" after the first user, was built in big numbers—as many as 2,200 for US railroads over the next 40 years—and is even today in large-scale production for the Chinese Railways. The original examples were interesting in that they were "tandem" compounds with each high-pressure cylinder sharing a common piston rod with the low-pressure ones. The problems of attending to the packing of the glands between each pair of cylinders were assisted to some extent by providing a small crane on each side of the smoke-box. Perhaps this illustrates part of the reason why Santa Fe's locomotive department, in their later and wiser days, never bought any locomotive, however huge, with more than two simple cylinders. Santa Fe in 1911 rebuilt 20 Class "900" 2-10-2s into 10 of the "3000" class 2-10-10-2 articulated locomotives which were then the most powerful in the world. They were not successful and were converted back to 2-10-2s in 1915, when, following changes in the management of the railroad, conversion was begun of all Santa Fe's compounds—numbering nearly 1,000 of many wheel arrangements—to simple propulsion.

The 2-10-2s had priority in this conversion and after

Below: The 3800 class 2-10-2s were the final development of the Santa Fe type on the Santa Fe Railway. Introduced in 1924, they had a tractive effort 30 per cent more than the 900 class.

World War II the fleet of this type was supplemented by more modern versions. As steam approached its end in the late-1940s nearly 300 of the 2-10-2s were still on the roster. Following improvements to the springing and side-control, and the dropping of compounding, the type had been found to be excellent workhorses for general heavy freight haulage, as well as on the helper duties for which the "900" class had originally been built. One "900" class (No940) was preserved in Johnson Park, Bartlesville, Oklahoma in 1954.

Above: The very first "Santa Fe" or 2-10-2 type supplied to the Atchison, Topeka & Santa Fe Railway in 1903 by Baldwin, No900 is seen a t San Augustme, Texas, in April 1951.

The locomotive was on static display and suffered over time as all such locos do from the effects of weather, floonding from the nearby river Caney, and vandalism. In 2008 the engine was restored by public subscription and is now on display at a much improved site.

KCS 1023 0-8-0 (1906)

SPECIFICATIONS
Gauge: 4ft 8 1/2 in
Tractive effort: 47,124 lb
Steam pressure: 210 psi
Driving wheels: 55 in

Originally built in the Alco, Pittsburg, Pennsylvania workshops in 1906 as a Consolidation Class E-3, 2-8-0 No88 for KCS RR. It was converted into a switcher in the Pittsburg, Kansas KCS shop and became a K-1 0-8-0 in 1925. It served on KCS until 1954 when it was put into storage and eventually wound up in Schlanger Park as a static display on September 17, 1955. It spent the next 50 years being vandalized and lost many of its more easily removable parts like the bell and whistle. When it finally sunk into the mud as the ties rotted it was rescued by The Heartlands Museum in Carona, Kansas,and transferred there in 2012 and has been systematically and skillfully rebuilt at great expense.

Above: The Kansas City Southern engine No1023 has been located in Schlanger Park, Pittsburg, Kansas since 1955, when the Kansas City Southern railroad sold it to the City of Pittsburg. This image was taken in April 2011 in Schlanger Park. In February 2012, the engine was moved to the Heart of the Heartlands Railroad Museum in Carona, Kansas.

Below: No1023 in her heyday on switching duties.

UP No618 2-8-0 (1907)

SPECIFICATIONS
Gauge: 4ft 8 1/2 in
Cylinders: (2) 22 x 30 in
Driving wheel: 57 in
Steam pressure: 200 psi
Tractive effort: 43,350 lb
Weight: 212,800 lb

No618 is a 2-8-0 consolidation type locomotive which has 2 unpowered truck wheels followed by eight powered wheels and no wheels under the cab. At one time the Union Pacific Railroad rostered over 1,000 such locomotives. They were all–rounders, used for freight, passenger service and switching alike.This is because the consolidation design is very versatile as well as powerful. No618 could pull a freight train of 60 cars on a level track and could tackle steep gradients with similar loads if double headed. No618 was part of a three loco order for the Oregon Short Line (OSL) railroad built by the Baldwin Locomotive Works of Eddystone, Pennsylvania in July 1907. The OSL was a subsidiary of the Union Pacific Railroad. The three locos were given OSL road numbers 1066-1068. Our loco was No1068.The OSL ran from North Utah into Idaho,Wyoming and Montana.

It survived when UP went bankrupt in the 1890s due to its separate status.

The Oregon locomotives were numbered into the UP system in 1915, No1068 becoming UP No618. At the time however they retained the OSL script on the tender.The locos were classified as C-57 class- C for consolidation and 57 for the driver diameter.

No618 retired from UP service in 1958 and was donated to the Utah State fair as a static exhibit. The loco was transferred to the Promontory Chapter NHRS in 1969 and leased to the Wasatch Mountain Railway in 1970 where it remained until 1990. During that time it was in full operational order between 1970-76 and 1986-1990.It was acquired by the Heber Valley Historical railroad in 1992 and once again restored to full service in 1995.

The locomotive was undergoing boiler test and repairs at the beginning of the 2014 season and an appeal was launched to restore it to full running for the 2015 season.

Left: UP 618 with its Vanderbilt tender, note the OSL logo on the back.

No112 NWP 4-6-0 (1908)

SPECIFICATIONS
Gauge: 4ft 8 1/2 in
Construction date: 1908
Weight: 137,800 lb

Northwestern Pacific No112 was built by the American Locomotive Company at Schenectady, New York, in 1908 for the newly formed Northwestern Pacific Railroad. The Northwestern Pacific Railroad (NWP) is a regional railroad in northern California. It was founded in the late 1800s as a joint venture between the Southern Pacific and the Santa Fe, and began operations in 1907. In 1908, the NWP received 4-6-0 steam engine No112 from Alco. The locomotive's green paint is unusual, as by then most steam locomotives had been painted black. No112 served the NWP until 1962.

It worked on freight trains and in the 1914 completion of the line connecting Tiburon, in Marin County, and Eureka, in Humboldt County. An early event in its history was a 1913 fall through a wharf at Tiburon and into San Francisco Bay. It was soon fished out, renovated and put back to work. As larger locomotives were purchased by the NWP, No112 was used in secondary service, running on branch lines in different part of the railroad making itself useful as a maid of all work until eventual retirement in 1962.

It was donated to the Pacific Coast Chapter of the Railway & Locomotive Historical Society for preservation. Renovated at the Bethlehem Shipyards in San Francisco in 1965, No112 was then repainted in a representation of the two-tone green paint scheme worn by several first class NWP passenger engines in the late 1920s. In 1969 NWP No112 was fortunately donated to the California State Railroad Museum where it remains.

Above: No112 in service with the NWP.

Left: No112 on the roundhouse turntable at the California State Railroad Museum.

AT & SF Class 1300 4-4-6-2 (1909)

SPECIFICATIONS
Gauge: 4ft 8 1/2 in
Tractive effort: 53,700 lb
Axleload: 58,963lb (26.8t)
Cylinders, HP: (2) 24 x 28 in
Cylinders, LP: (2) 28 x 28 in
Driving wheels: 7 5 in
Heating surface: 4,756 sq ft
Superheater: 323 sq ft
Reheater: 728 sq ft
Steam pressure: 200 psi
Grate area: 52.7 sq ft
Fuel (oil): 4,000 US gall
Water: 12,000 US gall
Adhesive weight: 268,0 16 lb
Total weight: 612,192 lb
Length overall: 104 ft 10 3/4 in

It has been related how the great Atchison, Topeka & Santa Fe Railway exchanged straightforward locomotives for fairly complex ones and did not find the change satisfactory. The further solution was then to move from the fairly complex to some very complex machinery indeed. One of the first results of this policy was this really quite extraordinary 2-4-6-2 compound Mallet design, of which two prototype locomotives (Nos. 1300-1) were supplied by Baldwin in 1909. They were intended for fast passenger traffic and to this end had the largest driving wheels ever used on a Mallet, plus a very low center of gravity. The long, thin boiler was also complicated, with a corrugated firebox and plate stays rather than the usual staybolts. The boiler barrel was in two sections with a separate forward chamber which acted as a feed-water heater.

One modest but satisfactory complication, at this time becoming standard, was a superheater, but Santa Fe had to go one stage further and have a reheater as well. This feature is similar in form to a superheater, except that it is arranged to apply extra heating to the steam after it leaves the high-pressure cylinders and before going to the low-pressure engine. A small gain in thermal efficiency is achieved by its use, but, in contrast to normal superheating, in general railroads did not find the extra cost of maintaining reheaters worthwhile.

Aside from all the complications, the main problem with such locomotives as these was that the Mallet layout, as then being built, was a little too flexible for comfort at fast speeds. Violent oscillations were liable to occur and some very specific restrictions to the degree of freedom of the front engine had first to be devised before these articulateds were suitable for anything but slow slogging freight trains.

In 1915, Nos. 1300 and 1301 had their front engines and front boiler stages removed, and so with comparative ease they returned to service as simple two-cylinder 4-6-2s. At last their inherent high-speed capability could be used. Incidentally, after finding out that the two 2-4-6-2s were not going to work, Santa Fe had some 2-6-6-2 articulated locomotives with amazing flexible boilers that were even more complicated. This time, the front stage of the boiler was fixed to the leading chassis and was actually hinged to the rear part. Both concertina connections and ball-and-socket joints were tried, but to no avail.

Below: Complicated and shortlived, AT&SF's extraordinary Class 1300 compound Mallet.

GNR Class H4 4-6-2 (1909)

SPECIFICATIONS
Gauge: 4 ft 8 1/2 in
Tractive effort: 35,690 lb
Axleload: 55,400 lb
Cylinders: (2) 23 1/2 x 30 in
Driving wheel: 73 in
Heating surface: 3,177 sq ft
Superheater: 620 sq ft
Steam pressure: 210 psi
Grate area: 53.3 sq ft
Fuel: 28,000
Water: 8,000 gall
Adhesive weight: 151,200 lb
Weight: 383,750 lb
Length: 67 ft 6 in

The Great Northern Railway was the northernmost US transcontinental route to be completed between the Mississippi River and the Pacific Ocean. It was a personal enterprise on the part of a legendary railroad tycoon called Jerome Hill. He had no government assistance and in the face of universal prophesies of disaster the GN was driven through to the Pacific coast in three short years. Even Hill himself was not present when the last spike— a plain iron one— was driven without ceremony on September 18, 1893, for he had his sights on a day in December 1905, when his superbly de-luxe "Oriental Limited" would cross the continent to connect with his own steamship, the SSMinnesota, en route to China and Japan.

The GN section of the route ran 1,829 miles from St Paul to Seattle, and before all-steel rolling stock arrived in the 1920s, 4-6-2s were adequate as haulage units, with some assistance in the mountains. But it must not be thought that in the days when timber-bodied cars was used on the "Oriental Limited" the train lacked anything in the way of luxury for its riders.

Delivery from the Baldwin Locomotive Works in 1909 of 20 new superheated "H-4" class Pacifics coincided with running the "Oriental Limited" through from Chicago via the Burlington's route to St Paul. A further 25 came from Lima in 1913. Superheating was soon to become universal and this class was an early example of the technique. The arrangement used involved elements in the smokebox rather than in flue tubes and so was less efficient but still very creditable. The "H4s" burnt coal but some were later converted to oil-burning.

All the elements had come together of the style of steam locomotive that would sweep the world. Cast frames, cylinders integral with the smokebox saddle, wide firebox, superheater, compensated springing with three point suspension, inside-admission piston valves and Walschaert valve gear proved an unbeatable combination. The "H4s" had long lives of 40 years and over. Few alterations were made over the years, although boosters were fitted in the 1940s to some of the class to cope with wartime loads. Alas, none have been included amongst the handful of GN steam locomotives preserved for posterity.

Above: Although no H-4 locomotives survived GNR No1355, Class H-5, 4-6-2 is preserved and currently being cosmetically restored at Sioux City, Iowa at the former Milwaukee Road roundhouse.

NCO No9 4-6-0 (1909)

SPECIFICATIONS
Gauge: 3 ft 0 in
Cylinders: (2) 16 x 20i n
Driving wheels: 44 in
Steam pressure: 180 psi
Water: 5,000 gall
Tractive effort: 17,800 lb
Weight: 165,150 lb
Length: 53 ft 11 in

NCO No9 was one of four small oil-burning 3ft gauge ten-wheelers delivered by Baldwin between 1907 and 1911. They pursued a busy but uneventful existence for 15 years or so until their new owner Southern Pacific widened the gauge. However, SP had another narrow-gauge line not far away which had been effectively in its hands since 1900. The Carson & Colorado Railway ran from Mound House, Nevada, nearly 300 miles southwards to Keeler, California and the best of the NCO engines were moved there after 1928. By 1943 all but the southernmost 70 miles of the C&C had been abandoned or widened.

The route of the Nevada-California-Oregon Railway was constructed between 1880 and 1912. It ran 235 miles, generally northwards, from Reno, Nevada, via Alturas in California to Lakeview, Oregon, incorporating the earlier Nevada & Oregon and Nevada & California railroads as well as other lines. Most of the mileage came into the hands of Southern Pacific by purchase in 1926 and was quickly altered to standard gauge.

The remaining narrow-gauge tracks ran through the remote Owens Valley which, though once prosperous, had become desolate following the abstraction of its water to supply the city of Los Angeles. Even so, once World War II was over, the fame of the slim gauge steam operations attracted hundreds of railfans who came both individually and by special excursions. Two of the 4-6-0s, Nos 8 and 9, plus an ex-NCO 2-8-0 (No18) soldiered on down the years. No.9 was the last steam locomotive to run, not only on the narrow-gauge but on the whole Southern Pacific system. This was in 1959 during a period of standby duty beginning in 1954 when a narrow-gauge diesel locomotive arrived. In 1960 the line was finally abandoned. An unconventional feature of all three locomotives was the semicircular tender, partitioned for oil and water supplies. Otherwise Nos. 8 and 9 followed normal turn-of-the-century practice. No18 (supplied to NCO in 1911) was more modern, having outside Walschaert valve gear.- It says enough of the fame of the line that three of its steam locomotives have survived—No.8 at Carson City, Nevada, No18 at Independence, California and No9 at Bishop, California.

Below: Baltimore & Ohio compound 0-6-6-0 No2400 was the first Mallet -type articulated locomotive in the US.

Boston & Maine No410 0-6-0 (1911)

SPECIFICATIONS
Manufacturer: American Locomotive Company (ALCO), Manchester, NH
Type: Six-wheeled steam switcher
Weight: 74 tons

No410 was built in 1911 by the American Locomotive Company Manchester, NH for the Boston & Maine Railroad. It is one of only two remaining B&M 6 wheel switcher locomotives. At one time the B&M would have employed over 200 such engines. Switchers of this type were often used to ply between the vast networks of mill ,factory and yard trackage around Lowell in the golden years of steam. Their relatively compact dimensions, light weight (around 70 tons), and short wheelbase enabled them to handle the sharp curves and rough track often encountered in these industrial areas.

B&M used this switcher for almost four decades before it was sold in June 1950 to H.E.Fletcher where it was used in his granite quarry in Westford, MA until 1979 when it was moved to the B&M workshops at Iron Horse Park, Billerica. There was a unsuccessful attempt to cosmetically restore the loco for a proposed museum along the Boston Fort point Channel in the early 1990s.It languished for a while until in 1993 it was moved to Lowell, MA, where both the National Park Service and the Boston and Maine Historical Society have cosmetically restored it for display. Boston & Maine No410 holds the distinction of being the largest steam locomotive in the state of Massachusetts.

Right: No410 at its present site at Dutton Street, downtown Lowell, MA.

NATIONAL STREE
410
G-11-a
B-7-93-3

Class C-5a No1208 (1911)

SPECIFICATIONS
Gauge: 4 ft 8 1/2 in
Cylinders: (2) 24 x 30in
Driving wheel: 63 in
Steam Pressure: 185 psi
Tractive effort: 43,130 lb

The Class C-5a was a 2-8-0 Consolidation type locomotive built by the American Locomotive Company in 1912 and was used for freight by the Chicago, Milwaukee, St. Paul and Pacific Railroad (often referred to as the Milwaukee Road) (reporting mark MILW). From its introduction in 1866 and well into the early twentieth century, the 2-8-0 design was considered to be the ultimate heavy freight locomotive. The 2-8-0's forte was starting and moving "impressive loads at unimpressive speeds" and its versatility gave the type its longevity. The practical limit of the design was reached in 1915, when it was realized that there was no further development possible with a locomotive of this wheel arrangement. No C class locomotives survived and they were all phased out between 1945 and 1954.

The Milwaukee Road was a Class I railroad that operated in the Midwest and Northwest of the United States from 1847 until 1980, when its Pacific Extension was embargoed through the states of Montana, Idaho, and Washington. The eastern half of the system merged into the Soo Line Railroad on January 1, 1986. The company went through several official names and faced bankruptcy several times in that period. The railroad no longer exists as a separate entity, but much of its track continues to be used by its successor and other roads, and is commemorated in buildings like the historic Milwaukee Road Depot in Minneapolis, Minnesota.

Below: Class C-5a No1208 takes on water at Sioux Falls, South Dakota in August 1940.

No15 East Broad Top RR 2-8-2 (1911)

Narrow gauge was always rare east of the Mississippi and 45 years ago the only such system of any size still operating was the 33-mile 3ft gauge East Broad Top Railroad, with headquarters at Orbisonia, Pennsylvania. The EBT was chartered in 1856, completed in 1874 and it spent its life doing mainly what it was built to do; that is, carrying coal from mines towards its far end to an interchange with the Pennsylvania RR at Mount Union.

It was a well-arranged operation; narrow gauge was no handicap because the coal had to be unloaded and reloaded at the Mount Union cleaning and grading plant anyway. Carload freight of other kinds was dealt with by replacing standard-gauge trucks with narrow-gauge ones for the ride over the EBT. Its long survival was due to these factors, as well as to the more fundamental point that coal-hauling was what railroads were first built for and were best at doing.

Above: East Broad Top 2-8-2 No15 heading North out of Orbisonia, PA.on June 4th 2006.

The EBT was also rare in that, by the early 1950s, it was still all-steam with a fleet of six reasonably modern Baldwin 2-8-2s which had been supplied one at a time over the years 1911 to 1920. All were of similar appearance but the three later ones were larger and had piston instead of slide valves. Alas, in 1955 coal sales were at very low ebb, and so it was no surprise that in April 1956 the last coal train ran and all these wonders seemed at an end.

Now the idea of preserving an operating commercial railroad as a tourist attraction had come to Pennsylvania in 1959, when the little Strasburg Railroad was saved from extinction in this way. Now, luckily, the man who bought the East Broad Top for scrap was a railfan. Moreover, Nick Kovalchick had (as *Trams* magazine put it in 1960) "wanted to run a railroad ever since his parents couldn't afford to buy him a toy train years ago."

The new EBT, although rather remote in situation, ran tourist trains northwards out of Orbisonia each weekend from June to October and daily during July and August up to 2011.The future of the railroad is now uncertain.

Above: Back in 1952, there was still hard work to be done by the Baldwin 2-8-2s. No16 heads a train of empty hoppers at the Mount Union coal washery.

SPECIFICATIONS
Gauge: 3 ft 0 in
Tractive effort: 27,600 lb
Axleload: 29,700 lb (13.5t).
Cylinders: (2) 19 x 24 in
Driving wheels: 48 in
Heating surface: 1,676 sq ft
Superheater: 357 sq ft
Steam pressure: 150 psi
Grate area: 36 sq ft
Adhesive weight: 108,000 lb
Weight: 244,750 lb
Fuel: 18,000 lb
Water: 5,000 gall
Length: 64 ft 10 in

ATSF 3000 Class 2-10-10-2 (1911)

SPECIFICATIONS
Gauge: 4 ft 8 1/2 in
Cylinders: HP (2) 28 x 32 in
Cylinders: LP (2) 38 x 32 in
Driving wheels: 57 in
Steam pressure: 225 psi
Tractive effort: 109,113 lb
Weight: 616,000 lb

The first 2-10-10-2s were constructed in 1911 by the Atchison, Topeka and Santa Fe Railroad at a cost of $43,880 each. This class of ten locomotives was built from existing, more conventional, 900 & 1600 class 2-10-2s with new front engines and tenders from Baldwin. They were used between Bakersfield and Barstow and up to San Bernardino. The boilers could not sustain pressure and after ten years, the locomotives were remanufactured into 20 3010 class 2-10-2s. The tenders survived intact for many years used behind some of the 2-10-2s. This wheel arrangement was rare. Only two classes of 2-10-10-2 locomotives have been built; the Atchison, Topeka and Santa Fe Railway's 3000 class, and the Virginian Railway's class AE.

Known as Santa Fe Mallets, these are among the few 2-10-10-2s built. Ten in number, the engines were cobbled together by the Santa Fe using existing 2-10-2 engine units united with low-pressure engineer units supplied by Baldwin. Baldwin also designed and built the unusual "turtleback" tender that probably was designed to improve rearward vision.

Although the 3000 class locomotives appeared to have exceedingly long boilers, the forward section of the boiler in front of the rear set of cylinders actually contained first a primitive firetube superheater for further heating the steam before use; the steam was carried forward from the boiler proper by outside steam pipes as shown in the photograph. Also in this space was a reheater for the high-pressure exhaust before it was fed to the forward low-pressure cylinders.

In front of that, there was a feedwater heater, a space where cold water from the tender could be warmed before being injected into the water proper. This worked similarly to the boiler itself; the firetubes passed through the feedwater.A distinctive ribbed firebox trailed the long boiler. Like other Santa Fe Mallets, the actual tube length was relatively short, but a "reheater" ahead of the forward tube sheet and boiler joint was supposed to maintain steam heat as it traveled forward to the HP cylinders. It added 2,659 sq ft, but was nowhere nearly as efficient as later superheater designs.

Overall, the design was unsuccessful and the engines were converted back to 2-10-2 simple-expansion locomotives in 1915-1918.

Left: Santa Fe demonstration trains Winfield Ks

No1352 2-8-2 (1912)

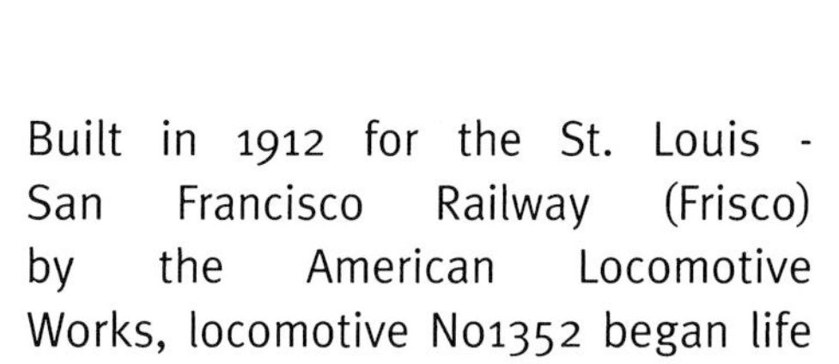

SPECIFICATIONS
Gauge: 4 ft 8 1/2 in
Cylinders: (2) 26 x 36 in
Driving wheels: 63 in
Steam pressure: 195 psi
Tractive effort: 53,355 lb
Weight: 322,600 lb

Built in 1912 for the St. Louis - San Francisco Railway (Frisco) by the American Locomotive Works, locomotive No1352 began life as 2-8-0 Consolidation type No1321. Short on cash and in need of more modern steam motive power during World War II, the Frisco rebuilt the engine into a 2-8-2 Mikado type locomotive, complete with stretched boiler, modified tender, Coffin feed water heater, and other modern steam improvements aimed at increasing efficiency and improving horsepower. After being retired by the Frisco in the late 1950's, the engine spent time in numerous states. No1352 eventually found its way to a long disused engine house in rural Illinois. The American Steam Railroad purchased the locomotive from this location, mostly disassembled, in 2008. The St. Louis San Francisco Railway No1352 2-8-2 steam locomotive is now the American Steam Railroad's locomotive project. They will be restoring the locomotive to operation in the Historic B&O Roundhouse in Cleveland Ohio, not only for the sake of running train trips, but to use as an educational tool. This will allow the public to see an example of the magnificent machines that were once the backbone of American industry and commerce.

Below: Here's #1352's sister, Frisco locomotive #1351, in Collierville Tennessee. This picture gives you an idea of what the locomotive will look like when it is back together.

Southern Pacific No2353 4-6-0 (1912)

SPECIFICATIONS
Gauge: 4 ft 8 1/2 in
Cylinders: (2) 22 x 28 in
Driving wheel: 63 in
Steam pressure: 210 psi
Tractive effort: 38,400 lb
Weight: 208,000 lb
Length: 82 ft 5 1/2 in

No2353 is one of 10 heavy 4-6-0 T-31 class "ten wheel" steam locomotives built by the Baldwin Locomotive Works for the Southern Pacific Railroad in August 1912.The locomotive was delivered to Southern Pacific in October and the boiler was changed as soon as 1917. In 1927, No2353 was leased to the San Diego & Arizona line, and later returned to Southern Pacific in 1939, serving in the San Francisco Bay Area. No2353 was retired from service on 18 January 1957 and displayed for the next 29 years at the California Mid-Winter Fairgrounds in Imperial, California.

In 1984, the Mid-Winter Fair's operator donated No2353 to the Pacific Southwest Railway Museum, with physical transfer of the locomotive occurring in the summer of 1986. Over the next ten years, volunteers restored No2353 to working order, with its first public appearance under steam happening on 2 March 1996.However the need for extensive boiler repairs caused No2353 to be withdrawn from service again in 2001.Since then the loco has been on static display at the museum.

Southern Pacific No2353 has at least one surviving classmate Southern Pacific No2355 has been on static display in Mesa, Arizona's Pioneer Park since 1958. Efforts to cosmetically restore this loco have been underway since at least 2008

Right: Southern Pacific No2353 is preserved at the Pacific Southwest Railway Museum on Depot Street , Campo, CA

Below: Southern Pacific Railroad No2353, a 4-6-0 oil burning steam locomotive built by the Baldwin Locomotive Works in 1912. The restored unit was photographed on March 7, 1999 in operation at the Pacific Southwest Railway Museum in Campo, California over the original San Diego & Arizona Railway mainline, powering one of the museum's period demonstration trains.

2353
2353

No1031 K-15 Class 4-6-0 (1913)

SPECIFICATIONS
Gauge: 4 ft 8 1/2 in
Cylinders: (2) 20 x 26 in
Driving wheels: 64 in
Steam Pressure: 200 psi
Tractive effort: 27,600 lb

No1031 is one of twenty-five Ten Wheeler type (4-6-0) locomotives built for the Atlantic Coast Line by Baldwin in 1913. It has 64in drivers and 20in x 26in cylinders. With a 44.1 sq ft grate, 181 sq ft firebox and total heating surface of 2,449 sq ft, including 400 sq ft superheating, it operated at 200psi delivering 27,600 lbs tractive effort.

The locomotives quickly earned the nickname "Copperheads," because they originally had polished copper rims around the tops of their smokestacks. Primarily used to haul both freight and passenger services in the Fayetteville, Rocky Mount and Wilmington, NC, district they nevertheless worked as far afield as Richmond, VA, and Jacksonville, FL. As dieselization gathered pace on the ACL, many were transferred to serve on smaller subsidiary lines. For most of its life, No1031 operated mainly out of Florence,

SC, hauling freight and passenger trains. It was taken out of service in 1952, but was then lent it to the Atlantic Coast Line subsidiary, the Virginia & Carolina Southern. Later it was transferred to the East Carolina Railway.

In 1959, it was placed on open air display behind the Florence, SC, passenger station but, after continuing deterioration in its condition, the City of Florence donated it to the North Carolina Railroad Museum in 1994. Two years later, it was cosmetically restored to its 1940s appearance.

B & SR No8 2-4-4T (1913)

SPECIFICATIONS
Gauge: 2 ft
Tractive effort: 10,072 lb
Axleload: 21,340 lb
Cylinders: (2) 12 x 16 in
Driving wheels: 35 in
Steam pressure: 180 psi
Fuel: 3,000 lb (1.4t).
Water: 1,000 gall
Adhesive weight: 38,800 lb
Weight: 69,700 lb
Length: 34 ft 7 3/4 in

This tiny locomotive from a tiny railway was nevertheless a pathfinder. It seems she was the first in the world to be restored for use as an instrument of pleasure after being withdrawn from normal commercial use. The Baldwin Locomotive Works in 1913 delivered No7 to the 2ft gauge Bridgton & Saco River Railroad up in Maine. The locomotive followed the Forney style with the addition of a lead truck, making a 2-4-4T type, and had slide valves actuated by Walschaerts valve gear, vacuum brakes and an unsuperheated boiler. These features all helped by their simplicity to give this elegant little iron foal qualities of usefulness and reliability, leading to a long life and finally to survival beyond the end of the age of steam.

In 1930 the 35-mile B&SR had become the 27-mile Bridgton & Harrison, but traffic was miniscule and by the end of the decade abandonment was clearly not far off. On days when the train was not operating regularly, the manager would put on a special for a few dollars to satisfy visiting railfans. One group even went some way towards raising fundsto buy the line. In the event, a scrap merchant put in a bid which the fans could not match.

In 1941 No8 (plus sister No.7 and a number of pasenger and freight cars) was bought by a cranberry grower from Massachusetts called Ellis D. Atwood. As the war ended, Atwood began building a railroad on his farm. It was formed as a circuit 5 1/2 miles in length, the dykes between the cranberry bogs providing a ready-made alignment on which he and his men could spike down the secondhand rails he had bought.

The idea was that the line should provide essential transport for the estate and only on high days and holidays be a pleasure line for himself and his friends. But it was not to be. People for miles around started to clamour for invitations, and soon enough the idea of opening the Ellis D. Atwood Railroad (Edaville for short) to the public was born. Monday, Apnl 7, 1947 was the day when the golden spike was ceremonially driven, since when Edaville has been systematically redeveloped to the detriment of the original locos and buildings.

Shay Locos B-B-B (1913)

SPECIFICATIONS
Gauge: 3 ft
Tractive effort: 36,150 lb
Axleload: 19,666 lb (8.9t)
Cylinders: (3) 11 x 12 in
Driving wheels: 32 in
Heating surface: 881 sq ft
Superheater: 189 sq ft
Steam pressure: 200 psi
Grate area: 23 sq ft
Fuel (oil): 1,200 gall
Water: 3,000 gall
Adhesive weight: 118,000 lb
Weight: 118,000 lb
Length: 50 ft 2 in

The story of the Shay is also the story of the Lima Locomotive Works of Lima, Ohio. Before 1880 it was just the Lima Machine Works, but in that year they built their first locomotive, a strange steam-driven flat car to the designs of a veteran logging man called Ephraim Shay. The first Shays had vertical boilers but later examples had locomotive-type ones. These were offset to one side of the center-line in order to balance the two- or three-cylinder in-line steam engine with vertical cylinders mounted on the other side. This drove the axles via longitudinal shafts fore and aft, universal joints and bevel gears. The basic Shay had two axle trucks but further powered trucks driven in the same way could added at the rear. Three-tri Shays were common and four truck ones were also built.

Because it had all wheels driven and because there was the maximum amount of flexibility. along its wheelbase, and a because of its simplicity and robustness, the Shay pulled useful loads and held to the rails on crazy temporary tracks in forests which were being felled. It was also ideal for any steeply-graded and sharp curved railroad. Over the next 65 years a total of 2,771 were sold, most of them in the first 35 or so. Their popularity was a clear indication of their usefulness and efficiency. The last one was delivered in 1944 by which time Lima had become a prestigious supplier of giant high-powered locomotives.

A valuable but by no means obvious feature was the short, fat heavily-tapered boiler barrel which minimised the change in water level at the firebox end for different inclinations. This meant that Shay could run off and on to 10 per cent (l-in-10) grades without trouble. Their flexibility meant they could negotiate 76° curves with serious kinks and wildly varying cross-levels without derailing. In addition, they were available straight out of Lima catalog in all gauges and sizes from 320,000lb total weight down to 50,000lb or less. They could burn coal, oil or forest waste and were easy and cheap to maintain in primitive workshops. Perhaps one should add that, flat out at 12mph or so, they sounded like the "Overland Limited" setting alight the prairies at 90mph.

No10: a 3 ft narrow gauge three-truck Shay steam locomotive constructed

Below: Bloedel Stewart & Welch steam locomotive No1 (Class B Shay) at BC Forest Discovery Centre in Duncan, British Columbia.

Above: Shay No10 fueling up at the Yosemite Mountain Sugar Pine Railroad. Built by the Lima Locomotive Works of Lima OH. in March of 1928 for the Pickering Lumber Corporation's West Side Lumber Company's operations in Tuolumne, CA. The largest narrow gauge Shay built. Purchased from West Side Lumber Co. in 1967 by Yosemite Mountain Sugar Pine Railroad to become the only steam locomotive on the old Madera Sugar Pine Company track at that time.

for the Pickering Lumber Company. The locomotive was completed on March 2, 1928 by the Lima Locomotive Works of Lima, Ohio and later acquired by the West Side Lumber Company in 1934. No10 burns oil, with a capacity to hold 1,200 gallons of oil and 3,420 gallons of water. This locomotive is reputedly the largest narrow gauge Shay locomotive ever constructed.

No15: also a 3 ft narrow gauge three-truck Shay steam locomotive, was originally constructed as the No9 for Norman P. Livermore & Company, of San Francisco, California, and soon thereafter sold to the Sierra Nevada Wood & Lumber Co.

who in 1913 had one of the largest logging railroads with over 60 miles of main route based at Toulomne, California. The locomotive was completed on May 20, 1913 by the Lima Locomotive Works of Lima, Ohio. No15 burns oil, with a capacity to hold 1,000 gal of oil and 2,000 gallons of water. In 1917, No15 was acquired by Hobart Estate Co. as their No9. In 1938, No15 was given its current number when purchased by the Hyman-Michaels Co., operating out of San Francisco. The West Side Lumber Company purchased No15 only a year later. When the West Side shut down in the 1960s, a tourist operation, the West Side & Cherry Valley, acquired the No15. After hauling tourists for a number of years, the locomotive sat on static display in Tuolumne, California, until the Yosemite Mountain Sugar Pine RR acquired it in 1988.

Shays still run on several tourist railroads, notably at Cass, West Virginia; Tacoma, Washington State; Georgetown, Colorado; and at the Yosemite Mountain Sugar Pine RR .

The Yosemite Mountain Sugar Pine Railroad (YMSPRR) is a historic 3 ft narrow gauge railroad with two operating steam train locomotives located near Fish Camp, California, in the Sierra National Forest near the southern entrance to Yosemite National Park. Rudy Stauffer organized the YMSPRR in 1961, utilizing historic railroad track, rolling stock and locomotives to construct a tourist line along the historic route of the Madera Sugar Pine Lumber Company.

NYC No6721 0-6-0 (1913)

SPECIFICATIONS
Gauge: 4 ft 8 1/2 in
Cylinders: (2) 21 x 28 in
Driving Wheels: 57 in
Steam pressure: 180 psi

No6721 was built by Alco in December 1913 as #54075.

One of only a handful of surviving New York Central steamers, this B-11k class 0-6-0 is on display outside Union Station in Utica, NY. Although the 0-6-0 wheel arrangement was more popular in Europe than in North America, a large number of them were used in the USA and Canada. The last locomotives built for a US railroad were 0-6-0 switchers from the Roanoke shops built for the Norfolk & Western Railroad.

L-1 Class Soo Line 2-8-2 Mikado (1913)

No1003 was built in 1913 by the American Locomotive Company in March 1913. It was used by the Soo Line until retirement in 1954, when it went into serviceable storage in Gladstone, Michigan as part of the railroad's strategic reserve. Soo Line No1003 is a restored 2-8-2 Mikado type steam locomotive of the Minneapolis, St. Paul and Sault Ste. Marie Railway ("Soo Line") L-1 class. It is occasionally operated on the major railroads of the American Upper Midwest.

SPECIFICATIONS
Gauge: 4 ft 8 1/2 in
Cylinders: (2) 28 x 30 in
Driving wheels: 63 in
Steam Pressure: 170 psi
Tractive effort: 53,947 lb
Weight: 290,000 lb

In December 1959, the railroad donated the locomotive to the city of Superior, Wisconsin, where it was put on public display. The locomotive was sold partially disassembled in 1994 to Wisconsin Railway Preservation Trust (WRPT), an organization whose goal was to return the locomotive to operations. WRPT raised $250,000 for the locomotive's restoration. No1003's first run after restoration under its own power occurred on October 27, 1996, when it steamed up the Duluth, Missabe and Iron Range Railway's Proctor Hill. It performed a few more test runs before its first public excursion in 1997. In 1998 it ran the "triple-header" excursion with Northern Pacific No328 and Soo Line No2719.

The locomotive made its final journey under its FRA-mandated 15-year boiler certificate on November 13, 2010. But shortly afterwards, the operators raised funds to have the engine overhauled and certified for another 15 years of operation. No1003 returned to service in September of 2012.

Class F-16 4-6-2 (1913)

No461 was one of four Pacific type F-16 class locomotives delivered to the Chesapeake & Ohio Railroad in June 1913 by the Baldwin Locomotive Works.. Baldwin duplicated Alco's existing design in grate area, firebox size, heating surface, bore and stroke, and adhesion weight.

The Pacific Type became one of the most prolific and common steam locomotive designs during the first two decades of the 20th century and was by far the most widely used for passenger service. The 4-6-2's large drivers and high tractive efforts of the time made them ideal for such operations where they could regularly cruise at speeds over 70 mph.

SPECIFICATIONS
Gauge: 4 ft 8 1/2 in
Cylinders: (2) 27 x 28 in
Driving wheels: 74 in
Steam pressure: 185 psi
Tractive effort: 43,376 lb
Weight: 290,000 lb

However there were some differences from the Alco design- 16" diameter piston valves served the cylinders, firebox heating surface included 30.8 sq ft of arch tubes and as a concession to the C&O's hilly profile, driver diameter was decreased by 5".

Baldwin's specs reveal that the C&O expected the F-16s to pull a 692 ton train consisting of:

- 1 steel express and 1 steel postal car, each weighing 113,200 lb
- 1 steel combined car of 130,300 lb
- 1 steel coach of 135,200 lb
- 1 diner of 157,700 lb
- 5 steel sleepers weighing 147,000 lb each

The ruling grade over which the trains were to keep time was 60 ft/mile (1.1%) over a 13 1/2 mile stretch from milepost 291 to Allegheny, Virginia. Up this stretch, the trains were to average 24 mph. It was generally considered that from their construction in 1913 to their scrapping in 1951-1952, the F-16s served well." At first they hauled flatland express trains, later taking the Charlottesville-Newport News and Ashland-Louisville sections. Over the years, the locomotives were fitted with automatic stokers.

After World War II, the class entered heavy-duty local service such as the daily Ashland-Elkhorn run and the Columbus-Toledo "accommodation train."

Left: Chesapeake & Ohio engine No461 is a Vanderbilt tender-equipped medium Pacific. Seen here at the Washington, D.C. terminal, this engine is classed F-16 by the railroad and was built by the Baldwin Locomotive Works in 1913. Specifications include 27x 28in cylinders, 74 in driving wheels, 185 psi boiler pressure, 290,000 lbs. without tender, and produces 43,376 lbs tractive effort.

Pennsylvania Railroad K-4 4-6-2 (1914)

SPECIFICATIONS
Gauge: 4ft 8 1/2 in
Cylinders: (2) 27 x 28 in
Driving wheel: 80 in
Tractive effort: 44,460 lb
Steam pressure: 205 psi
Weight: 308,890 lb
Length: 83 ft 6 in

The Pennsylvania Railroad's K-4 class 4-6-2 "Pacific" (425 built 1914–1928, PRR Altoona, Baldwin) was their premier passenger-hauling steam locomotive from 1914 through the end of steam on the PRR in 1957. Nos 5400–5474 were built by Baldwin, while all others were constructed at the PRR's Juniata Shops numbered as shown below.

YEAR	QUANTITY	ROAD NUMBERS
1914	1	1737
1917	41	Assorted numbers
1918	111	3667-3684, 5334-5349 plus assorted numbers
1919	15	Assorted numbers for PRR Lines West
1920	50	3726-3775
1923	57	3800, 3801, 3805-3807, 3838-3889
1924	50	5350-5399
1927	92	5400-5491
1928	8	5492-5499

A number of K-4s locomotives had streamlining applied over the years, to varying degrees. All were later removed, restoring the locomotives to their original appearance. Locomotive #3768 was clad in a shroud designed by famed industrial designer Raymond Loewy in February 1936. This was a very concealing, enveloping streamlined casing which hid most of the functionality of the steam locomotive, leading to its nickname of "The Torpedo" by train crews. At first, the locomotive was not painted in standard Dark Green Locomotive Enamel (DGLE) but instead in a bronze color. It was later refinished in DGLE. A matching tender ran on unusual six-wheel trucks. Like most streamlined steam locomotives, the shrouds impeded maintenance and the covers over the wheels were later removed. For a time, the locomotive was the preferred engine for the Broadway Limited.

Right: The K-4 in streamlined guise.

Below: The original Baldwin works photo of the K-4 class.

Attempts were made to replace the K-4s, including the K-5 and the T1 duplex locomotive, but none was really successful, and the K-4s hauled the vast majority of express passenger trains until replaced by diesel locomotives. The K-4s was not powerful enough for the heavier trains it often pulled from the mid-1930s onward, so they were often double or even triple headed. This was effective, but expensive—several crews were needed. The PRR did have the extra locomotives, many having been displaced by electrification.

It is recognized as the State Steam Locomotive of Pennsylvania. On December 18, 1987, Pennsylvania Governor Robert P. Casey signed into law House Bill No1211 naming the PRR K-4 as the "official" state locomotive, according that title to both K-4 survivors, 1361 and 3750.

Below: No3750 is a firm favorite locomotive among railfans and engineers alike, the Railroad Museum of Pennsylvania's K-4 locomotive is a fine example of the culmination of reliable, high speed, passenger steam locomotives.

Climax Locomotive (1915)

SPECIFICATIONS
Gauge: 4ft 8 1/2 in
Cylinders: (2) 9 x 12 in
Driving Wheels: 28 in
Boiler pressure: 160 psi
Length: 26 ft 10 in
Weight: 34,000 lb

The Climax locomotive was designed for maximum adhesion in situations like logging camps and was used extensively by lumber companys.

Its design was originally patented by Rush S. Battles in 1891. This had horizontal cylinders connected to the drive shaft through a 2-speed transmission located under the center of the boiler. The drive shaft passed just above the axle centers, requiring the use of hypoid bevel gears to transfer power to each truck. All the gearing was open, exposed to the elements. Battles' patent describes the core design that became the Class B Climax, and his patent illustrations show the name Climax emblazoned on the locomotive cab.

All Climax locomotives were built by the Climax Manufacturing Company (later renamed to the Climax Locomotive Works), of Corry, Pennsylvania. In addition, an agency and service facility was established in Seattle, Washington to sell and maintain locomotives for West coast buyers. Production began in 1888 and the last Climax locomotive was produced in 1928. Between 1000 and 1100 were built.

Many loggers considered the Climax superior to the Shay in hauling capability and stability, particularly in a smaller locomotive, although the ride was characteristically rough for the crew

Approximately 20 Climax locomotives survive in North America, of which about five are operational. The Corry Historical Museum in Corry, Pennsylvania has a Climax on display in its own exhibit room inside the museum, with the locomotive sitting on a section of track. The museum is open from 14:00 – 16:00 on weekends from Memorial Day to Labor Day, and admission is free.

Two Climax locomotives are preserved in Canada, both at the BC Forest Discovery Centre in Duncan, British Columbia. Shawnigan Lake Lumber Co. No2 is a 25 ton Class B locomotive, and was built in 1910 as shop number 1057. Hillcrest Lumber Co. No9 was built to a larger, 50 ton Class B design in 1915, and is Climax shop number 1359.

Triplex 2-8-8-8-2 Erie Railroad (1915)

SPECIFICATIONS
Gauge: 4ft 8 1/2 in
Tractive effort: 160,000 lb
Axleload: 69,813 lb
Cylinders, HP: (2) 36 x 32in
Cylinders, LP: (4) 36 x 32in
Driving wheels: 63 in
Heating surface: 6,886 sq ft
Superheater: 1,584 sq ft
Steam pressure: 210 psi
Grate area: 90 sq ft
Fuel: 32,000 lb
Water: 10,000 gallons
Adhesive weight: 761,600 lb
Weight: 853,050 lb
Length: 105 ft 1 in

By 1905 the Mallet articulated was here to stay but there was still a need for greater tractive effort for helper service on particularly steep grades. Limited drawbar strength made very high tractive efforts irrelevant for pulling trains, but for pushing at the rear the use of really substantial power was possible. The Erie Railroad's "Triplex" or triple-Mallet articulated was a product of this thinking, intended to save locomotive expense in working heavy tonnage up Gulf Hill in the Susquehanna Division of the Erie Railroad. Baldwin offered the Erie a solution in the form of a compound Mallet with three engine units.

In fact, the locomotive was a 2-8-8-0 Mallet with an 0-8-2 steam-powered tender attached. The front and rear units were driven by four low-pressure cylinders, while the center unit had two high-pressure cylinders. All six cylinders had the same bore and stroke. The leading unit exhausted up the chimney in the normal way, while the rear engine had its own separate exhaust at the back of the tender. The exhaust steam passed through a feed-water heater beneath the tender en route.

Nominal tractive effort was substantially greater than that of Union Pacific's "Big Boy" and the weight on the drivers of these four locomotives (three were built for the Erie and one slightly different for the Virginian Railroad) was the greatest of any reciprocating steam locomotive type ever built, so that part of the objective was achieved. Alas, those responsible failed to match this with adequate steam generating capability. Since only low speeds were envisaged, the boiler was in principle big enough. However, only half the exhaust steam (and that at very low pressure) was available at the blastpipe and this seems to have been insufficient to produce enough smokebox vacuum to ensure good steaming for the locomotive.

On test, the first "Triplex" (named Matt. H, Shay after one of the Erie's senior engineers) created a world record by hauling solo a 250-wagon train weighing 17,600 tons over a line which was generally uphill and included 0.9 per cent (1-in-l 10) adverse gradients. The distance was 23 miles and the average speed was 13 1/2 mph. In service, though, there were difficulties and the Erie locomotives were withdrawn in 1925.

Below: 7776 remarkable Erie Railroad "Triplex" compound Mallet articulated 2-8-8-8-2 Matt. H. Shay, built by Baldwin in 1914. Three of these giants were built and were intended for helper service on the Erie's Gulf Hill grade. The center unit took high-pressure steam from the boiler, while the two outer units were the low pressure ones.

MacDermot 4-6-2 Overfair Railroad (1915)

SPECIFICATIONS
Gauge: 1 ft 7 in
Tractive effort: 3,700 lb
Axleload: 5,544 lb
Cylinders: (2) 8 x 9 in
Driving wheels: 26 in
Heating surface: 443 sq ft
Steam pressure: 200 psi
Grate area: 8 sq ft
Fuel: 1,000 lb
Water: 300 gallons
Adhesive weight: 15,120 lb
Weight: 24,080 lb
Length: 28 ft 1 in

The unlikely motivation for these delightful mini-locomotives was construction of the Panama Canal. To celebrate its completion in 1915, the Panama-Pacific-International Exposition was mounted at San Francisco. A wealthy and mechanically-minded young man, Louis MacDermot, secured a franchise to run an ambitious steam-operated line called the Overfair Railroad some 2 1/2 miles long covering the length of the exposition site.

MacDermot made all his own drawings and had built machine-and erecting-shops in the grounds of his family's home in Oakland, California. Here an 0-6-0T and four 4-6-2s were built. Sixty cars with seats for 1,000 passengers were built, and there was also a freight train to be used for advertising purposes. Three trains were intended, but in the event only two were required. Allowing for one 4-6-2 in reserve, three were sufficient and the fourth was never quite completed. The locomotives were perfect replicas of the classic passenger power of the day except that—quite incongruously—full-size air pumps were a rather too conspicuous feature of the left-hand side. The relatively massive coal-fired boilers produced saturated steam which was fed to conventional piston-valve cylinders. Walschaert valve gear was used, all the mechanism being one-third full size. It is a tribute to the elegant simplicity of the steam locomotive that there needed to be no compromise with correct functioning because of the reduction in size.

In December 1915, after 10 months of operation, the exposition and its railway closed leaving MacDermott financially embarrassed. MacDermot operated one of his locos in poor condition at Alameda Zoo in 1939.

MacDermot died in 1945, but all five of his superb locomotives still exist.They were stored on MacDermot's estate and would not run again until the 1980s. Al Smith had previously worked for the Southern Pacific Railroad and decided to purchase the locomotives from an auction because of his love of railroads. In 1979 he began laying the rails along Scotts Creek with the help of volunteers. Today the railroad is maintained and run by the Swanton Pacific Railroad Society which is connected with Cal Poly. It attracts many visitors, especially railfans, who volunteer to support its on-going operation.

Above: One of the three 19in gauge 4-6-2s built by Louis MacDermotin 1915 for the Panama-Pacific Exposition at San Francisco, Note how a seated engineer drives these locomotives, based on one-third of full-size practice

Right: The trains still run at the Swanton Pacific Railroad Company in California.

1914

1913

H Class Pennsylvania Railroad 2-8-0 (1916)

SPECIFICATIONS
Gauge: 4ft 8 1/2 in
Cylinders: (2) 26 x 28 in
Driving wheels: 62 in
Steam pressure: 205 psi
Tractive effort: 53,197 lb
Weight: 247,500 lb

No7688, an H-10 class built in 1915 by the Lima Locomotive Works for the Pennsylvania Railroad has been preserved by the Railroad Museum of Pennsylvania. It joins two earlier examples of the H class at the museum. No7688 was added to the National Register of Historic Places in 1979 as Consolidation Freight Locomotive No7688. The Pennsylvania Railroad's class H-8, H-9s and H-10s steam locomotives were of the 2-8-0 "Consolidation" type, the last three classes of such built by the railroad. The three classes differed only in cylinder diameter and thus tractive effort, each subsequent class increasing that measurement by an inch. The first H-8 was built in 1907 and the last H-10 in 1916; within a few years they were replaced on heavy freight assignments by 2-8-2s and 2-10-0. They became the railroad's standard light freight locomotive, replacing all other class H 2-8-0s, and a number remained in service until the end of PRR steam locomotive operation in 1957. 968 class H-8 of various subclasses were constructed, along with 274 class H-9s and 273 class H-10s. A number of H-8 locomotives were rebuilt to H-9 specification.

Class H-10s was built primarily for PRR Lines West, and featured a typical Lines West tender with sloping side coal boards at the top, to enable a bigger load of coal to be carried.

Some locomotives of this type were leased to the PRR-owned Long Island Rail Road, becoming the primary freight-hauling type on that system.

7688

CN No3254 S-1-b 2-8-2 (1917)

Canadian National 3254 is a preserved Canadian National class S-1-b 2-8-2 type steam locomotive. It is a part of the operating fleet at the Steamtown National Historic Site in Scranton, Pennsylvania.

No3254 was built by the Canadian Locomotive Company in Kingston, ON, in 1917. It started life as No2854 on the Canadian Government Railways but, when consolidated into the government owned Canadian National, it was renumbered .The locomotive weighs nearly 280,000 lbs, has 63" drivers and 27in x 30in cylinders. With a 56 1/2 sq ft grate, 235 sq ft firebox and total heating surface of 4,155 sq ft (including 804 sq ft superheating), it operated at a boiler pressure of 180 psi delivering 53,115 lbs tractive effort. The tender weighs 177,100 lbs light with a 9000 gallon and 14 ton coal capacity. No3254 was last shopped by Canadian National at its works in Allendale, ON, in February 1958. It appears to have retired soon after and been placed in storage.

SPECIFICATIONS
Gauge: 4ft 8 1/2 in
Cylinders: (2) 27 x 30 in
Driving wheels: 63 in
Steam pressure: 180 psi
Adhesive weight: 209,970 lb
Water capacity: 9000 gall
Tractive effort: 53,115 lb
Weight: 277,550 lb

In 1961, Willis F. Barron of Ashland, Pennsylvania, bought the locomotive intending to operate it on a Reading Railroad branch that served Ashland but, by the time he had moved No3254, the Reading had already abandoned and dismantled the branch.

In 1982, Barron sold No3254 to the Gettysburg & Northern Railroad, which operated track between Gettysburg and Mount Holly Springs, Pennsylvania.

Below: No3254 is seen here at the restored Delaware, Lackawanna & Western Railroad roundhouse. Built in 1902 as a forty-six stall, full circle roundhouse, it was modernised in 1937. Much of the structure was torn down in the 1950s and only three original sections of the roundhouse have survived.

No1630 2-10-0 Decapod Frisco (1918)

No1630 was built in 1918 by the Baldwin Locomotive Works for use in Russia as a class Ye locomotive. However, it, along with approximately 200 other locomotives, remained in the United States, due to the inability of the Bolshevik government to pay for them, following the Russian Revolution. No 1630 was converted from 5 ft Russian track gauge to 4 ft 8 1/2 in standard gauge. The St. Louis – San Francisco Railway used it as a mixed traffic engine. The locomotive was then sold to Eagle-Picher, who used it to haul lead ore from a mine to their smelter. In 1965, the locomotive was donated to the Illinois Railway Museum, in Union, Illinois, where it began operating in 1973. 1630 was taken out of service in 2004, and after more than six years undergoing repairs and a federally mandated rebuild, it was returned to operational condition on October 30, 2013. On Memorial Day weekend 2014, the locomotive returned to excursion service.

SPECIFICATIONS
Gauge: 4ft 8 1/2 in
Cylinders: (2) 24 x 28 in
Tractive effort: 47,453 lb
Steam pressure: 180 psi
Weight: 210,000 lb
Length: 71 ft 0 in

Class Q-3 2-8-2 (1918)

SPECIFICATIONS
Gauge: 4 ft 8 1/2 in
Locomotive Weight: 292,000 lbs
Driving wheels: 64 inches
Cylinders: (2) 26 x 30 in
Tractive effort: 53,800 lbs

During World War I, the federal government took control of the nation's railroads and formed the United States Railroad Administration (USRA) to efficiently mobilize troops and supplies. The USRA oversaw the mass production of standardized locomotives and operations of all privately owned railroads. Consisting of representatives from ALCO, Baldwin Locomotive Works, and Lima Locomotive Works, the USRA Locomotive Committee designed over 1,800 locomotives using the best of current technology. Although many railroads resented the USRA's control, the organization streamlined the railroad industry and made advances for railroad labor by increasing wages and decreasing the workday to eight hours. USRA control ended on March 1, 1920 but its durable locomotives continued to have a lasting influence on the railroad industry.

Constructed in just 20 days by Baldwin Locomotive Works, the B&O No4500 was the first USRA locomotive produced under federal management. The No4500 was equipped with the latest technology of its time, including a superheater and stoker. The weight of the versatile locomotive was considered "light" by most standards, yet it was quite powerful.

In the later years of its life, the No4500 operated on the B&O's Ohio, Newark, St. Louis, and Ohio River divisions. In 1957, the No4500 was renumbered as No300 to make room on the B&O roster for four-digit diesel locomotives. That same year, the No300 retired from service, and was sent to the Baltimore & Ohio Railroad Museum. There it was restored to its original number. In 1990, the No4500 became a National Historic Mechanical Engineering Landmark.

H-60 Class Nickel Plate Road (1918)

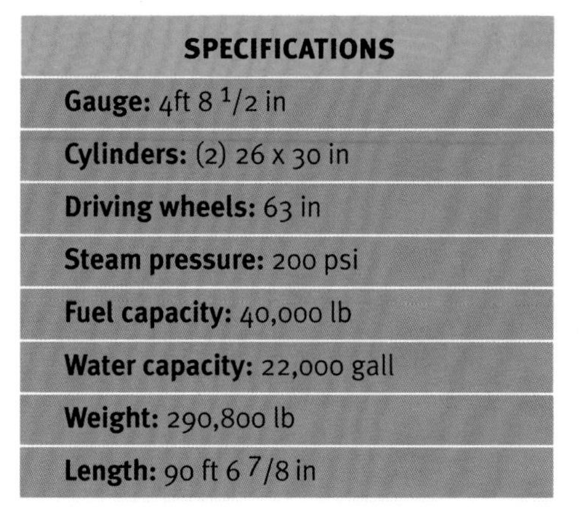

SPECIFICATIONS
Gauge: 4ft 8 1/2 in
Cylinders: (2) 26 x 30 in
Driving wheels: 63 in
Steam pressure: 200 psi
Fuel capacity: 40,000 lb
Water capacity: 22,000 gall
Weight: 290,800 lb
Length: 90 ft 6 7/8 in

Nickel Plate Road No587 is a USRA Light 2–8–2 steam locomotive built in September 1918 by the Baldwin Locomotive Works for the Lake Erie and Western Railroad as its number 5541. In 1923, the LE&W was merged into the New York, Chicago and St. Louis Railroad, ("Nickel Plate Road"), and allocated 587 as its new number in 1924. NKP No587 is generally referred to as a Baldwin locomotive. However, its supporting truck and cylinders are actually from another Lima Locomotive Works (LLW) engine when the original cylinders failed. No587 is the best remaining representation of the Mikado 2-8-2 locomotive style originally designed and built as part of the World War I rearmament program.

No587 served on the NKP railroad for 37 years on the route from Indianapolis to Michigan city. The locomotive remained relatively unchanged from its original design and operated until March 1955 when it was retired.

In September 1955, No587 was donated to the city of Indianapolis and put on display in Broad Ripple Park, Indianapolis, Indiana. Prior to being put on display, the locomotive's original tender was switched with another NKP steam engine No639, because the tender on 639 was in need of repair and 587's original tender was in good mechanical condition. No587 was originally equipped with the 16-ton, 10,000 gallon tender commonly used behind USRA 2-8-2s, but in the 1930s, it received a larger 16RA tender used on many NKP engines. This tender carried 19 tons of coal and 16,500 gallons of water. It is identifiable by having a six-wheel truck under the coal bunker and a four-wheel truck under the water cistern.

In 1934, Lima Locomotive Works delivered 25 22RA tenders to the NKP for Mikados. These tenders were nearly identical to those behind the Berkshires (2-8-4) built by Lima.

In 1955, 2-8-2 No639 was shopped with a 22RA tender on which the stoker was inoperable, and the railroad switched tenders to keep No639 running. No587 was displayed in Indianapolis's Broad Ripple Park with the larger 22RA tender in 1955. No639 was retired in 1957 and displayed in Bloomington, IL, with No587's 16RA tender

No587 remained in Broad Ripple Park until 1983 until it was determined that the locomotive was a good project for restoration. The locomotive was then loaned to the Indiana Transportation Museum.

From 1983 to September 1988 the Indiana Transportation Museum leased a work area at Amtrak's Beech Grove Shops. During restoration the museum was surprised to find that when the welds holding the fire box doors closed (for safety purposes) were removed there were still ashes in the ashpan. This indicated that the locomotive was simply pulled from active service and stored until being donated to the city of Indianapolis.No587 was operated by the Indiana Transportation Museum and is considered its crown jewel.

In 2003, the 587's Federal Railway Administration's (FRA) operating permit expired. This is due to FRA requirements that all boiler tubes and flues on steam locomotives be replaced every 15 years, or 1472 days of operation (whichever comes first).

Currently, No587 is undergoing its second restoration dependent on funding and available volunteer efforts.

No800 2-10-10-2 Class AE (VGN) (1918)

SPECIFICATIONS
Gauge: 4 ft 8 1/2 in
Tractive effort: 176,600 lb
Axleload: 61,700 lb
Cylinders HP: (2) 30 x 32 in
Cylinders, LP: (2) 48 x 32 in
Driving wheels: 56 in
Heating surface: 8,605 sq ft
Superheater: 2,120 sq ft
Steam pressure: 215 psi
Grate area: 108.7 sq ft
Fuel: 24,000 lb
Water: 13,000 gall
Adhesive weight: 617,000 lb
Total weight: 898,000 lb
Length overall: 99 ft 6 in

Until 1918, Mallet articulated steam locomotives with more than 16 driving wheels had been fairly conspicuous by their lack of success. The Virginian Railroad, which faced serious haulage problems in the Appalachian Mountains, had been persuaded by Baldwin to have a Triplex 2-8-8-8-4, similar to those which were a failure on the Erie. Although attempts were made to give the new locomotive a better steam-raising capacity so as to satisfy the vast appetite of the three sets of machinery, it was not long before she was divided in two. So with one new boiler the VGN got two new locomotives, a 2-8-8-0 and a 2-8-0, both of which gave good service, but not in the way intended.

Nevertheless, the problem of hauling VGN's immense coal drags down to tidewater at Norfolk, Virginia, which first involved climbing the notorious 2.11 per cent (1 -in-47) incline from the main collection point at Elmore, West Virginia, to Clark's Gap summit, remained unsolved. In 1917, with swollen wartime traffic round their necks, the management decided to have another go and ordered a batch of huge 2-10-10-2s from the American Locomotive Co. The dialog between builder and customer was no doubt made more meaningful by the traumatic experiences of the recent past.

Be that as it may, the results were excellent. Small tenders made the "800s" less impressive than they actually were but their vital statistics were huge. For

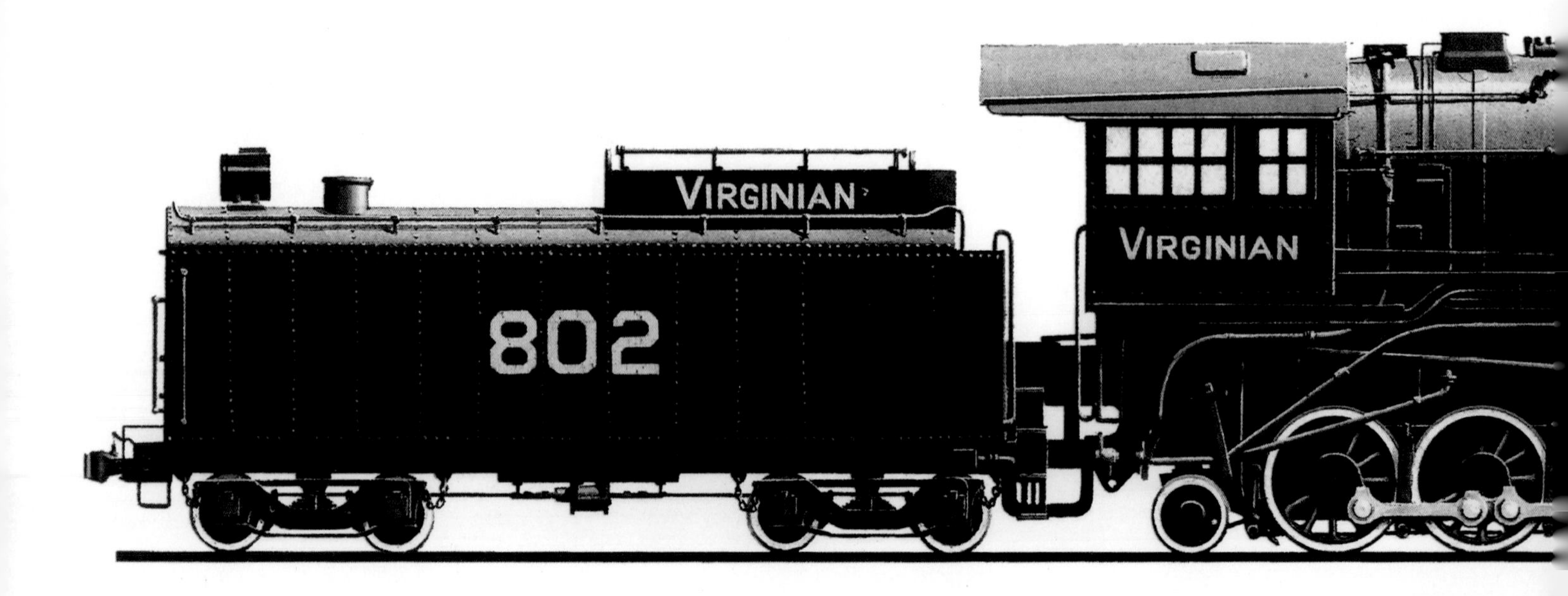

example, 4ft diameter low-pressure cylinders were the largest ever used on a locomotive. Their adhesive weight was 14 percent greater than that of a Union Pacific "Big Boy" and the tractive effort 40 per cent greater when live steam was admitted to the low-pressure engine at starting. Naturally, power output was much lower, as one might expect from a machine intended for low-speed operation, but these iron mammoths did all that was expected of them and gave 30 years of good service to their owners. When working on their intended task, it was customary to have a 2-8-8-2 at the head of a 5,500 ton train, well able to handle this load other than on the 2.11 percent (1-in-47) grade. Two 2-10-10-2s then pushed from the rear and the whole caravan moved upgrade noisily but steadily at some 5 1/2 mph. It must have been one of the greatest sights and sounds in railroading.

The 800s were interesting in that the high-pressure cylinders had conventional piston valves, but the huge, low pressure ones used old-fashioned slide valves. These were quite adequate for the lower temperatures involved on the L.P. side, while being easier to keep streamtight.

They are also thought to be the only successful class of locomotives in the world with as many as 20 driving wheels. Certainly the Mallet principle was never carried further than with these giants, although the size and power of articulated locomotives with few coupled wheels would in the end be even greater. It is perhaps true to say that the problem of Clark's Gap was ameliorated rather than solved by the "800s". A few years later a real solution was found when this hilly stretch of line was the subject of an electrification scheme. The 2-10-10-2s were then given useful but less heroic work to perform elsewhere on the system and survived until the 1940s. None of the Class AEs survived.

In the early hours of 1st April 1941, AE-class loco No800 was hauling an eastbound coal train out of Roanoke, Virginia. On approaching Stewartsville, 12 miles out, the boiler exploded killing the crew.

It is thought that the strainers on the hoses of the tender water outlets were partially blocked causing the explosion.

Below: What remains of the boiler of No800 following the fatal explosion at Stewartsville, April 1, 1941.

No24 2-6-2 (1919)

The Sandy River's last locomotive (and the fifth 2-6-2 on the line) was No24 delivered by the Baldwin Locomotive Works in 1919. The tender tanks on the Sandy River line were typically 84in wide (even that was rather wide for a 24in gauge) but the boss's careless handwriting led to Baldwin building one 8ft 4in wide. Not surprisingly, this overgenerous water cart overbalanced and had to have a slice taken out of its middle. But this tiny hunk of locomotive engineering was very soundly built—Walschaert valve gear, an unsuperheated boiler and slide valves making an excellent combination for ease of maintenance. "Piston valves wear out, slide valves wear in" is the saying, whilst a little extra fuel consumption would hardly be noticed on such a small machine.

Above: 2-6-2 No.24, the last locomotive supplied to the Sandy River & Rangeley Lakes RE.

The Prairie or 2-6-2 wheel arrangement was first tried in any quantity around the turn of the century. It bore the same relationship to the 2-6-0 Mogul type as the 4-4-2 Atlantic bore to the 4-4-0 American standard. You could say that it was a Mogul with a much larger firebox, and hence more power without additional tractive effort. So it was eminently suitable for lines in the "wide open spaces" of the mid-west, hence the name Prairie. But intense competition and a rapidly expanding economy meant a continuous search for increased productivity, and so the Prairie was quickly superseded by larger 4-6-2 Pacifics and 2-8-2 Mikados. So for main-line use the 2-6-2 was quickly eclipsed. It did however, find a niche on the short lines of the US railroad system.

SPECIFICATIONS
Gauge: 2 ft 0 in
Tractive effort: 10,085 lb
Axleload: 5,400 lb
Cylinders: (2) 12 x 16 in
Driving wheels: 33 in
Steam pressure: 170 psi
Fuel: 6,000 lb
Water: 2,000 gall
Adhesive weight: 42,000 lb
Weight: 91,000 lb
Length: 44 ft 7 in

The craze for narrow gauge railroads that swept the state of Maine at the same time resulted in construction of seven little narrow-gauge railroads, four of which were later consolidated into one system, which then became little only in respect of distance between the rails. This was the Sandy River & Rangely Lakes Railroad—a 46-mile main line with 60 miles of branches plus 16 locomotives and hundreds of freight cars.

No24 survived until the Sandy River line was abandoned in 1935. A railfan actually bought her for $250 but, alas, problems of storage and finance forced him two years later to let the engine go for scrap. But memories of No.24 linger on in model form, while the railroad itself surely sets a record in the number of pages of print published per mile of railroad abandoned.

E-1 Class No506 2-10-2 (1919)

SPECIFICATIONS
Gauge: 4 ft 8 1/2 in
Cylinders: (2) 27 x 32 in
Driving wheels: 57 in
Steam pressure: 200 psi
Tractive effort: 71,200 lbs
Weight: 352,000 lb

No506 is the first of ten E1 Class Santa Fe type (2-10-2) locomotives built for the Duluth Missabe & Northern by Alco in 1919- road numbered as 506-515). They were ordered following the successful introduction of six E Class 2-10-2s built by Baldwin for the railroad in 1916. E No502, one of that first order, is at the St. Louis Museum of Transportation Yard.

The E1s had a larger tender with a capacity for 10,000 gallons of water rather than 9,000.

Like the E, the tender on the E1 also came with a "dog house" for the brakeman These were fitted on some tenders until 1937, when the law mandated a separate seat for brakemen in the locomotive cab.

The driving wheels are slightly smaller than the E class, 57" on the E1 versus 60" on the E, a little more suited to heavy drag freight. The E1s cost $56,795 each and could haul trains of one hundred 50 ton ore cars.

Growing demands because of WWI necessitated the purchase of the ten additional Santa Fes. The locomotives joined the Duluth Missabe & Iron Range roster when the DM&N was absorbed by that railroad in 1938.The engine is now preserved at The National Railroad Museum of Green Bay, Wisconsin.

Left: E class No502 is one of six Santa Fe type (2-10-2) locomotives built by Baldwin in 1916 and designed for heavy freight on the Duluth, Missabe & Northern. To help it negotiate sharp curves, the center drivers are "blind," with no flanges.

Frisco No4003 2-8-2 (1919)

SPECIFICATIONS
Gauge: 4ft 8 1/2 in
Cylinders: (2) 26 x 30 in
Driving wheel: 63 in
Steam pressure: 200 psi
Tractive effort: 54,700 lb
Weight: 303,000 lb

Frisco No4003 is a 2-8-2, Mikado type, standard gauge steam railway locomotive built by the American Locomotive Company in 1919 as a standard USRA light Mikado for the Pennsylvania Railroad. The PRR, for unknown reasons, rejected 33 of 38 locomotives in the order. The USRA reassigned 23 of them (road numbers 4000-4007 and 4017-4031) to the St. Louis – San Francisco Railway, better known as the "Frisco." The Frisco also received 10 sisters from the Indiana Harbor Belt Railroad (road numbers 4008-4016 and 4032), making 33 in all.

Originally, the Frisco told the USRA that they were not interested in the locomotives, either. However, the railroad saw how good they were, and eventually decided to accept them. To prepare them for service, the Frisco modified them, fitting them with cast trailing trucks and boosters. Additionally, most of the Mikados that the Frisco received from the USRA, including No4003, had their cab roofs raised for additional headroom.

No4003 was built by ALCO in March 1919 in their Schenectady, New York, shops as USRA Engine #20008. It was ALCO construction number 60946, and cost $53,619 to build. After the locomotive had been rejected by the PRR, it was purchased by the Frisco in August 1919, and placed in service hauling freight trains between Fort Smith, Arkansas, and Monett, Missouri. No4003 was also used for occasional passenger service and would be one of only three ALCO-built Mikados in service on the Frisco.

The 2-8-2 Mikado design, like Engine No4003, was a good choice on the line between Fort Smith and Monett, which was one of the Frisco's major branch lines. Mikados were relatively large locomotives, which meant that they could haul general purpose freight trains of between 3,000 and 5,000 tons. It was a design that was also well-suited to hauling passenger trains on steep grades, something that would have been found on the Fort Smith-Monett line as it wound its way through the Boston Mountains.

The freight trains that No4003 hauled likely carried fruit and zinc among other products. Monett, at least during the first half of the twentieth century, was the shipping center of the berry-growing region in that part of Missouri. The fruit was brought to Monett by truck and wagon from the surrounding farms where it was then crated and loaded onto freight cars and shipped. By 1941, approximately 1,000,000 quarts of berries were shipped from the region. In addition, nearby Joplin was the center of a zinc mining region in Southwest Missouri, and some of the zinc was brought to Fort Smith where it was refined in two smelters that existed on the northeast side of town. The smelters' location on Midland Boulevard was located adjacent to the Frisco line as it entered Fort Smith from Van Buren.

By 1940, the Frisco, like many American railroads, began using diesel powered locomotives on their lines. Diesel locomotives are able to start a heavy train from a standstill more quickly than a steam locomotive can. Additionally, diesel locomotives are ready to work at any time, and spend much less time out of service for service and repairs than do steam locomotives. They can also travel greater distances without stopping for fuel. The many advantages of diesel

power would have been appealing to the Frisco, as they were to other railroads. The Frisco began an earnest effort in 1948 to switch to diesel locomotives, and, as a result, No4003 was retired by the Frisco in 1952. The last steam-powered train on the Frisco ran between Birmingham and Bessemer, Alabama, in February 1952.

Although it was retired in 1952, the Frisco kept No4003 until September 1954 when it was donated to the City of Fort Smith. The City of Fort Smith placed the locomotive in Kay Rodgers Park on Midland Boulevard where it remained until c. 2000. At that time, the City donated the locomotive to the Fort Smith TrolleyMuseum as long as the Museum paid for the cost of locomotive's move. Professional house movers were hired and the move from the Park to the Museum's grounds was completed c. 2000. No4003 currently resides at the Fort Smith Trolley Museum, and sits among other pieces of rolling stock on a spur off of the original Frisco rail line.

Frisco No4018 2-8-2 (1919)

No4018 is a class USRA Light 2-8-2 "Mikado" steam locomotive which operated for three decades hauling freight between Bessemer and Birmingham, Alabama on the St. Louis – San Francisco (Frisco)Railway. It went on display at the Alabama State Fairgrounds in 1952 and is one of only a few locomotives of its type that survive.

Called a "war baby" because it was part of the build-up of cargo capacity ordered through the United States Railroad Administration during World War I, Engine No4018 was constructed in October 1919 to a standardized USRA Light Mikado design by the Lima Locomotive Works of Lima, Ohio for the Pennsylvania Railroad. The engine is a 2-8-2 locomotive of the USRA standard Mikado type, inherited from Japan, and is coal-fired.

Engine No4018 was sold to the St. Louis and San Francisco Railway in 1923 and spent much of the next three decades carrying trains between Bessemer and Birmingham. The USRA Light 2-8-2 "Mikado" design itself was later improved upon by the Lima Locomotive Works evolving into the 2-8-4 Berkshire locomotive. Part of this evolution involved increasing the size of the Mikado's firebox. This larger fire box improved the engine's coal burning efficiency, however the additional weight facility adding the second wheel set to the trailing truck. Steam engines were phased out in favor of diesel locomotives in the mid-20th century and No4018 was, in fact, the last steam locomotive to operate on any part of the Frisco Railway and quite possibly the last to operate within the Birmingham metro area.

At the request of then-Birmingham mayor J. W. Morgan, the locomotive

SPECIFICATIONS
Gauge: 4ft 8 1/2 in
Cylinders: (2) 26 x 30 in
Driving wheel: 63 in
Steam pressure: 200 psi
Tractive effort: 54,700 lb
Weight: 303,000 lb

was saved from the scrap-yard and given a full cosmetic overhaul before making its final five-mile run to Birmingham Fairgrounds on February 29, 1952.

As the city of Birmingham planned a major redevelopment of the fairgrounds area in 2009, fund-raising began for moving the engine and tender to Sloss Furnaces National Historic Landmark. The move took place February 19–21, 2009. Since the move, a cosmetic restoration has been completed.

R-1 Class 2-10-2 (1919)

The Lehigh Valley Railroad needed more motive power to handle the surge during World War I and bought Seventy-six "Santa Fe" type locomotives from the Baldwin Locomotive Works. Forty were delivered in 1917 and the other thirty-six came in 1919. These locomotives were designated as Class R-1 and given road numbers 4000 through 4075. They had 63in diameter drivers, 29in x 32in cylinders, a 200 psi boiler pressure, exerted 72,620 pounds of tractive effort and each weighed 370,000 pounds. The firebox with its 100 square foot grate area was designed to burn a combination of anthracite and soft coal.

These locomotives were bought for slow-freight services between Manchester, New York and Sayre, Pennsylvania, a distance of 88 miles with a ruling grade of .4%. Each of these R-1s replaced two "Consolidations." In this service they were rated at 4,000 tons and 6 1/2 hours.

After the war, as traffic diminished and the need for fast freight returned, the 2-10-2s proved too slow for main line traffic and the LV sold eleven of them to the Hocking Valley RR in 1920 and then another five in 1923. They were renumbered 275-294. Numbers 275-294 were scrapped between 1939 and 1949. Numbers 4060-4075 were sold to the Hocking Valley in 1920 and became HV numbers 130-145. The HV numbers 130-139 became C&O numbers 2950-2959 as our example shown here.The LV rebuilt twenty of them to be 2-8-2s. These remaining forty, numbers 4000 through 4039 were used until the end of steam.

There are no surviving LV 2-10-2 "Santa Fe" type locomotives.

Right: Sante Fe type No2954 in the service of the C&O in 1948.

SPECIFICATIONS
Gauge: 4ft 8 1/2 in
Cylinders: (2) 29 x 32 in
Driving wheel: 63 in
Steam pressure: 200 psi
Tractive effort: 72,620 lb
Weight: 370,000 lb

4300 Class 4-8-2 (1923)

The 4-8-2 or "Mountain" type was appropriately named; its origins are a nice illustration of the difference between tractive effort and power. Locomotives with a high tractive effort are often described as powerful, but this is misleading. The 4-8-2 was developed from the 4-6-2 but, whilst the extra pair of drivers meant that a higher tractive effort could be exerted, the power output—which depends on the size of the fire—had to remain limited because there was still only one pair of wheels to carry the firebox. For climbing mountains a high tractive effort is essential, but high *power* output only desirable. These things were relevant to the Southern Pacific Railroad, for their trains leaving Sacramento for the east had the notorious climb over the Sierras to face, from near sea level to 6,885ft in 80 miles.

So in 1923 SP went to the American Locomotive Co. of Schenectady for the first batch of 4-8-2 locomotives. The design was based on standard US practice, the one feature of note being the cylindrical so-called Vanderbilt tender. A booster engine was fitted, driving on the rear carrying wheels, and this could give an extra 10,000lb of tractive effort, provided the steam supply held out.

SP impressed their personality on the "4300"s by having them oil-burning and by their trade mark, the headlight mounted below centre on the silver-grey front of the smoke box. The 77 engines of the class were very successful, all the later ones being built in SP's own shops at Sacramento. Some of the earlier batches had 8-wheel tenders of lower capacity, instead of 12-wheel ones. None of the class has been preserved.

Above: Southern Pacific 4-8-2 No.4348 at Roseville Yard, California in May 1954.

Left: SP No4360 ready to leave San Francisco for San Jose with a commuter train.

SPECIFICATIONS
Gauge: 4 ft 8 1/2 in
Tractive effort: 57,100 lb
Axle load: 61 ,500 lb
Cylinders: (2) 28 x 30 in
Driving wheels: 73 1/2 in
Heating surface: 4,552 sq ft
Superheater: 1,162 sq ft
Steam pressure: 210 psi
Grate area: 75 sq ft
Fuel (oil): 4,700 gall
Water: 13,300 gallons
Adhesive weight: 246,000 lb
Weight: 611,000 lb
Length: 97 ft 9 in

Class O-1A 2-8-2 (1923)

SPECIFICATIONS
Gauge: 4 ft 8 1/2 in
Tractive effort: 58,090 lb
Axleload: 64,280 lb
Cylinders: (2) 27 x 30 in
Driving wheel: 64 in
Heating surface: 4,178 sq ft
Superheater: 769 sq ft ($71m^2$).
Steam pressure: 200 psi
Grate area: 59 sq ft
Fuel: 38,000 lb
Water: 10,000 gall
Adhesive weight: 233,850 lb
Weight: 502,780 lb
Length: 80 ft 9 in

Seventy Class "o-1" 2-8-2s came from Baldwin in 1910 and 1911, ever-growing traffic having outclassed the six-coupled power previously depended upon. A further 133 were supplied during the war years and immediately after, also by Baldwin. The o1 -As were a modestly modified version, the differences being hardly greater than those caused by alterations made from time to time to members of the original class. The Burlington also had some larger but less numerous 2-8-2s of classes o-2, o-3 and o-4, the latter being the USRA heavy Mikado. There were both coal- and oil-burners amongst them.

Herewith the typical 20th Century steam locomotive of North America! Supplied by the builder who made more steam locomotives than anyone else, with a wheel arrangement more common than any other, for a railroad of average size and situated in the geographical center of the sub-continent, in the median year of steam construction during the 20th Century. So please welcome Chicago, Burlington & Quincy Class o1-A 2-8-2 built by Baldwin in 1923.

Because the Burlington covered so much wheat-growing country in Illinois, Iowa, Nebraska and Kansas, the 2-6-2s, which had suited the company for so long (and which had become the most numerous type on the system), were known the world over as "Prairies." However, by the 1920s the Mikados had ousted the Prairies from their premier position.

The o-1 s and o-1 As had no original features, but their extreme ordinariness was just what was needed to handle the everyday freight traffic of prairie towns, where the tracks were never out of sight of a grain elevator. In the end it was only the force majeure of dieselisation that led their owners to dispense with their services. Even so, several stayed on the roster into the 1960s, while o-1 A No4960 in particular became famous as a performer on railfan trips well on towards the end of the decade.

Refurbished No4960 made its first official run on the Grand Canyon line in 1996. The Grand Canyon Railway commemorates the anniversary of its rebirth every September with a special roundtrip run to the Grand Canyon using steam locomotive No4960 fueled by recycled vegetable oil. The steam locomotive also makes several eight-mile trips on special days throughout the year with a single class of service in the train's historic 1923 Harriman cars.

H-23 Class 4-6-2 (1923)

No2719 was built in May, 1923 by the American Locomotive Works (ALCO) in Schenectady, New York. It was one of six H-23 Pacific class 4-6-2 steam locomotives built for the Soo Line (the Minneapolis, St. Paul and Sault Ste. Marie Railway). No2719 operated until the mid-1950s when it was used to haul the Soo Line's last steam-powered train, a June 21, 1959, round-trip excursion between Minneapolis, Minnesota, and Ladysmith, Wisconsin. It was brought out of retirement to haul the last steam trains on Soo Line's tracks in 1959. It is estimated that No2719 traveled more than 3 million miles during its time on the Soo Line. It was then was overhauled and put into storage. No2719 was then given to the City of Eau Claire, Wisconsin to be displayed in Carson Park.

On December 17, 2006, the 2719 was moved to the Lake Superior Railroad Museum. The museum operates the North Shore Scenic Railroad and 2719. After extensive work during the summer of 2007, the engine was test fired on August 24, 2007, and made a successful round trip test run from Duluth to Two Harbors, Minnesota, on August 25, 2007. Soo Line 2719 ran a regular excursion schedule 2007 - 2013. In May 2013, it met Milwaukee Road No261 for the first time. Soo

SPECIFICATIONS
Gauge: 4ft 8 1/2 in
Cylinders: (2) 25 x 26 in
Driving wheels: 75 in
Steam Pressure: 200 psi
Heating surface: 3,172 sq ft
Superheater: 1260 sq ft
Tractive effort: 36,833 lb
Weight: 281,080 lb
Length: 82 ft 6 1/2 in

2719 pulled special excursions for that weekend (National Train Day).

No2719's FRA boiler flue time was to expire on July 31, 2013, but its flue time was extended so that it could operate into late summer of 2013. It pulled its final excursion on September 14, 2013, afterwards No2719 was drained and moved deep into the Lake Superior Railroad Museum for display, with hopes of restoring it back to operation in the future. In October 2014, No2719 was thoroughly power-washed before being moved indoors for display.

Above: Soo Line No2719 hauling an excursion train on the North Shore Scenic Railroad, September 2009

CNJ 113 0-6-0 (1924)

Currently based in Minersville,PA, this fully operational steam loco was built by Alco Schenectady in June 1924.The engine was renovated and was steamed up on November 19th, 2010. No113 is one of only two surviving Central New Jersey locomotives.

SPECIFICATIONS
Gauge: 4ft 8 1/2 in
Cylinders: (2) 23 x 26 in
Driving wheel: 51 in
Steam pressure: 200 psi
Weight: 194,000 lb

No487 K-36 Class 2-8-2 (1925)

SPECIFICATIONS
Gauge: 3 ft
Cylinders: (2) 20 x 24 in
Driving wheels: 44 in
Steam pressure: 195 psi
Tractive effort: 36,200 lb
Weight: 143,850 lb
Length: 68 ft 3/4 in

No487 Class K-36 is another of the ten Mikado type (2-8-2) locomotives Baldwin built for the Denver & Rio Grande Western in 1925 (road numbered 480-489) at a cost of $27,950 each. The K-36s were designed to haul freight trains but were occasionally used on passenger trains

The product of nearly fifty years experience of mountain operations, they were the last narrow gauge engines bought by the railroad and were part of a general upgrading of its narrow gauge lines in the 1920s. Originally assigned to the Marshall Pass line between Salida and Gunnison, CO, as well as to helper service from Chama to Cumbres, they were later assigned to the Third Division out of Alamosa. Equipped with special valves to allow brake control between locomotives while double-heading, they became the workhorses of the narrow gauge railroad. The K-36s also worked on the Farmington Branch when traffic boomed in the 1950s with the development of the oil industry in the San Juan Basin. In 1955, the Farmington freight office handled more business than any other station on the D&RGW system. No487 was last operated by the D&RGW on 27th October 1967. Soon after, the railroad abandoned most of its narrow gauge lines. The locomotive underwent an overhaul starting in 1973 and returned to service on the Cumbres & Toltec in 1974.

K-36 2-8-2 (1925)

SPECIFICATIONS
Gauge: 3 ft 0 in
Tractive effort: 36,200 lb
Axleload: 39,558 lb
Cylinders: (2) 20 x 24 in
Driving wheels: 44 in
Heating surface: 2,107 sq ft
Superheater: 575 sq ft
Steam pressure: 195 psi
Grate area: 40 sq ft
Fuel: 16,000 lb
Water: 5,000 gall
Adhesive weight: 143,850 lb
Weight: 286,500 lb
Length: 68 ft 6 in

The Denver and Rio Grande Western K-36 class are ten 3.0 ft narrow gauge, Mikado type, 2-8-2 steam locomotives built for the Denver and Rio Grande Western Railroad (DRGW) by Baldwin Locomotive Works. They were shipped to the Rio Grande in 1925, and were first used along the Monarch Branch and Marshall Pass, but were later sent to the Third Division out of Alamosa. Of the original ten, four are owned by the Durango and Silverton Narrow Gauge Railroad (D&SNG) and five by the Cumbres and Toltec Scenic Railroad (C&TS). Of the class No485 fell into the turntable pit at Salida and was scrapped in Pueblo in 1955, with many parts being saved.

The locomotives' name of K-36 comes from two different sources. The K in the name comes from the locomotives' wheel arrangement (Mikado), and the 36 stands for 36,200 pounds of tractive effort.

For a long time, oil finds near Farmington, New Mexico, generated enough freight traffic to keep the DRGW open. At the same time the prospects were not good enough to justify widening the gauge to standard or to dieselise. It was remarkable that this delicate balance between modernisation and abandonment was maintained for so long. For this we all must be grateful, because when the end of freight operation came in 1968, railroad preservation had become fashionable and the states of Colorado and New Mexico bought the best 64 miles to run as a museum-piece

Above: A Denver & Rio Grande Western K-36 2-8-2 No488 takes on water at Alamosa, Colorado.

Below: Denver & Rio Grande Western K-36 2-8-2 No.481. Note the large snowplough on this classic narrow-gauge loco.

481
481

tourist railroad. The preserved section, known as the Cumbres & Toltec Scenic Railroad, runs from Antonito, Colorado to Chama, New Mexico. With the mountain trackage came 124 cars of various kinds and, most important of all, nine narrow-gauge 2-8-2s of classes K-36 and K-37.

When the 10 much larger K-36s came from Baldwin Locomotive Works in 1925 they were entirely standard products of their day, apart from exceptional size and power for the slim gauge. The boilers and cylinders were the same as those of a typical 2-8-0 made for standard-gauge. With outside frames and cranks, the distance between the cylinder center lines would be the same for the two gauges. Seven of the 10 built have been preserved, one on the Silverton line. They are rated to haul 232 tons on a 4 per cent (1-in-25) grade.

When further similar locomotives were needed a year or two later, the Rio Grande produced the "K-37" class "in-house" from some surplus standard-gauge 2-8-0s, new frames and wheels being the only substantial pieces of hardware required. The old boilers, cabs, tender bodies, cylinders and smaller fittings have survived, could be re-used. Although three have survived, boiler inspectors have been a little chary of certifying their boilers, dating from as early as 1902, for the hard work expected of locomotives working on the magnificent route of the Cumbres & Toltec Scenic Railroad.

P-1 Class 4-6-4 (1925)

SPECIFICATIONS
Gauge: 4 ft 8 1/2 in
Tractive effort: 44,244 lb
Axleload: 72,009 lb
Cylinders: (2) 26 x 28 in
Driving wheels: 80 in
Heating surface: 4,225 sq ft
Superheater: 1,051 sq ft
Steam pressure: 220 psi
Grate area: 71 sq ft
Fuel: 35,840 lb
Water: 12,000 gall
Adhesive weight: 1,196,390 lb
Weight: 582,680 lb
Length overall: 87 ft 5 in

The Wabash Railroad Company had headquarters in St. Louis and its tracks extended to Chicago, Kansas City, Omaha, Toledo and Detroit. With a route length of 4,000 miles, the Wabash was one of those medium-size lines which led a charmed life in spite of serving an area far too well provided with railroads. Its survival depended on the personal touch—that little extra bit of devotion on the part of the staff which made Wabash freight or passenger service just that critical amount better than its competitors. The personal touch was all that was available for the Wabash had for long been unable to afford much in the way of new equipment. Until dieselization, the work was all done with a steam fleet which by the end of the war had no units less than 14 years old and included a high proportion built before World War I.

An urgent need for passenger locomotives in 1943 led to the rather drastic step of rebuilding six K-5 class 2-8-2s into the road's first and only 4-6-4s, Class P-1. A further P-1 was produced in 1947 from a K-4 2-8-2. This remarkable piece of locomotive surgery was done in the Wabash Shops at Decatur, Illinois.

The original 2-8-2 locomotives were built by Alco in 1925 and, in accordance with the fashion of the moment, had a three-cylinder arrangement. The rebuilds had only two cylinders, three instead of four main axles, much larger driving wheels, two-axle instead of single-axle rear trucks, and roller bearings instead of plain ones. The boilers would also no doubt have needed renewal after 20 years service, so not too much of the originals could have been used. One might wonder about the need for smallish, but fast new passenger engines in the middle of a war. The reason seems to lie in the fact that many Wabash passenger trains were light yet the only locomotives available to work them (apart from 40-year-old Class H 4-6-2s) were the large M-1 and 0-1 4-8-2s and 4-8-4s supplied in 1930-31. Even these had 70in drivers— rather small for passenger services. So a batch of modest-sized roller-bearing 4-6-4s with high driving wheels, produced by rebuilding some older 2-8-2 engines in need of work anyway, would release bigger locomotives for freight and troop movements. At the same time the War

Production Board's ban on new passenger locomotives would remain unbroken and the maintenance burden would be relieved by the up-to-date features of the rebuilds. Hence, benefits all round in a totally satisfactory way and the "Wabash Cannonball" would run to time.

The P-1s were semi-streamlined in a particularly handsome way, while the striking blue and white livery suited such excellent trains as the Blue Bird and Banner Blue between St. Louis and Chicago. However, an obsolescent fleet of steam locomotives, expensive to run and to maintain, showed up badly against the diesel alternative. So all too soon after production of the last P-1, the Wabash was to become an all-diesel line. Regrettably, none of these fine locomotives has been preserved.

Above: All Aboard the "Cannonball"! A blue-and-white 4-6-4 of the Wabash Railroad at St. Louis, Missouri, in 1946.

A Class Northern Pacific Railroad (1926)

SPECIFICATIONS
Gauge: 4 ft 8 $^{1}/2$ in
Tractive effort: 61,600 lb
Axle load: 65,000 lb
Cylinders: (2) 28 x 30 in
Driving wheels: 76 in
Heating surface: 4,660 sq ft
Superheater: 1,992 sq ft
Steam pressure: 225 psi
Grate area: 115 sq ft
Fuel: 48,000 lb
Water: 12,500 gallons
Adhesive weight: 260,000 lb
Weight: 739,000 lb
Length: 105 ft 4 in

Above: One of the 1938 build of A-3s, No.2664 eases a freight train past the small depot at Manitoba Junction, Minnesota.

In the late 1920s the need for a larger firebox caused a number of railroads to come to the same conclusion at around the same time. The genesis of the 4-8-4 lay in the imbalance between possible tractive effort and grate area of its predecessor the 4-8-2. The Northern Pacific Railroad had a special problem in that its local coal supplies—known rather oddly as Rosebud coal—had a specially high ash content; hence the need for a big firebox and a four-wheel instead of a two-wheel truck at the rear. By a photo-finish the Northern Pacific's class "A" 4-8-4 was the first and hence the type-name Northern was adopted.

The Canadian National Railway, whose first 4-8-4 appeared in 1927 made an unsuccessful play for the name Confederation. Delaware, Lackawanna & Western put forward Pocono for their version. Other early members of the 4-8-4 Club—eventually to be over 40 strong in North America alone—were the Atchison, Topeka & Santa Fe.

The firebox was indeed big, measuring 13 $^{1}/2$ x 8 $^{1}/2$ ft, exceeding that of any other line's 4-8-4s. Northern Pacific themselves found their first Northerns so satisfactory they never ordered another passenger locomotive with any other wheel arrangement, and indeed contented themselves with ordering modestly stretched and modernised versions of the originals —sub-classes A-2, A-3, A-4 and A-5—right up to their last order for steam in 1943.

The originals were twelve in number and came from the American Locomotive Co of Schenectady. Apart from those enormous grates they were very much the standard US locomotive of the day, with the rugged features evolved after nearly a century of locomotive building on a vast scale. A booster fitted to the trailing truck gave a further 11,400lb of tractive effort when required at low speeds.

Above: Biggest of the giants— A-5 class 4-8-4 No.2685 starts "The Alaskan" out of Minneapolis in June 1954.

The next 4-8-4 to operate on NP was another Alco product, built in 1930 to the order of the Timken Roller Bearing Co to demonstrate the advantages of having roller bearings on the axles of a steam locomotive. This "Four Aces" (No1111) locomotive worked on many railroads with some success as a salesman. The NP was particularly impressed—not only did they buy the engine in 1933 when its sales campaign was over but they also included Timken bearings in the specification when further orders for locomotives were placed. On NP No1111 was renumbered 2626 and designated A-l.

Baldwin of Philadelphia delivered the rest of the Northern fleet. The ten A-2s of 1934 (Nos 2650-59) had disc drivers and bath-tub tenders, and the eight A-3s of 1938 (Nos 2660-67) were almost identical. The final two batches of eight and ten respectively were also very similar; these were the A-4s of 1941 (Nos 2670-77) and the A-5s of 1943 (Nos 2680-89). These last two groups may be distinguished by their 14-wheel Centipede or 4-10-0 tenders of the type originally supplied for Union Pacific. This final batch is the subject of the art-work above. The amount of stretching that was done may be judged from the following particulars . . .

Tractive effort: 69,800 lb
Axle load: 74,000 lb
Driving wheels: 77 in
Steam pressure: 260 psi
Fuel: 54,000 lb
Water: 25,000 gall
Adhesive weight: 295,000 lb
Weight: 952,000 lb
Length: 112 ft 10 in

Other particulars are the same as the A class.

Northern Pacific had begun well by receiving a charter from President Abraham Lincoln in 1864 to build the first transcontinental line to serve the wide north-western territories of the USA. Through communication with the Pacific coast was established in 1883. By the time the 4-8-4s began to arrive it had established itself under the slogan "Main Street of the North West," and connected the twin cities of St. Paul and Minneapolis with both Seattle and Portland.

The flag train on this run was the North Coast Limited, and the 4-8-4s assigned to it, after taking over from Chicago Burlington & Quincy Railroad power at St. Paul, ran the 999 miles to Livingston, Montana, without change of engine. This is believed to be a world record as regards through engine runs with coal-fired locomotives. No doubt it was made possible by using normal coal in a firebox whose ash capacity was designed for the massive residues of Rosebud lignite.

PS-4 4-6-2 Southern Railway (1926)

Sixty-four Ps-4 Class were built in total, the first twenty-seven by Alco between 1923 and 1924. Allocated road numbers Nos1366-1392). Derived from the USRA "heavy" Pacific type, but with smaller driving wheels (73in instead of 79in), they were amongst the largest locomotives built for the Southern, only slightly smaller than its Ts-1 Class Mountain (4-8-2) types. The engine weighed 300,000 lbs with a 195,600 tender weight with a capacity of 16 tons of coal and 12,000 gallons of water. With 27in x 28in cylinders, a 70.4 sq ft grate, 314 sq ft firebox, 4,578 sq ft heating surface (including 860 sq ft superheating) and operating at a boiler pressure of 200 psi, the locomotive delivered 47,535 lbs tractive effort.

No1401 was one of the second batch of thirty-seven Ps-4s delivered in 1926 road numbered 1393-1409, 6476-6482 and 6688-6691), thirty-two from Alco and five from Baldwin. At 304,000 lbs, they were slightly heavier than the first batch and, with a 7.5 sq ft grate and total heating surface of 4,594 sq ft, including 905 sq ft superheating, had slightly more heating surface, although this made no appreciable difference to the 47,535 lbs tractive effort. They were also mated with larger tenders weighing 261,600 lbs with a capacity of 16 tons of coal and 14,000 gallons of water.

The history of the Southern Railway's Pacifies began in World War I, when the United States Railroad Administration, which had taken over the railroads for the duration, set out to design a standard set of steam locomotives to cover all types of traffic. One of these was the so-called USRA "heavy" 4-6-2. Based on this design, the American Locomotive Company built the first batch of 36 Class "Ps-4" 4-6-2s in 1923.

In 1925 President (of Southern Railway) Fairfax Harrison, visited his line's namesake in England and was impressed with its green engines. He determined that his next batch of 4-6-2s would make an equal if not better showing. He naturally chose a style very similar to the English Southern

Region except that a much brighter green was used together with gold—the small extra cost paid off quickly in publicity. Colored locomotives were then quite exceptional in North America. A little later the earlier batch of locomotives appeared in green and gold also.

The 1926 batch of 23 locomotives had the enormous 12-wheel tenders illustrated here, in place of the USRA standard 8-wheel tenders on the earlier engines, and a different and much more obvious type (the Elesco) of feed water heater involving the large transverse cylindrical vessel just in front of the smokestack. Some locomotives from each batch had the Walschaert's gear, others had Baker's. A final batch of 5 came from Baldwin in 1928. These had Walschaert's valve gear and 8-wheel tenders of large capacity. All were fitted with mechanical stokers.

Of the 64 locomotives built, 44 were allocated to the Southern Railway proper, 12 to subsidiary Cincinnati, New Orleans & Texas Pacific and 8 to the Alabama Great Southern, although "Southern" appeared on the tenders of all.

The Ps-4 class was the last steam passenger locomotive type built for the Southern and they remained in top-line express work until displaced by diesels in the 1940s and 1950s. No1401 is preserved and is superbly displayed in the Smithsonian Museum, Washington, D.C.

SPECIFICATIONS
Gauge: 4 ft 8 1/2 in
Tractive effort: 47,535 lb
Axle load: 61,000 lb
Cylinders: (2) 27 x 28 in
Driving wheels: 73 in
Heating surface: 3,689 sq ft
Superheater: 993 sq ft
Steam pressure: 200 psi
Grate area: 70.5 sq ft
Fuel: 32,000 lb
Water: 14,000 gall
Adhesive weight: 182,000 lb
Weight: 562,000 lb
Length: 91 ft 1 1/2 in

UP 9000 4-12-2 (1926)

The crossing of the Continental Divide by the Union Pacific Railroad was a major challenge for this iconic railroad. As traffic increased in weight it called for outstanding locomotive power to work over the grades involved. In the 1920s, traffic was being handled by 2-10-2S and 2-8-8-0 compound Mallets. The latter had adequate tractive effort but speeds above 25mph were not then possible with the Mallet arrangement, while the 2-10-2s had limited adhesion.

Above: Ready for its next attempt at the formidable Sherman Hill (40 miles of l-m-66), Class 9000 No9013 waits at Cheyenne in May 1953.

The idea of a 12-coupled engine was made possible by using the then new lateral-motion device developed by the American Locomotive Company. This arrangement enabled a long-wheelbase locomotive to negotiate sharp curves. Very few 12-coupled classes of steam locomotive then existed. In fact the first 4-12-2 when it appeared from Alco in 1926 was well over double the weight of its nearest 12-coupled rival. The three-cylinder arrangement was adopted to spread the thrust of the pistons which it was thought would unbalance the engine. Tests indicated that No.9000 could take the same higher speeds and with much lower coal consumption. Production was put in hand at once and by 1930 the world's first and last class of 4-12-2s totalled 88.

Some of the 9000s more unusual features did cause some trouble. Much was hoped for from the conjugated motion for the center-cylinder which, being rugged, simple and accessible, had apparently all the attributes. But, as also was found in Britain (its country of origin), the motion needed careful maintenance. There were two reasons—first, wear of any of the pins or bearings led to over-travel of the valve, and this in turn led to the middle cylinder doing more than its share of the work, often with dire results. The situation was particularly severe on the center big-end bearing. A few engines were provided with separate third sets of Walschaerts gear set between the frames to operate the valves of the inside cylinder. This was driven from a second return crank mounted on the crankpin of the fourth axle on the right-hand side of the locomotive. In order to encourage that 30ft 5in of fixed wheel-base to perform as a contortionist, the leading and trailing coupled wheels were allowed 1in sideplay either side of the center line. The first locomotive originally had the center (third) pair of driving wheels flangeless, but later examples had thin flanges. The sideplay was controlled against spring pressure, but no other devices such as spherical joints in the side rods were provided. The arrangements were entirely successful, so much so that the class as a whole would see steam out

SPECIFICATIONS
Gauge: 4 ft 8 1/2 in
Tractive effort: 96,650 lb
Axleload: 60,000 lb
Cylinders: (3) 2 x 27 x 32 in 1 x 27 x 51 in
Driving wheels: 67 in
Heating surface: 5,853 sq ft
Superheater: 2,560 sq ft
Steam pressure: 220 psi
Grate area: 108 sq
Fuel: 42,000 lb
Water: 15,000 gall
Adhesive weight: 355,000 lb
Weight: 782,000 lb
Length: 102 ft 7 in

a quarter of a century later, although they were soon to be displaced from prime assignments and to other parts of the system by some greatly improved Mallet-type articulateds—the "Challengers"—introduced a few years later.

Only one of these unique locomotives is preserved; this is No9004 at the Transportation Museum at Los Angeles.

Right: Front-end clutter adds to the impressive proportions of the prototype CJass 9000, built by A/co in 1926.

A-6 Class 4-4-2 (1927)

SPECIFICATIONS
Gauge: 4ft 8 1/2 in
Tractive effort: 41,360 lb
Axleload: 33,000 lb
Cylinders: (2) 22 x 28 in
Driving wheels: 81 in
Steam pressure: 210 psi
Grate area: 49.5 sq ft
Fuel (oil): 2,940 US gall
Water: 9,000 gall
Adhesive weight: 62,000 lb
Weight: 465,900 lb
Length: 78 ft 0 1/2 in

No3025 is an Atlantic type (4-4-2), one of thirteen originally built for the Southern Pacific by Alco in 1904 and classed A-3.The locomotive was converted to A-6 class in 1927 making it the last word as regards 4-4-2s on Southern Pacific.

The large 81in drivers were designed for speed, and it could reach speeds topping 100 mph. The locomotive hauled several crack passenger services in California, including the Daylight, the Starlight and the Lark.

An oil burner, with 20in x 28in cylinders, No3025 operated at a boiler pressure of 210 psi, delivering 24,680 lbs tractive effort. It still has its inside Stephenson link motion and was the first standard gauge locomotive to go on display at Travel Town Museum in Burbank in 1952 after being donated to the museum by the Southern Pacific.

The A-6s were produced in SP's own shops at Sacramento and Los Angeles by rebuilding four of the 51-strong A-3 class locomotives built by Alco and Baldwin between 1904 and 1908. They took the road numbers No.3000-3003 of older 4-4-2s of Class A-l which had started life as Vauclain compounds and had by then been scrapped. Seven other members of the A-3 class were given similar treatment, but were not reclassified. Being one of this latter group , No3025, is the sole survivor of the 4-4-2s.

J-3a Class 4-6-4 Hudson (1927)

Some locomotive wheel arrangements had a particular association with one railway; such was the 4-6-4 and the New York Central. Subsequent designs of 4-6-4s took over the type-name Hudson applied to these engines by the New York Central. In 1926 the Central built its last Pacific, of Class K-5b, and the road's design staff, under the direction of Paul W Kiefer, Chief Engineer of Motive Power, began to plan a larger engine to meet future requirements. The main requirements were an increase in starting tractive effort, greater cylinder power at higher speeds, and weight distribution and balancing which would impose lower impact loads on the track than did the existing Pacifics. Clearly this would involve a larger firebox, and to meet the axle loading requirement the logical step was to use a four-wheeled truck under the cab, as was advocated by the Lima Locomotive Works, which had championed engines with large fireboxes over trailing bogies under the trade name of Super Power. As the required tractive effort could be transmitted through three driving axles, the wheel arrangement came out as 4-6-4. Despite the Lima influence in the design, it was the American Locomotive Company of Schenectady which received the order for the first locomotive, although Lima did receive an order for ten of them some years later.

Classified J-la and numbered 5200, the new engine was handed over to the

SPECIFICATIONS
Gauge: 4 ft 8 1/2 in
Tractive effort: 41,860 lb
Axleload: 67,500 lb
Cylinders: (2) 22 1/2 x 29 in
Driving wheels: 79 in
Steam pressure: 265 psi
Heating surface: 4,187 sq ft
Superheater: 1,745 sq ft
Grate area: 82 sq ft
Fuel: 92,000 lb
Water: 18,000 gall
Adhesive weight: 210,500 lb
Weight: 780,000 lb
Length: 106 ft 1 in

owners on 14 February 1927. By a narrow margin it was the first 4-6-4 in the United States, but others were already on the production line at Alco for other roads. Compared with the K-5b it showed an increase in grate area from 67.8sq ft to 81.5sq ft and the maximum diameter of the boiler was increased from 84in to 87in The cylinder and driving wheel sizes were unchanged, so the tractive effort went up on proportion to the increase in boiler pressure from 200psi to 225psi. The addition of an extra axle enabled the total weight on the coupled axles to be reduced from 185,000lb to 182,000lb despite an increase in the total engine weight of 41,000lb.Improved balancing reduced the impact loading on the rails compared with the Pacific.

The engine had a striking appearance, the rear bogie giving it a more balanced rear end than a Pacific, with its single axle under a large firebox. At the front the air compressors and boiler feed pump were housed under distinctive curved casings at either side of the base of the smokebox, with diagonal bracing bars. The boiler mountings ahead of the cab were clothed in an unusual curved casing.

No5200 soon showed its paces, and further orders followed, mostly for the NYC itself, but 80 of them allocated to three of the wholly-owned subsidiaries, whose engines were numbered and lettered separately. The latter included 30 engines for the Boston and Albany, which, in deference to the heavier gradients on that line, had driving wheels three inches smaller thar the remainder, a rather academic difference. The B&A engines were classified J-2a, J-2b and J-2c, the suffixes denoting minor differences in successive batches. The main NYC series of 145 engines were numbered consecutively from 5200, and here again successive modifications produced sub-classes J-la to J-le. Amongst detail changes were the substitution of Baker's for Walschaert's valve gear; the Baker's gear has no sliding parts, and was found to require less maintenance. There were also changes in the valve setting.

From their first entry into service the Hudsons established a reputation for heavy haulage at high speeds. Their maximum drawbar horsepower was 38 percent more than that of the Pacifies, and they attained this at a higher speed. They could haul 18 cars weighing 1,400 tons at an average speed of 55mph on the generally level sections. One engine worked a 21-car train of 1,650 tons over the 639 miles from Windsor (Ontario) to Harmon, covering one section of 71 miles at an average speed of 62.5mph.

The last of the J-1 and J-2 series were built in 1932, and there was then a pause in construction, although the design staff were already planning for an increase in power. In 1937 orders were placed for 50 more Hudsons, incorporating certain improvements and classified J-3. At the time of the

introduction of the first Hudson, the NYC, were chary of combustion chambers in fireboxes because of constructional and maintenance problems, but by 1937 further experience had been gained, and the J-3 incorporated a combustion chamber 43 long. Other changes included a tapering of the boiler barrel to give a greater diameter at the front of the firebox, raising of the boiler pressure from 225 psi to 275psi (later reduced to 265psi), and a change in the cylinder size from 25 x 28in to 22 1/2 x 29in The most conspicuous change was the use of disc driving wheels, half the engines having Boxpok wheels with oval openings, and the other half the Scullin type with circular openings. The final ten engines were clothed in a streamlined casing designed by Henry Dreyfus. Of all the streamlined casings so far applied to American locomotives, this was the first to exploit the natural shape of the locomotive rather than to conceal it, and the working parts were left exposed. Many observers considered these to be the most handsome of all streamlined locomotives, especially when hauling a train in matching livery. Prior to the building of the streamlined J-3s, a J1 had been clothed in a casing devised at the Case School of Science in Cleveland, but it was much less attractive than Dreyfus' design, and the engine was rebuilt like the J-3s; while two further J-3s were given Dreyfus casings for special duties.

The J-3s soon showed an improvement over the J-1 s both in power output and in efficiency. At 65mph they developed 20 per cent more power than a J-1. They could haul 1,246 ton trains over the 147 miles from Albany to Syracuse at scheduled speeds of 59mph and could reach 60mph with a 1,840 ton train. The most celebrated train of the NYC was the renown 20th Century Limited. At the time of the building of the first Hudsons this train was allowed 20 hours from New York to Chicago. This was cut to 18 hours in 1932 on the introduction of the J-le series, and in 1936 there was a further cut to 16 1/2 hours. Aided by the elimination of some severe service delays, and with J-3 power, the schedule came down to 16 hours in 1938, which gave an end-to-end speed of 59.9mph with 990 tons trains, and with seven intermediate stops totalling 26 minutes. On a run with a J-3 on the Century, with 1,036 tons, the 133 miles from Toledo to Elkhart were covered in a net time of 112 1/2 minutes, and the succeeding 93.9 miles from Elkhart to Englewood in 79 1/2 minutes, both giving averages of 70.9mph. A speed of 85.3mph was maintained for 31 miles with a maximum of 94mph The engines worked through from Harmon to Toledo or Chicago, 693 and 925 miles respectively. For this purpose huge tenders were built carrying 45 tons of coal, but as the NYC used water troughs to replenish the tanks on the move, the water capacity was by comparison modest at 18,000 gallons.

Eventually the engines allocated to the subsidiaries were brought into the main series of numbers, and with the removal of the streamlined casings in post-war years, the NYC had 275 engines of similar appearance numbered from 5200 to 5474. It was the largest fleet of 4-6-4 locomotives on any railway, and constituted 63 per cent of the total engines of that wheel arrangement in the United States.

Although the Hudson had their share of troubles, they were generally reliable, and the J-3s ran 185,000 to 200,000 miles between heavy repairs, at an annual rate of about 110,000 miles.

After World War II the Niagara 4-8-4s displaced the Hudson from the heaviest workings, but as that class numbered only 25 engines, the Hudsons still worked many of the 150 trains daily on the NYC booked at more than 60mph start-to-stop. Despite rapid dieselization the engines lasted until 1953-6, apart from an accident casualty.

Class 3450 4-6-4 AT&SF (1927)

SPECIFICATIONS
Gauge: 4 ft 8 1/2 in
Tractive effort: 43,223 lb
Axleload: 77,510 lb
Cylinders: (2) 24 x 28 in
Driving wheels: 74 in
Steam pressure: 220 psi
Heating surface: 5,088 sq ft
Superheater: 2,080 sq ft
Grate area: 98.5 sq ft
Fuel (oil): 7,000 gall
Water: 21,000 gall
Adhesive weight: 198,300 lb
Weight: 343,500 lb
Length: 100 ft 3 in

No3450 is the one of ten Hudson type (4-6-4) locomotives built for the AT&SF by the Baldwin Locomotive Works in 1927, road numbered 3450-3459. They were the first of the type bought by the Santa Fe, who ordered six more in 1937 (Nos3460-3465) completing its roster of sixteen. Only two have survived: as well as No3450, No3463 is undergoing restoration by Sustainable Rail International.

Some of the Class 3450s were later assigned to passenger and freight service on the Santa Fe's Valley Division, working the 270 miles up the San Joaquin Valley from Bakersfield through Fresno to Oakland, CA.

Retirements started in the early 1950s and all of the 3450s were gone by 1956 after running up some impressive mileage. For example when it was retired in 1953, No3450 had accumulated more than 2.4 million miles in service. It then went into storage until it was donated to the Southern California Chapter Railway and Locomotive History Society by the Santa Fe in 1955.

As built, No3450 weighed 343,500 lbs, 198,300 lbs on its 74" drivers. With 25 x 28in cylinders, an 88 sq ft grate, 268 sq ft firebox, and total heating surface of 5,088 sq ft, including superheating, it operated at a boiler pressure of 220 psi delivering 44,223

Left: A view of No3450's modified 79 inch driving wheels. These were added when the class was rebuilt in 1937.No3450 is preserved at the Society's museum in the Los Angeles County Fairgrounds at Pomona, California. It is not in operational condition but is preserved in good condition as a static exhibit.

lbs tractive effort.

The class was typical of its day and included coal-burning as well as oil-burning examples, they demonstrated very clearly that this great railway had finally thrown aside the thought of anything with a hinge in the middle or more than two cylinders. Not that big power was needed for much of Santa Fe's work; there are hundreds of miles of continuous level or near-level track between Chicago and Los Angeles as well as the more famous sections such as the ascents to the Cajon and Raton passes.

But it was considered that higher speeds could be run, and so in 1937 the class went into the shops for a rebuild.

The 3450s were substantially modified, cutting thirty-four tubes out of the boiler and adding a 28in combustion chamber, as well as 108 sq ft of thermic syphons to the firebox. This increased the firebox heating surface by over 25% to 338 sq ft.

Larger 79" disc type drivers were also fitted to the 3450s. The Walschaert valve gear remained unchanged.

The rebuilt locomotives weighed 352,000 lbs, 206,000 lbs on their drivers. The changes reduced the total heating surface to 4,349 sq ft, including 952 sq ft superheating. Tractive effort dropped slightly to 43,307 lbs, but the larger firebox much improved steaming capacity, higher boiler pressure of 230 psi and the new 79in drivers made for a faster locomotive, capable of speeds in excess of 100 mph. The modified tender weighing 396,246 lbs light, was a relatively late addition. It is one of the type built for the 3460 class and was mated to No3450 in 1952. It has a 20,000 gallon water and 7,000 gallon oil capacity. Both of the three axle trucks were equipped with Timken roller bearings. The rear axle has SKF bearings.

This was the period when Santa Fe's first streamline train—the now legendary "Super Chief"—was coming into service, and the idea of running other trains at faster speeds was important. Hence additions to the 3400s were 4-6-4s rather than 4-6-2s and they had relatively enormous driving wheels, 19 per cent more tractive effort than the 4-6-2s and considerably bigger grates. Larger tenders were also attached both to the new 4-6-4s as well as to some of the rebuilt 4-6-2s, and this helped by reducing time spent taking on fuel and water.

The first of the new locomotives was streamlined and became the only Santa Fe steam loco to be so treated. This was No3460, known colloquially as the "Blue Goose." The others well matched the standard heavyweight equipment used on such trains as the "Grand Canyon" and the "Santa Fe Chief," and could roll them at steamliner speeds over the long level miles of the midwest.

King George V King Class 4-6-0 (1927)

Great Western Railway (GWR) 6000 Class King George V is a preserved British steam locomotive.

No6000 was the first of the "King" Class, and was built in June 1927. It was shipped to the United States in August 1927 to feature in the Baltimore and Ohio Railroad's centenary celebrations. During the celebrations it was presented with a bell and a plaque, and these are carried to this day. This led to it being affectionately known as "The Bell."

The bell carries the inscription:

Presented to
Locomotive King George V
by the
Baltimore and Ohio Railroad Company
in commemoration of its
centenary celebration
September 24 - October 15, 1927

SPECIFICATIONS
Gauge: 4 ft 8 1/2 in
Cylinders: (4) 16 1/4 x 28 in
Driving wheels: 78 in
Steam pressure: 250 psi
Tractive effort: 40,300 lb
Weight: 199,400 lb
Length: 68 ft 2 in

No6000 was designed for express passenger work. With the exception of one Pacific (The Great Bear), they were the largest locomotives the GWR built. The Class was named after kings of England, beginning with the reigning monarch, King George V, and going back through history. Following the death of King George V, the highest-numbered engine was renamed after his successor; and following the abdication of the latter, the next-highest engine was also renamed after the new King.

After returning from the US it was allocated to Old Oak Common locomotive depot, in London but was moved to Bristol in 1950. It was subsequently allocated to Old Oak Common again in 1959, to be withdrawn by the Western Region of British Railways in December 1962 after covering 1,910,424 miles.The locomotive was officially preserved; being restored to main line running order and based at Bulmer's Railway Centre in Hereford and in 1971 became the very first steam engine to break the mainline steam ban that had been in place since 1969. Its restoration to main line service and subsequent operation is often credited with opening the door for the return of steam to the main lines of the UK. After years of running a costly overhaul was declined since classmate King Edward I had been restored for mainline operation. Subsequently King Edward II has been returned to working order after a lengthy restoration reducing any chance of No6000 being restored. The running engines have had their original chimneys, cabs and safety valve bonnets cut down to allow running on the modern railway, with its much deeper ballast and consequently reduced clearance under bridges. No6000 is now displayed in its original condition with full-height fittings at the National Railway Museum in York.

Above: Great Western Railway (GWR) 6000 Class No6024 King Edward I see here at Didcot Oxfordshire UK in 2011 is a preserved steam locomotive. No6024 ran from 1930 to 1962 for the Great Western Railway and latterly British Railways hauling express passenger services.

No3751 4-8-4 Northern (1927)

SPECIFICATIONS
Gauge: 4ft 8 1/2 in
Cylinders: (2) 30 x 30 in
Steam pressure: 210 psi
Driving wheels: originally 73 in later 80 in
Tractive effort: 66,021 lb
Weight: 423,000 lb

Santa Fe No3751 is a 4-8-4 Northern class steam locomotive that was originally owned and operated by the Atchison, Topeka and Santa Fe Railway. Built in 1927 by the Baldwin Locomotive Works, No3751 was Baldwin's and the Santa Fe railway's first 4-8-4. It had a 5 chime freight whistle mounted on it. Tests showed that No3751 was 20% more efficient and powerful than Santa Fe's 4-8-2 No3700 class, which at the point was Santa Fe's most advanced steam locomotive. In 1936, the engine was converted to burn oil. Two years later, the locomotive was given a larger tender able to hold 20,000 gallons of water and 7,107 gallons of fuel oil. No3751 was also present at the grand opening of Union Station in Los Angeles on May 7, 1939 pulling the Scout, one of Santa Fe's crack passenger trains as it arrived from Chicago. It was the first steam locomotive to bring a passenger train into Los Angeles Union Pacific Terminal.

In 1941, along with other 4-8-4s, No3751 received major upgrades including: 80 inch driving wheels, a new frame, and Timken roller bearings on all axles. That same year, it achieved its highest recorded speed at 103 mph. It continued to be a very reliable working locomotive until 1953, when it pulled the last regularly scheduled steam powered passenger train on the Santa Fe to run between Los Angeles and San Diego on August 25, this was its last run in revenue service. After that, it was stored at the Redondo Junction, California roundhouse in Los Angeles for four years before it was officially retired from the roster by the railroad in 1957, and in 1958 it was placed on display in San Bernardino. Today it is located in the Central City East neighborhood of Los Angeles, California and is listed on the National Register of Historic Places.

Right: Acheson, Topeka and Santa Fe No3751 leads an employee special westbound through Streator, Illinois, in 1992.

Above: No3751 steamed into San Diego on Sunday, June 1, 2008, its first visit since it hauled the last steam powered passenger train out of San Diego in 1953. Since then the old 1915 vintage San Diego station had not seen a steam powered train of any kind until the triumphant return of the first 4-8-4 built for the Santa Fe in 1926 as the first "Northern" class engine on the railroad.

3751

3751

G-1 Class 4-6-2 (1928)

SPECIFICATIONS
Gauge: 4ft 8 1/2 in
Driving wheels: 69 in
Boiler pressure: 210 psi
Cylinders: (2) 22 x 28 in
Tractive effort: 35,156 lb

G-1 Class locomotive was built for the Gulf, Mobile and Northern by the Baldwin Locomotive Works of Eddystone, Pennsylvania in 1928 as the first of two Pacifics ordered. The engine later became Gulf, Mobile and Ohio 580. It was retired in 1950.

It passed through various hands and different steam preservation railroads until in 1983, it was sold to Andrew J. Muller, Jr. to power tourist trains on the newly formed Blue Mountain and Reading Railroad based out of Temple, PA. Numbered as 425,the locomotive made many runs on the 26 miles branchline, as well as a few trips on the mainline. No425 was later joined by a second purchase a T-1 class No2102 in 1987. The Blue Mountain & Reading became much larger with the purchase of nearly 300 miles of former Conrail track throughout the early 1990s. The railroad was renamed to Reading, Blue Mountain and Northern (often shortened to Reading & Northern) merging in 1995. Having more track gave 425 and 2102 much greater scope to roam, and the engines became based out of the railroad's own headquarters of Port Clinton, Ohio.

On June 12, 1992 the 425 was painted dark blue. Inspired by the Reading Company's own blue painted Pacific (not, contrary to popular belief, the Central Railroad of New Jersey's Blue Comet), the locomotive's distinctive color made it stand out. The suggestion to paint 425 in dark blue was made to Andrew Muller by former Reading Company Engineer Charles W. Kachel who was often seen at the controls. The locomotive was a featured guest at the Steamtown National Historic Site Grand Opening in July 1995, and made a number of excursions out of Scranton. The locomotive returned to Port Clinton in late 1996. 425's last excursion was the Tamaqua Fall Fest on October 13, 1996. From 1997 - 2008 is when steam operations took an extended break.

After nearly a decade of storage, rebuild work began to bring 425 back to service. Following two years of restoration, the Pacific made its first operation under steam in December 2007 in a partially repainted appearance. Another test run was done on May 11, 2008 where the engine appeared in a new lighter blue color and an above-centered headlight. It made its return to excursion service in June 2008 on a round trip from Port Clinton to Jim Thorpe, a run it would make often. The RBMN's new star made many trips to Jim Thorpe and other locations over the next three years, with employee runs, tourist trains on the Lehigh Gorge Scenic Railway, and a featured attraction of the 2010 NRHS Convention.

Above: No425 in her dark blue /black former color scheme.

Below: Fully operational No425 leaves Jim Thorpe on a railfan excursion.

G-5s 4-6-0 Long Island RR (1928)

SPECIFICATIONS
Gauge: 4 ft 8 1/2 in
Tractive effort: 41,328 lb
Axleload: 63,000 lb
Cylinders: (2) 24 x 28 in
Driving wheels: 68 in
Steam pressure: 220 psi
Heating surface: 2,855 sq ft
Superheater: 613 sq ft
Grate area: 55 sq ft
Fuel: 24,000 lb
Water: 8,300 US gall
Adhesive weight: 178,000 lb
Weight: 409,000 lb
Length: 70 ft 5 1/2 in

Ninety of these fine 4-6-0s, designated Class G-5s were built in 1923, 1924 and 1925 in the road's own Juanita shops. The design owed a great deal to the E-6s 4-4-2 class of 1910, thereby avoiding some of the expense of new tooling and patterns. An exception was the driving wheels which were 15 per cent smaller than those of the E-6s. Incidentally, the "G" and the "E" stood for the respective wheel arrangements, while 's' stood for "superheated."

These handsome locomotives were specially designed for the Commuter traffic between New York City and the suburbs on Long Island. This was run by the Long Island Railroad, who were owned by the Pennsylvanian RR, whose basis of existence relied solely on the daily passengers. Many railroads used superannuated main-line locomotives for this purpose, but the Pennsylvania Railroad was not one of them, being then in the fortunate position of being able to afford the proper tools for the job.

The only important feature which was not standard North American practice was the Belpaire firebox. Invented by a Belgian engineer in 1864, the idea was to increase the

Above: The Long Island RR G-3 coped with the increasing commuter traffic between the two wars. Locomotive No39 was built in 1929.

Right: Long Island Rail Road Class G-5s 4-6-0 No38 makes a fine display of smoke at Kings Park, New York, one day in 1950.

surface area of the firebox and its internal volume, whilst keeping plenty of room for water to circulate round it. The simplicity of the round-top firebox was lost because the change of section of the boiler made construction more complex. Further more, a good deal of additional boiler stays were required, but improved steam generation was thought to make the extra costs worthwhile. Belpaire fireboxes were common in Europe but few other North American railroads used them.

Between 1924 and 1929, a fleet of 31 G-5s locomotives, identical to those on the Pennsy proper except that the keystone numberplate on the smokebox was originally replaced by a plain circular one, were built specially for the subsidiary railroad. Most of the shorter-distance Long Island traffic was operated electrically, so the new locomotives took the longer distance runs to the outer extremities of the island. When diesels took over outside the electrified areas in the 1950s, the G-5s were retired. No5741 is preserved in the Railroad Museum of Pennsylvania at Strasburg.

Right: PRR No5741 - Selected by the PRR itself for preservation upon its retirement, currently on permanent static display in the Railroad Museum of Pennsylvania.No5741 was added to the National Register of Historic Places in 1979 as Freight Locomotive No5741.The G-5s were the primary passenger locomotive on the Long Island Railroad until the end of steam operations in the mid-1950s, sharing the duty with K-4s's and other G-5's leased from the PRR, as well as diesels such as the Alco RS3 that ultimately replaced them. The G-5s locomotives on Long Island differed slightly from those produced for the PRR, with a larger tender typically used by the K-4s.

5741

Penn RR K-5 Class 4-6-2 (1929)

SPECIFICATIONS
Gauge: 4ft 8 1/2 in
Cylinders: (2) 27 x 30 in
Driving wheels: 80 in
Steam pressure: 250 psi
Tractive effort: 54,675 lb
Weight: 327,560 lb

The K-5 class locomotive was not proof of the old adage that bigger is best! Built in 1929, it was an experiment to see whether a larger Pacific than the standard K-4 was viable.Two prototypes were built, #5698 at the PRR's own Juniata Works, and #5699 by the Baldwin Locomotive Works. Although classified identically, the two locomotives differed in many aspects, as detailed below. They were both fitted with a much fatter boiler than the K-4s, but dimensionally similar to those of the I-1s 2-10-0 "Decapods". Most other dimensions were enlarged over the K-4s as well; the exceptions being the 70 square feet grate area and the 80 in driving wheels.

Because the K-5 was more powerful than the K-4 its factor of adhesion was far less good than the K-4 as it had little more weight on driving wheels (and thus adhesion). Factors of adhesion below 4 are often considered undesirable for steam locomotives, and the K-5 design did prove to be rather less sure-footed because of it. For this reason, 4-8-2 "Mountain" and 4-8-4 "Northern" designs with more driving wheels (and thus a greater allowable weight on drivers within the same axle load limit) were generally considered preferable for locomotives as powerful as the K-5.

Both K5 locomotives were given a 130-P-75 tender carrying 12,475 gallons of water and 22 tons of coal. Surprisingly for such large locomotives built at such a late date, both were equipped for hand firing. Both were fitted with Worthington-pattern feedwater heaters, power reverse, unflanged main driving wheels, and both used nickel steel boiler shells. As built, both carried their bell on the smokebox front, hung below the headlight; this arrangement was common on other roads but at the time unique on the PRR.

The two locomotives were originally assigned to haul Philadelphia to Pittsburgh through trains. After the electrification reached Harrisburg, they were reassigned to the twisting Harrisburg–Baltimore route. Finally, they were reassigned to Pittsburgh–Crestline trains.

Both locomotives were cosmetically restyled after World War II with the standard front-end treatmen given to most K-4s locomotives: a sheet-steel drop coupler pilot, higher-mounted headlight, and turbo-generator mounted on the smokebox front for easier access. They lost their unique bell placement at this time for a location equally unique for the PRR: under the generator service platform on the front of the smokebox.

Both locomotives were scrapped.

Below: The original Baldwin Works photo of the K-5 class is one of very few images of the type still surviving.

F-2a Class 4-4-4 (1929)

SPECIFICATIONS
Gauge: 4ft 8 1/2 in
Tractive effort: 26,500 lb
Axleload: 61,000 lb
Cylinders: (2) 17 1/4 x 28 in
Driving wheels: 80 in
Heating surface: 2,833 sq ft
Superheater: 1,100 sq ft
Steam pressure: 200 psi
Grate area: 55.6 sq ft
Fuel: 27,000 lb
Water: 8,400 gallons
Adhesive weight: 121,000 lb
Weight: 461,000 lb
Length: 81 ft 2 1/2 in

In 1936 the Canadian Pacific Railway introduced four trains which were announced as a High-Speed Local Service. In each case the formation consisted of a mail/express (packages) car, a baggage-buffet and two passenger cars. By North American standards they counted as lightweight, the weight being 200 tons for the four-coach train.

Most American railroads would have found some hand-me-down locomotives discarded from first-line passenger service to work them, but that was not the CPR way. They ordered five new 4-4-4 steam locomotives, designated the "Jubilee" type, from the Montreal Locomotive Works to work these trains—although spoken of as streamlined, they are better described as having a few corners nicely rounded. Running numbers were 3000 to 3004.

The new services for which this equipment was ordered comprised the "Chinook" in the West between Calgary and Edmonton (194 miles in 315 minutes including 22 stops) and the international "Royal York" between Toronto and Detroit (229 miles in 335 minutes with 19 stops) and two others between Montreal and Quebec. F-2a s were certainly magnificent. They had such sophisticated features as mechanical stokers, feed-water heaters and roller bearings. One feature that was important for operation in Canada was an all-weather insulated cab, able to provide comfortable conditions for the crew in a country where the outside temperature could easily drop to minus 40°F.

A further series of similar and slightly smaller 4-4-4s, numbered from 2901 to 2929, were built in 1938, designated class F-la. The second series was easily recognizable by the drive on to the rear coupled axle, instead of on to the front axle as with the F-2a. Nos 2928 and 2929 of this later series are preserved at the Canadian National Railway Museum at Delson, Quebec.

Above: No3002 of the original and larger batch of F-2a s at Toronto, April 1952 (and left) CPR's smaller 4-4-4, Class F-la No.2928, now preserved.

P-7 Class 4-6-2 Baltimore & Ohio (1928)

The Baltimore & Ohio Railroad has used Pacifics for its principal passenger trains since 1906. For the start of through running by B&O locomotives from Washington to Jersey City in 1927, the eighth class of 4-6-2 was placed in service; these 20 Class P-7s were over 50 per cent more powerful than the original Class P of 20 years earlier.

This was also the year in which the B&O celebrated the centenary of the granting of its charter with an ambitious "Fair of the Iron Horse" at Baltimore. Influenced by English engine King George V which was exhibited at the fair the P-7 4-6-2s took names from US presidents, but were painted a similar shade of green. It was many years since any B&O engine had been painted anything else but black. The twenty-first P-7, named President Cleveland and completed in 1928, was an experimental locomotive with a water-tube firebox and camshaft-driven Caprotti pattern poppet valves, similar in principle to those used in automobile practice. The latter were not a success and were altered to the conventional Walschaert pattern the following year. Unlike the earlier engines, all of which came from Baldwin, No5320 was built in the B&O's own Mount Clare shops. Incidentally, the names were allocated in historical order beginning with No.5300 President Washington, one loco sufficing for both presidents named Adams.

The water-tube firebox, while not sufficient of an improvement to displace convention, did remain on the locomotive until 1945. In 1937 one other P-7 No5310 President Taylor was also treated in this way, with equally inconclusive results. These two locomotives were designated P-9 and

Below: No5300 at the Baltimore and Ohio yards in the 1970s.

SPECIFICATIONS
Gauge: 4ft 8 1/2 in
Tractive effort: 50,000 lb
Axleload: 68,000 lb
Cylinders: (2) 27 x 28 in
Driving wheels: 80 in
Heating surface: 3,782 sq ft
Superheater: 950 sq ft
Steam pressure: 280 psi
Grate area: 70.2 sq ft
Fuel: 39,000 lb
Water: 11,000 gallons
Adhesive weight: 201,000 lb
Weight: 544,000 lb
Length: 87 ft 10 1/2 in

Above: P-7 class 4-6-2 No5308 of the Baltimore & Ohio. The class engines were named after presidents as influenced by the English fashion of naming locomotives for famous people. No5308 is President Tyler.

P9b respectively when running in this condition.

From 1937 to 1940 No5310 was streamlined, painted blue and renamed The Royal Blue for working the train of that name. Because of the difficulty of allocating a specific locomotive to a particular train, the rest of the class were also painted blue and this matched the colour of new B&O trains then being put into service. Soon afterwards, the whole class ceased to carry names and in 1946 a further batch of four was streamlined for running the "Cincinnatian," following a rebuild which included provision of cast locomotive beds, roller bearings and bigger 12-wheel tenders.

Above: In 1937 P-7 No5301 was streamlined and painted Royal Blue for working the train of the same name.

CP Class H-1b 4-6-4 No2816 (1930)

Canadian Pacific No2816, named the Empress, was built by the Montreal Locomotive Works in December 1930. Classified as a 4-6-4 H-1b No2816 is the only non-streamlined H-1 Hudson remaining (the other four remaining are the semi-streamlined Royal Hudsons). Locomotive No2816 was one of ten H-1b-class (the "H" meant the 4-6-4 wheel configuration, the "1" was the design number and the "b" meant it was the second production run) 4-6-4 Hudson built .It was first assigned to the line between Winnipeg and Fort William, Ontario. Later, it was transferred to service between Windsor, Ontario, and Quebec City, and finally it ran a commuter train between Montreal and Rigaud, Quebec. It made its last service run on May 26, 1960, after more than 2 million miles in active service. In 1963, the locomotive was sold to Monadnock, Steamtown & Northern Amusements Corp. Inc. (AKA: Steamtown, USA), which evolved into the Steamtown National Historic Site in 1986.

When Steamtown USA moved from Bellows Falls, Vermont, to Scranton, Pennsylvania, in the 1980s, No2816 made the trip with other engines. When the National Park Service took over from the Steamtown Foundation, No2816 also passed to the NPS, now Steamtown National Historic Site, and the NPS decided to divest itself

SPECIFICATIONS
Gauge: 4 ft 8 1/2 in
Driving wheels: 75 in
Length: 91 ft 1 in
Weight on drivers: 194,000 lb
Weight 360,000 lb
Locomotive and tender combined weight: 658,000 lb
Fuel type: Coal (Converted to burn oil during restoration)
Water capacity 14,000 gall
Steam pressure: 275 psi
Cylinders: (2) 22 in x 30 in
Power output: 4,700 hp
Tractive effort: 45,300 lb

Below: Canadian Pacific Railway 2816 at Steamtown in Vermont.

Below: CP No2816 steams through Carmangay, AB on October 20th 2006.

of foreign locomotives. In 1998 Canadian Pacific purchased No2816 after hearing of its availability from the crews who were running the Royal Hudson No2860, who had been looking for parts for No2860 and were offered the entire locomotive. It was moved in train from Scranton to Montreal via Binghamton and Albany, New York, before being shipped cross country to the BC Rail steam shops in Vancouver for restoration. The locomotive was completely stripped down and rebuilt, "the most thorough rebuild undertaken on a steam locomotive in North America since the end of their era" according to *CPR News*. The restoration team was able to use over 800 technical drawings of CPR H-lb class locomotives from the Canada Science and Technology Museum to completely restore No2816 to its 1950s appearance and to its original specifications.

T-1 Texas Class Locos 2-10-4 (1930)

SPECIFICATIONS
Gauge: 4ft 8 1/2 in
Cylinders: (2) 29 x 34 in
Driving wheels: 69 in
Steam pressure: 270 psi (max)
Tractive effort: 95,106 lb
Weight: 566,000 lb

The Chesapeake & Ohio Railroad designed a new class of freight locomotive to satisfy its needs for motive power. They tested an Erie Berkshire then basically stretched the Erie design adding one set of drivers, creating a 2-10-4 that was both powerful and fast.

The C&O ordered forty of its new design "Texas" type locomotives from the Lima Locomotive Works in 1930. These locomotives were designated as Class T-1 and were assigned road number 3000 through number 3039. They had 69" diameter drivers, 29" x 34" cylinders, a 265 psi boiler pressure, exerted 95,106lb of tractive effort and each weighed 566,000 pounds. The locomotives were equipped with a trailing truck booster that exerted 15,275 pounds of tractive effort. This basic design was used for most 2-10-4s constructed after 1930.

The firebox was 645.3 square feet. The boiler's nine foot diameter was packed with the most heating surface of any 2-cylinder locomotive ever built. The evaporative surface was 6,624 square feet and the superheater added 3030 giving it a combined heating surface of 9,654 square feet.

The C&O ran most of it "Texas" types from Russell, Kentucky to Toledo, Ohio with a few used in eastern Virginia. A typical train pulled by one of the T-1s would consist of 160 loaded cars of coal comprising a trailing load of 13,500 tons.

A weak point common to long-wheelbase freights was the difficulty in maintaining a proper counterbalancing scheme as the drivers wore unevenly in service. In the latter part of their careers, this class rode roughly and pounded the track to the point that a special gang stood by at the bottom of one long grade to repair the damage.

125 slightly modified 2-10-4s of the basic Chessie design were built by Pennsylvania RR during WW II as the J-1 class. These later engines benefited from cast-steel frames with integral cylinders and are regarded by many as the finest steam locomotives ever operated by the PRR.

There are no surviving examples of the C&O Class T-1 locomotives. All were scrapped in the early 1950s.

Above: The PRR version of the Texas locomotive was classified as J-1 class.

L. F. Loree 4-8-0 (1933)

No1403, L. F. Loree, 4-8-0 was designed as a unique triple-expansion compound locomotive with four cylinders. Two low-pressure cylinders were placed in the conventional position at the front, while the single intermediate-pressure and single high-pressure cylinders were placed beneath the cab on either side. All four drove on to the second coupled axle. Rotary-cam poppet valves of the Dabeg pattern distributed steam between the four cylinders. Bleed and by-pass valves enabled live steam direct from the boiler to be admitted to all cylinders for starting. Interestingly this locomotive was the only one built in the depression year of 1933 for any US railroad.

SPECIFICATIONS
Gauge: 4ft 8 1/2 in
Tractive effort: 75,000 lb
Axleload: 86,075 lb
Cylinders, HP: (1) 20 x 33 in
Cylinders, LP: (1) 27 1/2 x 33 in
Cylinders, LP: (2) 33 x 33 in
Driving wheels: 63 in
Heating surface: 3,351 sq ft
Superheater: 1,086 sq ft
Steam pressure: 500 psi
Grate area: 75 sq ft
Fuel: 30,000 lb
Water: 14,000 gallons
Adhesive weight: 313,000 lb
Weight: 658,000 lb

No1403 was a technical achievement, showing an improvement in thermal efficiency which reduced coal consumption to less than half the amount a normal locomotive would burn on a given task. But once more the debits were even greater. For example, high temperature of the steam caused lubrication problems in the cylinders. Another problem was bearing failures in the main driving axleboxes, caused by taking the thrusts of all four cylinders on the same pair of wheels.

Even though Loree never took the steam locomotive beyond the Stephenson concept, the Delaware & Hudson pioneered several details which were also very important. The roller bearings fitted to the main axles of No1403 were the result of an early application of such a feature to a D&H 4-6-2 in 1927. The first locomotive in the world to have roller bearings in the big ends of the connecting rods was another of the same class. The D&H also pioneered the cast-steel locomotive bed. All these items became common-place in the final days of steam locomotive design .

Leonor F. Loree, President of the Delaware & Hudson Railway from 1907 to 1935, was the driving force behind these groundbreaking improvements to the design of the steam locomotive. The experiments began in 1923 with 4-8-0 No1400 Horatio Allen .The main thrust of Loree's ideas was that the boiler was designed for a higher pressure than normal. This was too high to be withstood by the flat surfaces of a normal firebox, and so a water-tube arrangement was substituted. The steam pressure was also too high to be used to advantage with simple cylinders, and so No1400 was arranged as a two-cylinder compound with a high-pressure cylinder on one side and a low-pressure unit on the other. As on some other D&H 2-8-Os there was also a booster engine on the rear tender truck, the axles of which were coupled by rods.

The new locomotive fulfilled its designer's promises by burning some 30 per cent less coal to do the same work, compared with a conventional 2-8-0. The penalty was an increased cost of maintenance which perhaps fractionally outweighed the savings of the reduced fuel consumption.

Class A 4-4-2 (1935)

The Milwaukee Road class A comprised four high-speed, streamlined 4-4-2 "Atlantic" type steam locomotives built by ALCO in 1935-37 to haul the Milwaukee Road's Hiawatha express passenger trains. They were among the last Atlantic types built in the United States, and certainly the largest and most powerful. The class were the first locomotives in the world built for daily operation at over 100 mph and the first class built completely streamlined, bearing their casings their entire lives. They were road numbered 1-4. The American Locomotive Company of Schenectady, New York, delivered the first two superb oil-fired and brightly colored streamlined 4-4-2s. They received road numbers 1 and 2. In service they earned this prime designation by demonstrating that as runners they had few peers. They could develop more than 3000 horsepower in the cylinders and achieve 110mph on the level. It says enough about that success of these locomotives that they were intended to haul six cars on a 6 1/2 hour schedule, but soon found themselves handling nine cars satisfactorily on a 6 1/2 hour one. These schedules included five intermediate stops and 15 permanent speed restrictions below 50mph.

SPECIFICATIONS
Gauge: 4ft 8 1/2 in
Tractive effort: 30,685 lb
Axleload: 72,500 lb
Cylinders: (2) 19 x 28 in
Driving wheels: 84 in
Heating surface: 3,245 sq ft
Superheater: 1,029 sq ft
Steam pressure: 300 psi
Grate area: 69 sq ft
Fuel (oil): 4,000 gall
Water: 13,000 gallons
Adhesive weight: 144,500 lb
Weight: 537,000 lb
Length: 88 ft 8 in

The design was unusual rather than unconventional; the tender with one six-wheel and one four-wheel truck, for instance, or the drive on to the leading axle instead of the rear one, were examples. Special efforts were made to ensure that the reciprocating parts were as light as possible—the high boiler pressure was chosen in order to reduce the size of the pistons — and particular care was taken to get the balancing as good as possible with a two-cylinder locomotive. Another class A (No3) was delivered in 1936 and a fourth (No4)in 1937.

Although partially supplanted by the larger F-7 "Hudsons" from 1937, they remained in top-flight service until the end. Locomotive No3 was taken out of service in 1949 and cannibalized for spares to keep the other three running until 1951. None survive.

Above left: The Chicago, Milwaukee and St. Paul Railway's "Hiawatha" express stopped near Milwaukee due to trouble. This version of the train was pulled by a streamlined steam locomotive.

Left: Class A No1 hauls the Chippewa Express near Deerfield, Illinois in 1939.

S-7 Class 0-10-2 Union Railroad (URR) (1936)

SPECIFICATIONS
Gauge: 4ft 8 1/2 in
Tractive effort: 108,050 lb
Axleload: 71,300 lb
Cylinders: (2) 28 x 32 in
Driving wheels: 61 in
Heating surface: 4,808 sq ft
Superheater: 1,389 sq ft
Steam pressure: 260 psi
Grate area: 85.2 sq ft
Fuel: 28,000 lb
Water: 12,000 gallons
Adhesive weight: 343,900 lb
Weight: 644,360 lb
Length: 83 ft 5 1/2 in

Above: S-7 No302 in original Union Railroad livery

Ten Union Railroad 0-10-2 steam locomotives were built in 1936–1939 by the Baldwin Locomotive Works. These were the only 0-10-2 locomotives ever built in the United States and this purchase gave the name "Union" to this type. The Union Railroad classified them "S-7" and road numbered them 300-309 , they bore the brunt of wartime steel production traffic.

Designed specifically for the switching and transfer duties on the Union

Left: S-7 survivor No304 on static display at Greenville, PA.

Railroad line owned by U.S. Steel, serving a number of plants in the area and connecting with six trunk line railroads, the humble switcher was normally one of a forgotten army of has-beens, occupied well behind the scenes in the drudgery of putting cars together to make trains. Those industries and few railroads which needed purpose-built switchers usually found the USRA 0-6-0 or 0-8-0 designs adequate for their needs, occasionally with minor adaptations. An occasional 0-10-0 could be found.

The Union Railroad connected steel plants in the Pittsburgh area, where the switcher was king. So, as one might expect, the King of the Switchers was a Union RR locomotive, holding the world record for every possible attribute for this category of motive power. A requirement for 100,000lb-plus tractive effort meant ten-coupled wheels plus a four-coupled booster. There was a requirement for a large firebox and plentiful steam-raising ability necessitated the trailing truck. (with mechanical stoker) to provide an adequate source of power which necessitated the trailing truck. To increase tractive effort still further, a booster engine was fitted to the leading tender truck. The unusual wheel arrangement was also a result of the turntable restrictions on the total wheel base.

Operation was only at low speed, thus a leading truck's stability was not required. The intent was to eliminate helper requirements on grades, and thus a locomotive larger than the Union's previous switchers and 2-8-0 "Consolidations" was needed. Upon dieselization of the Union in 1949, nine of the locomotives were sold to fellow U.S. Steel railroad the Duluth, Missabe and Iron Range, where they served until 1962. One of these locomotives survives as a static exhibit at Greenville, Pennsylvania.

Below: The locomotive No304 is a reminder of the bygone age of stem to a new generation of railfans.

F-7 4-6-4 Milwaukee Road (1937)

SPECIFICATIONS
Gauge: 4ft 8 1/2 in
Cylinders: (2) 23 1/2 x 30 in
Driving wheels: 84 in
Axle load: 72,250 lb
Tractive effort: 50,294 lb
Weight: 415,294 lb

Alco built and delivered six high-speed streamlined 4-6-4 "Baltic" or "Hudson" type steam locomotives to the Milwaukee Road in 1937-38.They were designated Class F-7 and road numbered No100-105.Their purpose was to haul the Milwaukee's Hiawatha express passenger trains.. Following on from the success of the road's class A 4-4-2s, the F-7s allowed the road to haul heavier trains on the popular Chicago–Twin Cities routes.

The F-7s are definite contenders for the fastest steam locomotives ever built, as they ran at over 100 miles per hour daily. One run in January 1941 recorded by a reporter for Trains magazine saw 110 mph achieved twice—in the midst of a heavy snowstorm. Baron Gérard Vuillet, a French railroading expert, once recorded a run between Chicago and Milwaukee where the locomotive reached 125 mph and sustained an average 120 mph for 4.5 miles However, the English locomotive LNER Class A4 No4468 Mallard is officially accepted to be the world's fastest, with a run recorded at 125.88 mph in 1938.

The Milwaukee F-7s are accepted as the fastest steam locomotives by a different measure—scheduled speed between stations. In 1939, shortly after they were introduced into passenger service, the Twin Cities Hiawatha schedule was modified such that the engines would need to run the 78.3 miles between Portage and Sparta, Wisconsin in 58 minutes—a start-to-stop average of 81 mph.

First-built No100 was also the first withdrawn from service, on November 10, 1949; last-built No105 was the final one in service, withdrawn August 10, 1951. None survived.

Right: The Hiawatha Express making smoke leaving the Twin Cities.

Below: Milwaukee Road class F-7 2nd generation Hiawatha locomotive streamstyled by Otto Kuhler is towed out of the Alco works.

Royal Hudson Class 4-6-4 CPR(1937)

SPECIFICATIONS
Gauge: 4ft 8 1/2 in
Tractive effort: 45,300 lb
Axleload: 65,000 lb
Cylinders: (2) 22 x 30 in
Driving wheels: 75 in
Heating surface: 3,791 sq ft
Superheater: 1,542 sq ft
Steam pressure: 275 psi
Grate area: 81 sq ft
Fuel: 47,000 lb
Water: 14,400 gallons
Adhesive weight: 194,000 lb
Weight: 659,000 lb
Length: 90 ft 10 in

Canadian Pacific built 45 of their famous 4-6-4s between 1937 and 1945.

The genesis of these fine locomotives lay in a wish to improve upon the class G-3 4-6-2s which before 1931 had been the top-flight power of the system, by increasing their steam-raising capacity a substantial amount. A fire-grate 23 per cent larger was possible if the 4-6-4 wheel arrangement was adopted and the boilers of the new locomotives were based on this. But in other ways, such as tractive effort or adhesive weight, the new locomotives were little different to the old. Their class designation was H-1 and the road numbers were 2800 to 2819.

The boilers had large superheaters and combustion chambers (the latter an addition to the firebox volume, provided by recessing the firebox tubeplate into the barrel), as well as front-end throttles which worked on the hot side of the superheater. This enabled superheated steam to be fed to the various auxiliaries. There were arch tubes in the firebox and, necessary with a grate of this size, a mechanical stoker.

The first effect of the new locomotives was to reduce the number of engine changes needed to cross Canada, from fourteen to nine. The longest stage was 820 miles from Fort William, Ontario, to Winnipeg, Manitoba; experimentally a 4-6-4 had run the 1,252 miles between Fort William and Calgary, Alberta, without change.

For five hectic months in 1931 the afternoon CPR train from Toronto to Montreal, called the "Royal York" became the world's fastest scheduled train, by virtue of a timing of 108 minutes for the 124 miles from Smith's Falls to Montreal West, an average speed of 68.9mph.

An interesting feature, later provided on one of the H-1s, was a booster engine working on the trailing truck. One of the problems of a 4-6-4 was that only six out of 14 wheels were driven; this was no detriment while running at speed but starting was sometimes affected by the limited adhesion. The extra 12,000lb of tractive effort provided by the booster came in very handy; the mechanism cut out automatically at 20mph

In 1937 was another batch of 30 Hudson type, Nos. 2820 to 2849 designated H- 1c, (the earlier ones had been delivered in two batches of ten, "H-1a" and "H-1b") which had not only softer lines as a gesture toward streamlining which was the vogue back then but also sported a superb colored livery, Very few mechanical changes needed to be made—although there were certain improvements or changes such as power-operated reversing gear, dome-less boilers and a one-piece cast locomotive frame, while boosters were fitted to five of the locomotives. A further ten 4-6-4s, designated H-1d were delivered in 1938, while the last batch of five H-1e, Nos.2860 to 2864 of 1940, differed from the others in being oil burners. All the H-les and five of the H-lds had boosters.

The last batch of 4-6-4s were intended to operate in the far west, between Vancouver and Revelstoke, British Columbia, where oil firing had been the rule for many years. After the war, when the big Canadian oil fields were being exploited, all the H-ls operating over the prairies were also converted. This was made easier by the fact that it was

customary to allocate a particular locomotive to a particular depot when they were built and they would then remain there for many years. This unusually stable approach to locomotive allocation also allowed the booster-fitted locomotives to be rostered for sections of line where their extra push was needed. For example, booster fitted H-1cs allocated to Toronto could take the 18-car 1,300-ton "Dominion" express up the Neys Hill incline on Lake Superior's north shore unassisted with booster in operation; otherwise a helper engine would have been an obvious necessity.

Like other lines which had excellent steam power, well maintained and skillfully operated, the Canadian Pacific Railway was in no hurry to dieselise and, in fact, it was not until 1956 that the first 4-6-4 was scrapped. Four Royal Hudsons have been preserved (Hudson No2816 is not streamlined and thus is not "Royal" but is often mistakenly referred as a Royal Hudson.

No2839 Operated in the 1970s and 1980s. Now at the Nethercutt Collection and Museum, Sylmar, California. Good cosmetic shape, displayed outside.

No2850 The locomotive that hauled the Royal Train and known as "The" Royal Hudson, served a long career until 1960 when she was retired and is now preserved at the Canadian Railway Museum at Delson/Saint-Constant, Quebec. Very good cosmetic and mechanical shape, displayed indoors.

No2858 Preserved at the National Museum of Science and Technology at Ottawa. Good cosmetic shape, displayed inside.

No2860 Squamish, BC. First CPR Hudson built as a Royal Hudson, one of the last five built. Operable.

Above: Royal Hudson No2839, once destined for a museum in eastern Canada, wound up being sold to a group of owners in Pennsylvania. After a restoration to full working order to full CPR livery (with Southern lettering), the engine was leased to the Southern Railway for their steam excursion program in 1979–1980, but was found that the locomotive was not powerful enough for their excursions. During her brief career with the Southern, 2839 earned the nickname "beer can" due to the Royal Hudson's cylindrical streamlined design. After being returned from the Southern, the engine was stored on the Blue Mountain and Reading Railroad before being stored near Allentown, PA. The Blue Mountain and Reading Railroad attempted to restore and run her on excursions, but ultimately 2839 was sold. After a series of owners, the engine was shipped on a flat car from Pennsylvania to the Nethercutt Collection in Sylmar, California, where it has been cosmetically restored and put on display outside the museum with a Pullman car.

Class I-5 4-6-4 NY, NH & H(1937)

SPECIFICATIONS
Gauge: 4ft 8 1/2 in
Tractive effort: 44,000 lb
Axleload: 65,000 lb
Cylinders: (2) 22 x 30 in
Driving wheels: 80 in
Heating surface: 3,815 sq ft
Superheater: 1,042 sq ft
Steam pressure: 285 psi
Grate area: 77 sq ft
Fuel: 32,000 lb
Water: 18,000 gallons
Adhesive weight: 193,000 lb
Weight: 698,000 lb
Length: 97 ft

The New York, New Haven & Hartford Railroad (called the New Haven for short) had its main line from New York to Boston. This had been electrified in stages, beginning as early as 1905 and reaching its greatest extent at New Haven, 72 miles from New York in 1914.

There remained 159 miles of steam railroad from there to the "home of the bean and the cod." Trains such as "The Colonial" or the all-Pullman parlor car express "The Merchants Limited" heavily overtaxed the capacity of the existing class I-4 Pacifies and, in 1936, after a good deal of research and experiment, ten 4-6-4s were ordered from Baldwin of Philadelphia.

These handsome engines were the first streamlined 4-6-4s in the USA to be delivered. They were also very much an example to be followed in that firstly, the desire to streamline was not allowed to interfere with access to the machinery for maintenance.

The I-5 class with disc driving wheels, roller bearings and Walschaert's valve gear went into service in 1937. Road numbers were 1400 to 1409. They certainly met the promise of their designers in that they showed a 65 per cent saving in the cost of maintenance compared with the 4-6-2s they replaced and, moreover, could handle 16-car 1100-ton trains to the same schedules as the Pacific could barely manage with 12.

Another requirement was met in that they proved able to clear the 1 in 140 (0.7 per cent) climb out of Boston to Sharon Heights with a 12-car 840-ton train at 60mph But, alas, the I-5s were never able to develop their no doubt formidable high speed capability because of a rigidly enforced 70mph speed limit. For this reason and because the line was infested with speed restrictions, the schedule of the "Merchants Limited" never fell below 171 minutes including two stops, representing an average of 55mph.

None of the I-5s is preserved.

5001 Class 2-10-4 AT & SF (1938)

The 2-10-4 type got the name Texas' from a group of locomotives built by Lima for the Texas & Pacific Railroad in 1925. Yet the Atchison, Topeka & Santa Fe Railway had one earlier than that. In 1921 a Santa Fe type (2-10-2) had its rear truck replaced by a two-axle one. The railway got its second 2-10-4 (and its first designed as such) nine years later when new power was needed to hurtle vast freights across Kansas, Oklahoma, Texas and New Mexico along the southern and more easily graded of the two routes that ran southwest from Kansas City and joined at Belen, New Mexico, on the way to Los Angeles.

SPECIFICATIONS
Gauge: 4ft 8 1/2 in
Tractive effort: 108,960 lb
Axleload: 81,752 lb
Cylinders: (2) 30 x 34 in
Driving wheels: 74 in
Heating surface: 6,075 sq ft
Superheater: 2,675 sq ft
Steam pressure: 310 psi
Grate area: 121 sq ft
Fuel (oil): 7,100 gall
Water: 24,500 gall
Adhesive weight: 371,600 lb
Weight: 1,002,700 lb
Length: 123 ft 5 in

We have seen how Santa Fe's locomotive department promised itself, after traumatic experiences, never again to order locomotives with more than two cylinders. Within this limitation their own 2-10-2 type was the favorite, but fast running on moderate gradients needs a high power output in relation to tractive effort. This means a high heat input, hence a large grate and a two-axle rear truck to carry its weight making a 2-10-2 into a 2-10-4. Because of the Depression the unremarkable prototype 2-10-4 No.5000 delivered in 1930-built by Baldwin like almost all Santa Fe locomotives—remained a singleton for eight years, although it delivered the goods in a literal sense. In the meantime, however, the specification had grown to give for the first and only time in the history of steam a big-wheeled high-speed locomotive with as many as ten coupled wheels. Another way of getting speed was to cut out the need for fuel and water stops and

the huge 16-wheel tender which resulted from this thinking brought the fully-loaded weight to over a million pounds, another record for a two-cylinder steam locomotive. More than 100,000lb of nominal tractive effort is also confined to a handful of examples.

These racing mammoths did all that was expected of them and more, being capable of developing over 6,000hp in the cylinders. This applied particularly to the second wartime batch road numbered 5011 to 5035, which had roller bearings as well as larger tenders from the start.

Dieselization came early to the Santa Fe, the first freight units arriving in 1940. Whilst the 2-10-4s showed up well against them, the older steam locomotives then in the majority even on Santa Fe—did not. Moreover, in the west there were major problems in finding enough good water for steam locomotives; diesels eliminated much unproductive hauling in of water supplies. So, by the early 1950s, modern 4-8-4s and 2-10-4s were the only steam locomotives operating, and even these had ceased by August 1957.

Below: Santa Fe Class 5000 No5000 is a Texas type locomotive (2-10-4). It is currently on outdoor display next to Queen's Garden in downtown Amarillo, TX, close to the junction of SE 2nd and Lincoln Avenue.

AC Class Cab-forward 4-8-8-2 (1938)

SPECIFICATIONS
Gauge: 4ft 8 1/2 in
Tractive effort: 124,300 lb
Axleload: 69,960 lb
Cylinders: (4) 24 x 32 in
Driving wheels: 63.6 in
Heating surface: 6,470 sq ft
Superheater: 2,616 sq ft
Steam pressure: 250 psi
Grate area: 139 sq ft
Fuel (oil): 6,100 lb
Water: 22,000 gallons
Adhesive weight: 531,700 lb
Weight: 1,051,200 lb
Length: 124 ft 11 in

Above: Baldwin's works photo of No4159 shows the massive proportions of the class AC cab in front type locomotives.

The Mallet principle was tried in the form of two compound 2-8-8-2s (then the most powerful locomotives in the world) built by Baldwin in 1909. As much as 1,300 tons could be handled over The Hill' by these machines, but the many miles of snowsheds and tunnels caused difficulties for their crews in finding air to breathe. The 2-8-0s and 4-8-0s, previously used in multiple, would be spread out down the train, which helped: For some years oil fuel had been used, and this led to a suggestion for turning the locomotive round so that the cab came in front. The idea seems to have originated on a local Californian railroad called the North Pacific Coast, but was resuscitated by SP and Baldwin when, later the same year, 15 2-8-2s were built to this plan. They were otherwise very similar to the earlier 2-8-8-2s, and in fact the only significant change was due to the need for some assistance to persuade the oil to flow uphill a distance of some 100ft. This was done by making the tenders half-round in form, so enabling them to be pressurized with 5psi of air pressure. By 1913 there were 46 of these locomotives, most of which were later converted from compound to simple.

Another detail involved the arrangement of the Walschaert valve gear. Since the locomotives usually ran in what on a normal engine would be reverse gear, this normally meant having the die-block at the top of the curved link rather than in the best position, giving a direct drive at the bottom. By moving the position of the eccentric crank through 180°, the engines of the "cab-in-fronts" could have the best position of the motion in their normal direction of running.

Like other Mallets of the day, long before the idea of having a combustion chamber was evolved, the boiler of all these locomotives was in two parts. The rear section including the firebox was conventional, but in front was a separate section with its own fire tubes and tube plates, which acted as a feed-water heater. In due time modern single-stage boilers with a combustion chamber were substituted. One interesting device fitted to all the cab-in-fronts was a retractable longitudinal smoke-splitter which prevented the exhaust from impinging directly on to the vulnerable roofs of snowsheds.

By the end of 1913 three very similar batches of cab-in-front 2-8-8-2s had been built, designated classes MC-2,

Above: AC-4 No4154 at Roseville,California in 1954 on the upservice to Sacramento with a long and winding freight train.

MC-4 and MC-6, MC standing for "Mallet-Consolidation." Road numbers ran in chronological order from 4002 to 4048.

Twelve 'Mallet-Mogul' cab-in-front 2-6-6-2s (later altered to 4-6-6-2s) were built in 1911 for passenger work, but they were not found suitable for passenger speeds and so were relegated to freight after a few years. As well as becoming unstable, one problem with a fast-running compound Mallet was getting the huge volume of low-pressure steam in and out of big low-pressure cylinders quickly enough. This led

Above: AC class No4294 is preserved at the California State Railroad Museum.

to the high-pressure Mallet-type articulated, with four HP cylinders, as pioneered by the Chesapeake & Ohio.

After converting an MC-6 experimentally to simple working in 1927, SP in 1928 ordered a batch of 10 4-8-8-2s, classified (partly inappropriately) as "Articulated-Consolidation," or AC-4. SP then went on to have 185 more of these unique locomotives built by Baldwin, supplied in seven batches from 1929 to 1944. Locomotives of earlier "MC" classes were all converted to simple "AC" form, becoming classes AC-1, AC-2 and AC-3, between 1928 and 1937, while the original two conventional Mallets, Class MC-1 were converted to AC-1 form. Similarly, the "MMs" became Class AM-2 during the 1930s, making a final total of 257 cab-in-fronts on the Southern Pacific. They were used on several other lines besides the Overland Route.

The cab-in-fronts had one or two inherent weaknesses which caused problems from time to time. With the firebox at the leading end, the range of boiler water level for safe working was reduced when running uphill and thus steaming hard. There were more cases of boiler explosions than there should have been and this factor no doubt contributed. The other problem was that, because the furnace oil had such a long distance to flow, there was a tendency for the fire to go out and then relight explosively when the critical air-vapour mixture had been achieved. The dangers to the crews were accentuated by the totally-enclosed cabs. These and other combustion problems caused some difficulties and a few injuries, but in general the cab-in-fronts were very successful. High speed was not their thing, 55mph being the maximum their modest riding qualities would permit. In the 1940s, three cab-in-fronts (one in front, one in the middle and one out ahead of the caboose) would take a 5,000 ton load of perishables from Roseville Yard, 138 miles over "The Hill" to Sacramento in just under 8 hours. This compared with 16 hours using five locomotives hauling a lighter load before the Mallets came 30 years before. Now, 70 years on, diesel traction has effected further modest improvement to 7 hours.

Class E-4 4-6-4 C & NW (1938)

The Chicago and North Western Railway's Class E-4 comprised nine coal-burning streamlined 4-6-4 "Hudson" steam locomotives built in 1937 by Alco.

They were built to haul the road's famous *"400"* express passenger trains, but before they were even delivered the railroad's management decided that streamlined steam was the wrong direction and instead placed orders with General Motors Electro-Motive Division for new diesel locomotives. The displaced E-4s instead worked other trains until they were withdrawn from service in 1953.

The Chicago & North Western Railway had its own way of doing things; not for nothing did its trains run on the left-hand track, whereas most North American trains take the right. When, in 1935, the gloves came off for the fight between the Milwaukee, Burlington and C&NW companies for the daytime traffic between Chicago and the twin cities of St Paul and Minneapolis, the last-named was first into the ring with the famous "400" trains —named because they ran (about) 400 miles in 400 minutes. The C&NW stole this march over their competitors by running refurbished standard rolling stock hauled by a modified existing steam locomotive, instead of trains brand new from end to end.

Soon enough, though, the C&NW had to follow their competitors' example. They chose to copy the style of the Milwaukee's "Hiawatha" rather than the Burlington's diesel "Zephyr" and accordingly the American Locomotive Company was asked to supply nine high-speed streamlined 4-6-4s.

The new locomotives, designated E-4 and road numbered 4000 to 4008, were delivered in 1938, but in the meantime the C&NW management decided it had backed the wrong horse and went to General Motors Electro-motive Division for some of the first production-line diesel loco- motives. These took over the new streamlined "400" trains, leaving the new 4-6-4s to work the transcontinental trains of the original Overland Route, which the C&NW hauled from Chicago to Omaha.

Because of the arithmetic of design the basic physical statistics of the E-4 were very close to the Hiawatha 4-6-4s, yet it is very clear that lesser differences between the two meant two separate designs. So we have two classes of six and nine locomotives respectively, intended for the same purpose, built by the same firm at the same time, which had few jigs or patterns in common. Such was the world of steam railway engineering.

Amongst the advanced features of the E-4 may be mentioned Baker's valve-gear, oil

firing, roller bearings throughout and, particularly interesting, a Barco low water alarm. Boiling dry such a large kettle as a locomotive boiler is a very serious matter indeed and on most steam locomotives there is no automatic guard against the crew forgetting to look at the water-level in the gauge glasses.

It is particularly sad that, when the time came in the early 1950s for these handsome locomotives to go to the torch, none of them was preserved.

SPECIFICATIONS

- **Gauge:** 4ft 8 1/2 in
- **Tractive effort:** 55,000 lb
- **Axleload:** 72,000 lb
- **Cylinders:** (2) 25 x 29 in
- **Driving wheels:** 84 in
- **Heating surface:** 3,958 sq ft
- **Superheater:** 1,884 sq ft
- **Steam pressure:** 300 psi
- **Grate area:** 90.7 sq ft
- **Fuel (oil):** 6,000 lb
- **Water:** 20,000 gallons
- **Adhesive weight:** 216,000 lb
- **Weight:** 719,500 lb
- **Length:** 101 ft 9 1/2 in

Left: E-4 Class No4009 awaits the torch in May 1956.

Right: One of the Chicago and North Western Railway's E-4 class "400" fleet of locomotives lined up for coal and water at a coaling station in December 1942.

Mallard A-4 Pacific LNER UK 4-6-2 (1938)

No4468 Mallard is a London and North Eastern Railway Class A-4 4-6-2 Pacific steam locomotive built at Doncaster, England in 1938. It is historically significant as the holder of the world speed record for steam locomotives of 126mph.

The A4 class was designed by Sir Nigel Gresley to power high-speed streamlined trains. The wind-tunnel-tested, aerodynamic body and high power allowed the class to reach speeds of over 100 miles per hour although in everyday service it was relatively uncommon for any steam-hauled service in the UK to reach even 90 mph, much less 100. Mallard covered almost one and a half million miles before it was retired in 1963.

It was restored to working order in the 1980s, but has not operated since, apart from hauling some specials between York and Scarborough in July 1986 and a couple of runs between York and Harrogate/Leeds around Easter 1987. *Mallard* is now part of the National Collection at the United Kingdom's National Railway Museum in York. On the weekend of 5 July 2008, *Mallard* was taken outside for the first time in years and displayed beside the three other A4s that are resident in the UK, thus reuniting them for the first time since preservation. It departed the museum for Locomotion, the NRM's outbase at Shildon on the 23 June 2010, where it was a static exhibit, until it was hauled back to York on 19 July 2011 and put back on display in its original location in the Great Hall.

The locomotive is 70 ft long and weighs 165 tons, including the tender. It is painted LNER garter blue with red wheels and steel rims.

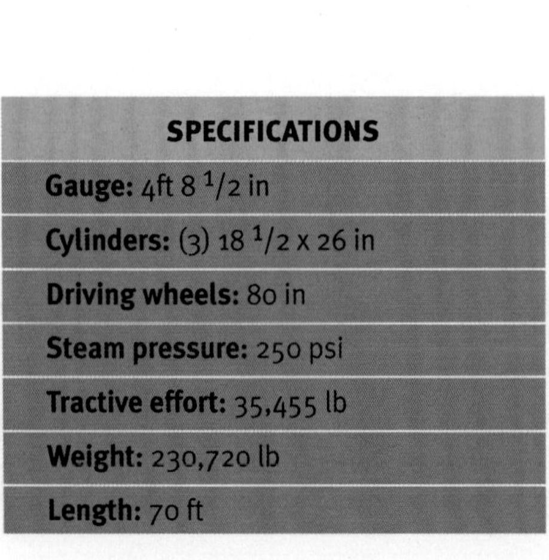

SPECIFICATIONS
Gauge: 4ft 8 1/2 in
Cylinders: (3) 18 1/2 x 26 in
Driving wheels: 80 in
Steam pressure: 250 psi
Tractive effort: 35,455 lb
Weight: 230,720 lb
Length: 70 ft

Right: Sir Nigel Gresley poses alongside London & North Eastern Railway locomotive No4498 named for him.Gresley chose a ''Big Engine'' policy for the LNER. This was especially suited to the London-Edinburgh mainline services. For other services, Gresley preferred to stay with well-tested pre-existing designs. This was particularly important because of the LNER's generally weak financial situation.

R-1 Class 4-8-4 Atlantic Coast Line (1938)

SPECIFICATIONS
Gauge: 4ft 8 1/2 in
Cylinders: (2) 27 x 30 in
Driving wheels: 80 in
Steam pressure: 275 psi (max)
Tractive effort: 63,900 lb
Weight: 460,270 lb
Length: 110 ft 11 1/4 in

Because of the growth of Florida tourism in the 1930s, the Atlantic Coast Line was experiencing a surge in its passenger business. To handle this heavy traffic, many of the main line trains were powered by 4-6-2 Pacifics pulling no more than 12 to 14 cars.

In 1937, the ACL received 12 new Class R-1 4-8-4s from the Baldwin Locomotive Works. They were assigned road numbers 1800 through 1811 and were immediately put into passenger service. These new 4-8-4s began to handle trains with as many as 21 heavyweight cars, eliminating the need for double-heading and running extra sections of many of the Richmond, Virginia to Jacksonville, Florida "Specials."

The Atlantic Coast line R-1's were built in 1938 by the Baldwin Locomotive Works for the road's lucrative passenger traffic between Richmond, Virginia and Jacksonville, Florida. The 1800's were the only "modern" steam locomotives owned by the ACL, and featured cast bed frames with cylinders and many accessories cast integrally, cast pilots, roller bearings on leading and driving axles, handsome Baldwin disk drivers, 275 PSIG boilers, Type "A" superheaters, Worthington SA feedwater heaters, front-end throttles, large fireboxes with combustion chambers and equipped with four thermic syphons, Elesco centrifugal steam separators, extensive mechanical lubrication, and massive 8-axle tenders. The engines featured jacketed smokeboxes and were painted in a sharp two-tone metallic gray/black paint scheme with silver striping and driver tires. The ACL herald on the tender was a separate embossed sheet metal disk which was attached to the tender, and which provided a dramatic change from earlier engines which merely had "Atlantic Coast Line" lettered on the tender.

These engines replaced double-headed USRA Pacifics on the road's passenger trains and their performance initially exceeded all expectations. During testing, no. 1800 accelerated a 20 car, 1500 ton passenger train (consisting of friction bearing heavy weight cars) from a dead stop to 70 miles per hour in 11 1/2 minutes and 11 miles. In passenger service, the R-1's made as high as 18,000 miles per month. The R-1's could hold to the scheduled running times of the fastest passenger trains with as many as 21 heavy weight cars. Initially, the maximum operating speed was limited to 80 mph, but (after the running gear balance was adjusted) their speed was later raised to 90 mph. Although 90 mph was the "official" speed limit for these engines, amateur observers often clocked them in excess of 100 mph on the ACL's mostly level and straight mainline.

The 1800's were well-designed steam locomotives and were nearly state-of-the-art for 1938, lacking only roller bearing axles on the trailing truck and tender axles. Their large fireboxes with combustion chambers and 4 Nicholson Thermic Syphons were among the largest in heating surface area of any 4-8-4 and undoubtedly made them prodigious steamers. One curiosity is Baldwin's use of relatively small piston valves for such large engines which seems to have been part of Baldwin's design philosophy at the time. The 1800's had only 12

Above: Atlantic Coast Line locomotive No1800 was the class leader among the 12 Class R-1 Steam Locomotives ,road numbered 1800-1811, purchased in 1938. These locomotive were heavy mainline steam power capable of hauling a 21 car heavyweight passenger trains at 80 mph.

inch diameter piston valves whereas even the ACL's much smaller Class P-5 Pacifics had 14 in valves. The R-1's also had somewhat smallish superheaters for their size (Type A, 1425 square feet). These two items would have restricted their "breathing" at high speed by impeding the steam flow into and out of the cylinders. Later 4-8-4 designs used 14 in piston valves with even longer valve travel and huge Type E superheaters to improve their high-speed performance. Nonetheless, the R-1s were known as outstanding performers on the ACL and their theoretical design deficiencies were not seen as drawbacks when they were in service.

O5A CB&Q (1939)

SPECIFICATIONS
Gauge: 4ft 8 1/2 in
Tractive effort: 67,500 lb
Axleload: 77,387 lb
Cylinders: (2) 28 x 30 in
Driving wheels: 74 in
Heating surface: 5,225 sq ft
Superheater: 2,403 sq ft
Steam pressure: 250 psi
Grate area: 106.5 sq ft
Fuel: 54,000 lb
Water: 18,000 gallons
Adhesive weight: 281,410 lb
Weight: 838,050 lb
Length: 105 ft 1 in

The Chicago, Burlington & Quincy Railroad, as its slogan says served "Everywhere West." In steam days its tracks went far beyond the relatively local implications of its name, serving St. Paul, Minneapolis, Kansas City, Omaha, Denver and St. Louis. In addition a north-south axis of wholly-owned subsidiary lines connected Billings, Montana, not far from the Canadian border, with Galveston, Texas, on the Gulf of Mexico.

In 1930 Baldwin supplied the Burlington with eight 4-8-4s intended for freight movement and classified 05. Subsequently in 1937 a further 13 of these giants were built. The locomotives were all coal-burners and not specially remarkable, although they did have Baker valve gear. Their one optional complication was a Worthington feed-water heater, as power reverse and mechanical stoker were essential for locomotives of such size and power. In the following years the Burlington's shops turned out 15 "Super 05s" designated 05A Although they were dimensional,

the same, a number of moder features were applied. There were "Box-pok" disc driving wheels in place of spoked, and Timken roller bearings to all axles as well as to the pins of the valve gear. Later on, some of the locomotives were equipped for oil-burning.

Generally, 4-8-4s are regarded as passenger power, but the 05s were classified as freight locomotives. One reason for this was that the CB&Q was the pioneer of diesel-electric streamlined trains and had been building up its fleet of such trains—the famous "Zephyrs"—ever since 1934. By 1939 all the principal routes were so operated by trains such as the "Denver Zephyr," "Twin Cities Zephyr" and others, so the need for powerful steam passenger locomotives was minimal. The other reason was that Burlington country was largely free from mountain grades. So a 4-8-4 had adequate adhesion for heavy freight haulage. The big firebox intrinsic to the type was needed for the high power output involved in running heavy long-distance trains at high speeds over straight alignments.

All except the first eight of the 36 05s and 05As were built by the Burlington in their own shops at West Burlington, Iowa.

The passing of steam left them completely in the hands of the two biggest of the big corporations. For a time after complete dieselisation the CB&Q kept an 05A (No.5632) in running order for special passenger trains.

Three surviving O5As are No5629 at the Colorado Railroad Museum, 5631 in Sheridan Wyoming , and 5633 at the Douglas Railroad Interpretive Center, Douglas Wyoming.

S-1 Pennsylvania Railroad 6-4-4-6 (1939)

SPECIFICATIONS
Gauge: 4ft 8 1/2 in
Cylinders: (2) 22 x 26 in
Driving wheels: 84 in
Steam pressure: 300 psi
Tractive effort: 71,900 lb
Weight: 608,170 lb
Length: 140 ft 2 1/2 in

The Pennsylvania Railroad S-1 class steam locomotive (nicknamed "The Big Engine") was a single experimental locomotive, the longest and heaviest rigid frame reciprocating steam locomotive ever built. The streamlined Art Deco styled shell of the locomotive was designed by Raymond Loewy for which he received U.S. Patent No2,128,490. The cost of the S-1 was $669,780.00, equal to $11,355,815 today. The S-1 was the only locomotive ever built with a 6-4-4-6 wheel arrangement. The six-wheel leading and trailing trucks were added, as the locomotive was too heavy for four-wheel units.

It was a duplex locomotive, meaning that it had two pairs of cylinders, each driving two pairs of driving wheels. Unlike similar-looking articulated locomotive designs, like the Mallet, the driven wheelbase of the S-1 was rigid.

The boiler for the S-1 was the largest built by the Pennsylvania Railroad Altoona Workshops, with 660 square feet of direct heating surface and 5001 in tubes and flues.

The S-1 was completed January 31, 1939 and was numbered 6100.

At 140 ft $2\,{}^{1}/_{2}$ in overall, engine and tender, No6100 was the longest reciprocating steam locomotive ever; it was too big for many PRR curves. Along with wheel slippage, this limited the locomotive's usefulness. In its brief service life it was restricted to the main line between Chicago, Illinois and Crestline, Ohio. It was assigned to the Fort Wayne Division and based at the Crestline enginehouse. No6100 hauled passenger trains such as The General and The Trailblazer on this route. Crews liked it, partly because of its very smooth ride. The great mass and inertia of the locomotive soaked up the bumps and the surging often experienced with duplex locomotives.

No further S-1 models were built as focus shifted to the T-1 class. The last run for No6100 was in December 1945 and the engine was scrapped in 1949.

Right: Industrial designer Raymond Loewy standing on one of his designs, the Pennsylvania Railroad's S1 steam locomotive.

FEF-2 CLASS 4-8-4 UP (1939)

Above: UP 844's 80 inch driving wheels and coupling arrangement represent the absolute zenith of steam locomotive engineering.

The FEF was a series of three steam locomotive types owned and operated by the Union Pacific Railroad. The classes were: FEF-1; FEF-2; FEF-3. "FEF" was an acronym for the wheel arrangement, "four-eight-four.

It began with 20 locos, road numbered 800 to 819, which were delivered in 1938.

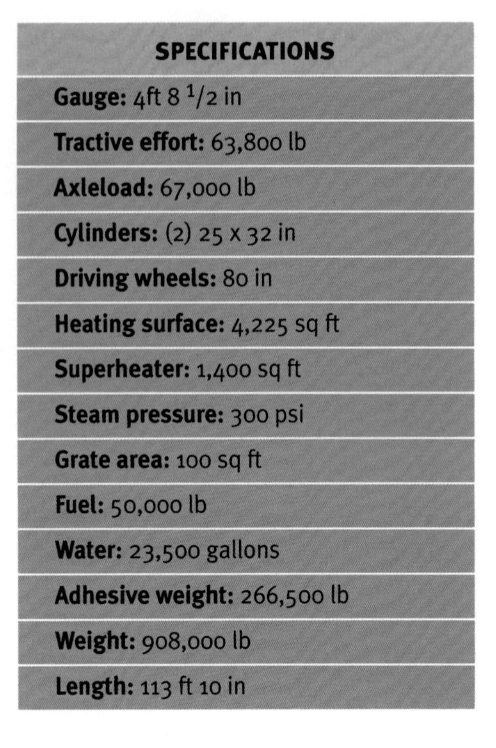

SPECIFICATIONS
Gauge: 4ft 8 1/2 in
Tractive effort: 63,800 lb
Axleload: 67,000 lb
Cylinders: (2) 25 x 32 in
Driving wheels: 80 in
Heating surface: 4,225 sq ft
Superheater: 1,400 sq ft
Steam pressure: 300 psi
Grate area: 100 sq ft
Fuel: 50,000 lb
Water: 23,500 gallons
Adhesive weight: 266,500 lb
Weight: 908,000 lb
Length: 113 ft 10 in

A further 15 (Nos.820 to 834) with larger wheels and cylinders as well as 14- wheel centipede tenders—instead of 12-wheel ones—came the following year and it is to these that the dimensions etc given above apply. This second batch was designated FEF-2, the earlier ones becoming class FEF-1.

A final batch of ten almost identical to the second one except for the use of some substitute materials, appeared in 1944. These were known as FEF-3s and were the last steam power supplied to UP. All the 800 s came from Alco.

The 800s as a whole followed the standard recipe for success in having two outside cylinders only, the simplest possible arrangement coupled to the 4-8-4 or Northern wheel arrangement being adopted as the conecrn originally felt regarding the suitability of eight-coupled wheels for very high speeds were found not to be justified. The negotiation of curves was made easier by the fitting of Alco's lateral motion device to the leading coupled wheels.

The basic simplicity of so many locomotives was often spoilt by their designers incorporating manufacturers'complicated accessories. The UP managed to resist most of them with the pleasing result that the locomotives had a delightfully elegant uncluttered appearance, unmarred by any streamline shroud. Instead benefiting from a thoroughly modern cast-steel locomotive frame, which replaced

many separate parts by one single casting. Another example was the use of a static exhaust steam injector instead of a steam-driven mechanical water-pump and feed water heater. A complication resisted by the UP was the provision of thermic syphons in the firebox; they held the view that on balance these quite common devices were more trouble than benefit. The speeds and forces involved in the drive train meant that current technology was taken beyond the then accepted limits but nonetheless the design of the rods and couplings were proven as well up to the job even though frequently moved at revolutions corresponding to running speeds above 100mph.The main principle of the new design was that the pulls and thrusts were transmitted from the connecting rods —and hence to three out of the four pairs

of wheels—by separate sleeve bearings instead of via the main crankpins in accordance with convention. The result was that separate knuckle-joints in the coupling rods were replaced by making the center pair of rods forked at both ends and combining the roles of crank-pins and knuckle-pins.

After World War II there was a period when coal supplies were affected by strikes and, in order to safeguard UP passenger operations, the 800s were converted from coal to oil burning; a 6,000 gallon tank was fitted in the bunker space. Otherwise only minor modifications were needed over many years of arduous service, a fact which is also much to the credit of the designers.

Normally the 4-8-4s were entrusted with the many expresses formed of the then conventional heavyweight stock, but the new engines' arrival on UP coincided with the introduction of diesel-electric streamline trains on much faster timings. In those early days the new form of motive power was not too reliable and 800 class locomotives

Above: Here is Union Pacific 844 at Lodgepole, Nebraska on 2 April 2010 as the locomotive is pulling a train eastbound. It has made a service stop in this small town in western Nebraska.

frequently found themselves replacing a multi-unit diesel at the head end of one of UP's important trains. They found no problem in making up time on the tight diesel schedules sufficient to offset extra minutes spent taking on water.

The last service passenger train hauled by an 800 was caused by such a failure; it occurred when in autumn 1958, the last one built took the "City of Los Angeles" over the last stretch of 145 miles from Grand Island into Omaha. No.844 gained time on the streamliner's schedule in spite of the crew's lack of recent experience with steam. A year later there came a time when all were out of service awaiting scrapping. Since then No.844 (renumbered 8444 to avoid confusion with a diesel unit) has been put back into service by a publicity-conscious Union Pacific and frequently performs for her fans.

H-8 Allegheny 2-6-6-6 (1941)

Sixty H-8 "Allegheny" class locomotives were built for the Chesapeake and Ohio Railway (C&O) between 1941 and 1948 by the Lima Locomotive Works. The "Allegheny" name refers to the C&O locomotives' job of hauling coal trains over the Allegheny Mountains.This was spurred on by the fight between competing roads, upon which cheap energy and therefore United States' industrial might depended. Just as the U.S. entered World War II in December 1941, the Lima Locomotive Works delivered to the Chesapeake & Ohio the first of the most super of all their celebrated 'super-power' designs. The design involved a unique 2-6-6-6 wheel arrangement and was totally modern and totally successful.

Though it was apparent to knowledgeable people in the railroad industry that steam locomotives might soon be replaced by diesel power, Lima and C&O set out to build the ultimate in high power steam locomotives and they succeeded. No diesel engine ever surpassed the output of these monsters, which were the heaviest steam locomotives ever constructed (by engine-only weight). The 3 axle trailing truck supporting the firebox was unusual, carrying over 190,000 lbs, allowing the huge firebox needed for the high power. As it turned out, steam locomotives continued in service almost another 20 years.

During the traumatic years which followed, 59 more were built, the last batch

SPECIFICATIONS
Gauge: 4ft 8 1/2 in
Tractive effort: 110,200 lb
Axleload: 86,350 lb
Cylinders: (4) 22 1/2 x 33 in
Driving wheels: 67 in
Steam Pressure: 260 psi
Superheater: Yes
Weight: 1,076,000 lb
Length: 130 ft 1 in

as late as 1948. The class was designated "H-8" and carried road numbers 1600 to 1659. A further eight were built for the nearby Virginian Railroad. Power output from a steam locomotive depends on the size of the fire and so the most important feature of the "H-8" was the huge deep firebox made possible by having that six-wheeled rear truck. The area of the grate was 11 per cent less than that of a Union Pacific "Big Boy," but the "H-8" firebox, not being situated above the rear driving wheels, was much deeper. Moreover, West Virginian coal was of better quality than that used on UP. Hence the H-8s could steam at higher rates, corresponding to a record drawbar-horsepower just short of 7,500. However, high horsepower involves high speed as well as a heavy pull, but it was some time before these vast machines—which took their name from the mountains they first conquered—had arrived in sufficient numbers to be used elsewhere and so prove their speed capabilities. About a third of the "Alleghenies" were fitted with steam connections for use on passenger work, on which their ability to reach 60mph was useful. On the climb from Hinton, West Virginia, eastward up to the 2,072ft summit at Allegheny tunnel, inclined at 0.57 per cent (l-in-175) two "H-8s'" one at each end, could manage an immense train of 140 cars weighing 11,500 tons. An important feature in this remarkable capability was the high adhesive weight upon which all hauling ability depends. A total of 254 tons was carried on the six driving axles, corresponding to an unprecedented axleload of well over 40 tons, a 37 per cent increase over that of their predecessors, the "H-7a" 2-8-8-2s. This was a world record for any major common-carrier railroad and only made possible because of Chessie's superb well-maintained heavy-duty permanent way. Even so, a great deal of track strengthening and tunnel enlargement had to be undertaken before the weight and bulk of the "H-8s" could be accommodated. The comparatively large driving wheels also helped in preventing problems with the weight of the loco digging in causing damage to the rails.

Right: A lost member of the H-8 class, No1630 in service on the C&O in the 1950s.

Below: Filling the 25,000 gallon water tank.

One H-8, the 1642, suffered a crown sheet failure and subsequent boiler explosion at Hinton, West Virginia, in June, 1953. The force of the explosion propelled the boiler endwise off the running gear, killing all three crew. While these locomotives had two sources of water for the boiler, a steam turbine pump fed Worthington hot pump and one injector, it is not known whether any were defective at the time of dispatch. According to the family of the locomotive's engineer, Wilbur H. Anderson, of Hinton, previous crews had complained of a faulty water level gauge. Anderson's widow, Georgia Anderson, was given $10,000 in compensation by the C&O

There are only two surviving Alleghenies, both housed indoors. One resides in The Henry Ford museum in Dearborn, Michigan (#1601), and the other at the B&O Railroad Museum in Baltimore, Maryland (#1604). Neither is in operational condition and they are likely to remain static displays given their incredible size and weight.

Southern Pacific No4449 GS-4 Class 4-8-4 (1941)

SPECIFICATIONS
Gauge: 4 ft 8 1/2 in
Cylinders: (2) 25 1/2 x 32 in
Driving wheels: 80 in
Tractive effort: 64,800 lb
Weight: 475,000 lb

Southern Pacific 4449 is the only surviving example of Southern Pacific Railroad's (SP) GS-4 class of steam locomotives. There is one other GS-class locomotive surviving, but it is a GS-6. The locomotive is a streamlined 4-8-4 (Northern) type steam locomotive. GS is abbreviated from "Golden State," a nickname for California (where the locomotive was operated in regular service), or "General Service." The locomotive was built by Lima Locomotive Works in Lima, Ohio, for SP in May 1941; it received the red-and-orange "Daylight" paint scheme for the passenger trains of the same name which it hauled for most of its service career. No4449 was retired from revenue service in 1956 and put into storage. In 1958 it was donated, by the railroad, to the City of Portland, who then put it on static display in Oaks Amusement Park, where it remained until 1974. It was restored to operation for use in the American Freedom Train, which toured the 48 contiguous United States for the American Bicentennial celebrations. As part of the American Freedom Train, the engine pulled a display train around the most of the United States. Afterwards, 4449 pulled an Amtrak special, the Amtrak Transcontinental Steam Excursion. After nearly two years on the road, 4449 was returned to storage in Portland, this time under protective cover and not exposed to the elements Since then, 4449 has been operated in excursion service throughout the continental US; its operations are based at the Oregon Rail Heritage Center in Portland, where it is maintained by a group of dedicated volunteers called Friends of SP 4449. In 1983, a poll of Trains magazine readers chose the 4449 as the most popular locomotive in the nation.

4449
4449
4449
4449

N-1 Class Pere Marquette 2-8-4 (1941)

SPECIFICATIONS
Gauge: 4ft 8 1/2 in
Cylinders: (2) 26 x 34 in
Driving wheels: 69 in
Steam Pressure: 245 psi
Tractive effort: 69,350 lb
Weight: 436,500 lb
Length: 101 ft 8 in

Left: Pere Marquette No1223 on permanent display in Grand Haven, Michigan. She is one of two surviving Pere Marquette 2-8-4 (Berkshire) engines.

Pere Marquette No1225 is a N-1 class 2-8-4 (Berkshire) steam locomotive built for Pere Marquette Railway (PM) by Lima Locomotive Works in Lima, Ohio in 1941. PM ordered this type of locomotive in three batches from Lima: class N in 1937 (PM road numbers 1201–1215), class N-1 in 1941 (numbers 1216–1227) and class N-2 in 1944 (numbers 1228–1239). No1225 cost $200,000 to build in 1941 ($3,206,787 in current dollars). No1225 is one of two surviving Pere Marquette 2-8-4 locomotives, the other being 1223 which is on display at the Tri-Cities Historical Society near ex-GTW coaling tower, in Grand Haven, MI.

The Pere Marquette Railroad used No1225 in regular service from the locomotive's construction in 1941 until the railroad merged into Chesapeake and Ohio Railway (C&O) in 1947. For the first part of its service life, 1225 was used to shuttle steel and wartime freight between Detroit, Saginaw, Flint and northern Indiana steel mills.

Class N-1 locos were renumbered to 2650–2661 after the merger. Part of the agreement, however, included the stipulation that locomotives that were acquired and fully paid for by PM would remain painted in PM livery after the merger. Although all the Berkshires received new numbers, only class N engines were repainted into standard C&O livery and renumbered. The majority of the class N locomotives were scrapped between 1954 and 1957, but class N-1s 1223 and 1225 both survived.

Slated for scrapping, No1225 was acquired by Michigan State University in 1957 and placed on static display. In 1971, work began to restore the locomotive to operation, an effort that culminated in its first excursion run in 1988. In 2010, when No1225 underwent its mandatory15 year inspection it was found that the firebox sheets had deteriorated to the point of needing replacement. That program was currently underway largely through small and large donations of funds and labor by the organizations supporters. Happily No1225 was back in service by late November 2013.

Below: The Berkshire had its biggest moment in 2004 when it was the star in the hit movie, The Polar Express, as the locomotive's blueprints where used to construct the digital version used in the movie. Even more, the locomotive's sounds and steam whistle were also incorporated into the movie.

Above: Lima- built class N-1 Berkshire locomotive No1220 at the clearing center Illinois 1949.

M-3 Yellowstone 2-8-8-4 DM&IR (1941)

Eight Class M-3 locomotives were built by Baldwin in 1941 for the Duluth, Missabe and Iron Range, which as the name implies was an outfit hauling iron ore from the mine to the port of Duluth, Minnesota on Lake Superior.Iron ore is heavy and the DM&IR operated long trains of ore cars, requiring maximum power. These locomotives were based upon 10 2-8-8-2s that Baldwin had built in the 1930s for the Western Pacific Railroad. The need for a larger, coal burning firebox and a longer, all-weather cab led to the use of a four-wheel trailing truck, giving them the "Yellowstone" wheel arrangement. They were the most powerful Yellowstones built, producing 140,000 lb of tractive effort, and had the most weight on drivers so that they were not prone to slipping.

These initial eight locomotives met or exceeded the DM&IR specifications so ten more were ordered (class M-4). The second batch was completed late in 1943 after the Missabe's seasonal downturn in ore traffic, so some of the new M-4s were leased to and delivered directly to the Denver & Rio Grande Western.The next winter the D&RGW again leased the DM&IR's Yellowstones as helpers over Tennessee Pass, Colorado and for other freight duties. The Rio Grande returned the Yellowstones after air-brake failure caused Number 224 to wreck on the Fireclay Loop. This was despite the Rio Grande's earlier assessment that these Yellowstones were the finest engines ever to operate there.

DM&IR's were the only Yellowstones to have a high-capacity pedestal or centipede tender, and had roller bearings on all axles. Some of the locomotives had a cylindrical Elesco feedwater heater ahead of the smoke stack, while others had a Worthington unit with its rectangular box in the same location.Only one Yellowstone was retired before dieselization took place on the Missabe; Number 237 was sold for scrap after a wreck. The rest of the 2-8-8-4s were retired between 1958 and 1963 as diesel locomotives took over.

Three of the eighteen built still

SPECIFICATIONS
Gauge: 4ft 8 1/2 in
Tractive effort: 140,000 lb
Axleload: 74,342 lb
Cylinders: (4) 26 x 32 in
Driving wheels: 63 in
Heating surface: 6,758 sq ft
Superheater: 2,770 sq ft
Steam pressure: 240 psi
Grate area: 125 sq ft
Fuel: 52,000 lb
Water: 25,000 gall
Adhesive weight: 565,000 lb
Weight: 1,138,000 lb
Length: 126 ft 5 in

survive and are on display: No227 at the Lake Superior Railroad Museum in Duluth, Minnesota, No225 in Proctor, Minnesota, and No229 in Two Harbors, Minnesota. DM&IR No227 is the best preserved of the three Missabe Yellowstones, thanks to the efforts of the Lake Superior Railroad Museum and members of the Missabe Railroad Historical Society.

Above: A Duluth, Missabe & Iron Range 2-8-8-4 pauses at Proctor, Minnesota.

Right: This impressive M-3 "Yellowstone" that welcomes you to Proctor as you drive up from Duluth was manufactured by the Baldwin Locomotive Works in 1941.It is one of eight locomotives in this class which DM&IR used in the Duluth-Proctor region.

Union Pacific "Big Boy" 4-8-8-4 (1941)

The Big Boy class initially comprised twenty 4-8-8-4 Class 1 locomotives delivered by Alco to Union Pacific in December 1941.They were road numbered 4000-4019.

The genesis of the "Big Boys" lay in the recovery of the US economy during the late-1930s from the Depression of 1929. In 1940 Union Pacific went to the American Locomotive Company (Alco) for a locomotive to handle yet heavier trains across the mountainous Wyoming Division between Cheyenne, Wyoming and Ogden, Utah. Out of Cheyenne going west the ruling grade was, until 1953, 1.55 per cent (l-in-65) on the notorious Sherman Hill, with a maximum elevation of 8,013ft at Sherman Summit. Eastwards out of Ogden, the crossing of the Wahsatch Mountains involved a stiff 60 miles uphill, much of it at 1.14 per cent (1 -in-88) as the rails climbed from 4,300ft altitude to 7,230ft at Altamont. The prime object of acquiring the new locomotives was to handle 3,600-ton trains unassisted on this latter section thus overcoming some of the limitations of the preceding 4-6-6-4 Challenger class locomotives. It was determined that the goals that Union Pacific had set for this new class of locomotive could be achieved by making several changes to the existing Challenger design, including enlarging the firebox to approximately 235 by 96 inches (about 155 sq ft), lengthening the boiler, adding four driving wheels and reducing the size of the driving wheels from 69 to 68 in. The Big Boys are articulated, as in the Mallet locomotive design. They were designed for stability at 80 miles per hour and were built with a wide margin of reliability and safety, as they normally operated well below that speed in freight service. Peak horsepower was reached at about 35 mph optimal tractive effort, at about 10mph. The locomotive without the tender was the longest engine body of any reciprocating steam locomotive in the world, hence the Big Boy name.

The engineering of the Big Boys was massive but wholly conventional. Such up-to-date features as cast-steel

SPECIFICATIONS
Gauge: 4ft 8 1/2 in
Tractive effort: 135,375 lb
Axleload: 67,750 lb
Cylinders: (4) 23 x 32 in
Driving wheels: 68 in
Heating surface: 5,889 sq ft
Superheater: 2,466 sq ft
Steam pressure: 300 psi
Grate area: 150 sq ft
Fuel: 56,000 lb
Water: 25,000 gall
Adhesive weight: 540,000 lb
Weight: 1,189,500 lb
Length: 132 ft 10 in

Right: UP "Big Boy" 4-8-8-4 at the head of freight extra X4019 conveying Pacific Fruit Express refrigerator cars through Echo Canyon, Utah. In accordance with UP practice, the train number of the extra is also that of the locomotive.

locomotive beds and roller bearings were adopted as a matter of course. All were coal burners except No.4005 which for a time burned oil. They are recorded as developing 6,290 horsepower in the cylinders, consuming some 100,000lb of water and 44,000lb of coal per hour whilst doing it. Towards the end of their lives, with the aid of the new line which reduced the grade to 0.82 per cent (1-in-122), the 4000s were taking 6,000-ton trains up Sherman Hill.

A dialog between UP's mechanical department and Alco's experienced design teams resulted in design work being completed in an amazingly short time of six months. Such was the confidence of UP management in the work done that they ordered the locomotives straight from the drawing board. The price was $265,174 each.

A further five were ordered and delivered in 1944, road numbered 4020-4024.

Yet even before the first one arrived at Omaha early in September 1941 , much preparatory work had to be done, including replacement of lighter rails with new 130lb/yd steel and new 135ft turntables at Ogden and Green River. Many curves had to be realigned, not so much because the locos could not get round them but rather that excessive overhang might mean contact with trains on adjacent lines. The front of the boiler swung out some 2ft sideways from the center of the track on a 10° radius curve. In fact, the maximum curvature that could be negotiated was as sharp as 20° and few locations on main tracks are as sharply curved as that. On good alignments the maximum speed of a Big Boy was about 70mph.

The last revenue train hauled by a Big Boy (No4015) ended its run on July 21, 1959. No4014 completed its last run earlier the same day at 1:50 in the morning.

Above: The impressive front end of Union Pacific articulated "Big Boy" 4-8-8-4 No.4002, built by the American Locomotive Company in 1941.

No4014 had traveled 1,031,205 miles for Union Pacific during its 20 years in service it was fully retired on December 7, 1961.

Most were stored operational until 1961, and four remained in operational condition at Green River, Wyoming until 1962. Of the 25 built, 8 were preserved at various locations around the United States. No4014 was donated by Union Pacific and is currently in Union Pacific's Steam Shop in Cheyenne, Wyoming, undergoing extensive restoration work which is intended to return the engine to operational status. When No4014 officially returns to service, it will displace UP 3985 as the largest and heaviest operational steam locomotive in the world.

Below: No4006 is one of twenty Class One Big Boy locomotives Road numbered 4000-4019 delivered in 1941 by Alco. It is now preserved at the St. Louis Museum of Transportation.

Challenger Class 4-6-6-4 (1942)

SPECIFICATIONS
Gauge: 4ft 8 1/2 in
Tractive effort: 97,400 lb
Axleload: 68,000 lb
Cylinders: (4) 21 x 32 in
Driving wheels: 63 in
Heating surface: 4,642 sq ft
Superheater: 1,741 sq ft
Steam pressure: 280 psi
Grate area: 132 sq ft
Fuel: 56,000 lb
Adhesive weight: 406,000 lb
Weight: 1,071,000 lb
Length: 1216 ft 11 in

The Union Pacific Challengers were a type of articulated 4-6-6-4 steam locomotive built by American Locomotive Company for the Union Pacific Railroad. A total of 105 of these locomotives were built between 1936 and 1943. The Challengers were nearly 122 ft long and weighed more than one million pounds. They operated over most of the Union Pacific system, primarily in freight service, but a few were assigned to passenger trains operating through mountain territory to California and Oregon. The locomotives were built specifically for Union Pacific and much of the experience gained later went into the design of the "Big Boy".

The name "Challenger" was given to steam locomotives with a 4-6-6-4 wheel arrangement. This means that they have four wheels in the leading pilot truck, which helps guide the locomotive into curves, two sets of six driving wheels, and finally four trailing wheels, which support the rear of the engine and its massive firebox. Each set of six driving wheels is driven by two steam cylinders. In essence, the result is two engines under one boiler.

In 1936 Union Pacific ordered 40 simple-expansion 4-6-6-4s with 69in driving wheels. They were road numbered from 3900 to 3939 .The leading bogie gave much better side control than a pony truck and the truck under the firebox assisted the fitting of a very large grate. The engines were distributed widely over the UP system and were used mainly on fast freight trains, but the last six of the engines were ordered specifically for passenger work. The most obvious difference between these earlier Challenger locomotives was the provision of much smaller 12-wheel tenders. Much of the coal which the UP used came from mines which the railroad owned.

In 1942 pressure of wartime traffic brought the need for more large engines and the construction of Challengers was resumed, a total of 65 more being built up to 1944. A number of changes were made, notably an enlargement of the grate from 108sq ft to 132sq ft cast steel frames in place of built-up frames, and an increase in the boiler pressure to 255psi accompanied by a reduction in cylinder size of one inch, which left the tractive effort unchanged. A less obvious but more fundamental change from the earlier engines was in the pivot between the leading unit and the main frame. In the earlier engines there were both vertical and horizontal hinges, but in the new engines, following the practice adopted in the "Big Boy" 4-8-8-4s, there was no horizontal hinge. The vertical hinge was now arranged to transmit a load of several tons from the rear unit to the front one, thus evening out the distribution of weight between the two sets of driving wheels, and thereby reducing the tendency of the front drivers to slip, which had

been a problem with the earlier engines. With no horizontal hinge, humps and hollows in the track were now looked after by the springs of each individual axle, as in a normal rigid locomotive.

All the engines were built as coal-burners, but in 1945 five of them were converted to oil-burning for use on passenger trains on the Oregon and Washington lines. Trouble was experienced with smoke obstructing the driver's view so these five engines were fitted with long smoke deflectors, and they were also painted in the two-tone grey livery which was used for passenger engines for a number of years.

Two examples survive today: Union Pacific 3985, used for excursion services by Union Pacific and Union Pacific 3977, on static display in North Platte, Nebraska.

Class T-1 4-4-4-4 PRR (1942)

The Pennsylvania Railroad's 52 T-1 class duplex-drive 4-4-4-4 steam locomotives, introduced in 1942 (2 prototypes) and 1945-1946 (50 production), were their last steam locomotives built and their most controversial. They were ambitious, technologically sophisticated, powerful, fast, and distinctively streamlined by Raymond Loewy. However, they were also prone to violent wheelslip both when accelerating and at speed, complicated to maintain, and expensive to run. Because of these problems the PRR decided in 1948 to place diesel locomotives on all express passenger trains, leaving unanswered the question of whether the T-1's flaws were solvable. It has been postulated that the wheel-slip problems were caused by the failure to properly train engineers transitioning to the T-1, resulting in excessive throttle applications, which in turn caused the wheel-slips on this very powerful locomotive. It was intended that the T-l s should have a full economic life before succumbing to diesels. In the event, the serious and intractable problems with them had the effect of accelerating dieselisation, and by the end of 1949 most of them were out of service.

SPECIFICATIONS
Gauge: 4ft 8 1/2 in
Tractive effort: 64,700 lb
Axleload: 69,000 lb
Cylinders: (4) 19 3/4 x 26 in
Driving wheels: 80 in
Heating surface: 4,209 sq ft
Superheater: 1,430 sq ft
Steam pressure: 300 psi
Grate area: 92 sq ft
Fuel: 85,000 lb
Water: 19,000 gall
Adhesive weight: 273,000 lb
Weight: 954,000 lb
Length: 122 ft 10 in

The story begins in the 1930s when there was a notable increase in the use of 4-8-4 locomotives in the United States, both-for freight and passenger service. There were, however, some problems with the very high piston thrust in these engines, and the resultant stresses in crank pins, while the balancing of the heavy reciprocating parts for high speeds also caused difficulties. All the problems could be solved, but R.P. Johnson, chief engineer of The Baldwin Locomotive Works suggested that they could be avoided by dividing the driving wheels into two groups in a single rigid frame, with separate cylinders for each, thus making the engine into a 4-4-4-4. Compared with the 4-8-4, piston loads were reduced, and it was easier to provide valves of adequate size but the rigid wheelbase was increased by the space required to accommodate the second set of cylinders. This increase was in itself sufficient to discourage some roads from further consideration of the proposal.

Despite the experience gained so far with their S-l, the PRR ordered two more duplex locomotives from Baldwin in July, 1940. The performance requirement was reduced to the haulage of 880 tons at 100mph and this could be met by a 4-4-4-4, with 80in driving wheels, and a grate area axle load was 69,250lb compared with 73,880lb of the S-l. The two engines, classified T-1 and road numbered 6110 and 6111 differed only in that 6111 had a booster. Apart from

the inclusion of certain PRR standard fittings, Baldwin was given a free hand in the design. Franklin's poppet valves were fitted at PRR insistence, as these had produced a notable increase in the power of "K-4" Pacific. Roller-bearings, light-weight motion, and disc wheels were amongst the modern equipment and the engine was clothed in a casing designed by Raymond Loewy, but quite different from that of No6100. They were delivered in April and May 1942. In 1944, No.6110 was tested on the Altoona testing plant and it produced a cylinder horsepower of 6,550 at 85mph with 25 per cent cut-off. In service the engines worked over the 713 miles between Harrisburg and Chicago, but despite these long runs they built up mileage slowly and spent an undue amount of time under repair. Slipping was again the main trouble, although in these engines 54 per cent of the total weight was adhesive.

At this point the PRR took a fateful step. Ignoring its old policy of testing and modifying a prototype until it was entirely satisfactory, it ordered 50 almost identical engines. Nos 5500-24 were built at Altoona and 5525-49 by the Baldwin Works and delivered between late 1945 and early 1946. With a shorter rigid wheelbase than the S-1 and a smaller maximum axle load, the Tls were allowed over the full steam-worked part of the PRR main line from Harrisburg to Chicago, and they worked through over the whole 713 miles. They took over all the passenger work on this route, including the 73.1 mph schedule of the Chicago Arrow over the 123 miles from Fort Wayne to Gary, and four other runs at more than 70mph At their best they were magnificent, with numerous records of 100 with 910-ton trains, including a pass-to-pass average of 100mph over 69 miles of generally falling grades with a load of 1,045 tons. They rode smoothly, and when all was well they were popular with the enginemen, but slipping remained a major hazard, not only slipping on acceleration but violent slipping of one set of wheels at high speed.

At this time the motive power department of the PRR was at a low ebb, both in equipment and in morale, and compared with the simple and well-known K-4 Pacifies, the T-l was overcomplicated, particularly its valve gear. Maintenance of the big engines proved to be a difficult job, and their appearances on their booked workings became less and less regular. The faithful K-4s were out again in force.

Various modifications were made to ease maintenance, mainly by the removal of parts of the casing, but one engine was rebuilt with piston valves. Eight engines had their cylinder diameter reduced in an attempt to reduce the tendency to slip but the problem was never solved. As time passed, the worsening financial state of the railroad led to the ordering of mainline diesels.

J-1a Class 2-10-4 (PRR) (1942)

Above: J-l class 2-10-4 with a freight train at Belleview, Ohio, in June 1956.

Soon after World War II began, it became apparent that heavier and more powerful locomotives would be needed to handle the surge in traffic on the Pennsylvania Railroad, due to the war effort.

With wartime restrictions on new locomotive designs imposed by the War Production Board the PRR was forced to borrow a design from another railroad. It chose the C&O's Class T-1 "Texas" type and the C&O loaned it a locomotive for evaluation. The PRR engineers prepared drawings based on the T-1 and the Altoona Shops built what would become North America's largest fleet of "Texas" type locomotives. In total the PRR's shops at Altoona, Pennsylvania would build 125 of these locomotives.

The C&O T-1 class of forty 2-10-4s were capable of moving 13,000-ton coal trains at a respectable speed over the less heavily-graded sections of their road. The T-ls were one of the Lima Locomotive Company's well-known "Super-power" designs and dated from 1930. The PRR classified their new acquisitions J-l and Jla, road numbered from 6150 to 6174 and 6401 to 6500. They were quite different in various fundamental ways from anything else in the stable but the Pennsylvania Railroad found them very good. For example, round-top fireboxes (instead of Belpaire with square corners), Baker valve gear and the provision of booster engines were almost unknown on the PRR. The modifications made to the C&O design were almost all superficial, but even so they changed the appearance almost beyond recognition. The 16-wheel tenders were of the unusual-looking standard PRR 210-F-84 type and that, combined with a cab having partially-rounded windows, took care of the C&O look at the rear. At the other end, solid rounded pilots with couplers normally dropped to a flush position, a high-mounted headlight and, of course, a smokebox numberplate cast in the form of the famous keystone herald in place of the "flying pumps," did the same for the front end.

SPECIFICATIONS
Gauge: 4ft 8 1/2 in
Tractive effort: 108,750 lb
Axleload: 83,116 lb
Cylinders: (2) 29 x 34 in
Driving wheels: 70 in
Heating surface: 6,568 sq ft
Superheater: 2,930 sq ft
Steam pressure: 270 psi
Grate area: 122 sq ft
Fuel: 60,000 lb
Adhesive weight: 377,800 lb
Weight: 984,100 lb
Length: 117 ft 8 in

No6325 Grand Trunk Western 4-8-4 (1942)

SPECIFICATIONS
Gauge: 4ft 8 1/2 in
Cylinders: (2) 26 x 30 in
Driving wheels: 73 in
Steam Pressure: 250 psi
Tractive effort: 59,000 lb
Weight: 403,000 lb

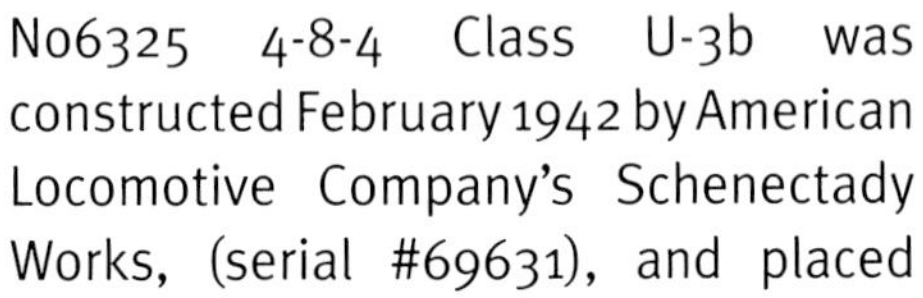

No6325 4-8-4 Class U-3b was constructed February 1942 by American Locomotive Company's Schenectady Works, (serial #69631), and placed in dual service with the Grand Trunk Western hauling both wartime materials and passenger trains between Detroit and Chicago.

In September 1946 this locomotive was chosen to pull President Truman's re-election campaign special, an assignment that would lead to No6325s eventual preservation in 1959 when it was put on outdoor display next to GTW's depot in Battle Creek, Michigan.

While No6325 languished in the spotlight its fellow locomotives continued racking up service miles until March 1960 when GTW officially dieselized its freight and passenger trains.

One other GTW 4-8-4 is saved, No6323 at the Illinois Railway Museum.

A dedicated group of local railroad enthusiasts undertook the project of rebuilding No6325 and bringing it back to full service, but enthusiasm and funds waned and by 1992 the 4-8-4 faced possible scrapping. The following year Jerry Jacobson purchased the loco and moved it to Coshocton, Ohio, for safe storage until he had the time and place to begin its rebuilding.

Finally six years later, and after nearly three years of reconstruction, No6325 steamed under its own power on July 31, 2001. The locomotive pulled numerous railfan trips and photographers' specials on Ohio Central rails, but in 2005 it was withdrawn from service with a worn driving axle bearing and has not operated since.

Below: No6325s usual place is in Age of Steam Roundhouse, Stall No.3, but it was separated from its tender and is undergoing minor repairs in the back shop.

Above: This is Northern No6323 in the Illinois Railway Museum Train shed.The Grand Trunk Western was one of the first US railroads to use 4-8-4 Northern type locomotives. As a subsidiary of the Canadian National Railway, however, they were known as "Confederations," the first 4-8-4 s having been delivered to the CN on the sixtieth anniversary of Canadian Confederation in 1927.

Y-6 Class 2-8-8-2 N&W (1942)

SPECIFICATIONS
Gauge: 4ft 8 1/2 in
Tractive effort: 152,206 lb
Axleload: 75,418 lb
Cylinders, HP: (2) 25 x 32 in
Cylinders, LP: (2) 29 x 32 in
Driving wheels: 58 in
Heating surface: 5,647 sq ft
Superheater:1,478 sq ft
Steam pressure: 300 psi
Grate area: 106 sq ft
Fuel: 60,000 lb
Water: 22,000 gall
Adhesive weight: 548,500 lb
Weight: 990,100 lb
Length: 114 ft 11 in

Above: Still working hard in 1951, Class Y-6 b No2181.

Thirty-five Y-6 Class locomotives were built at N&W's Roanoke shops between 1936 and 1940, sixteen Y-6as in 1942 and thirty Y-6bs from 1948 to 1952.

The Y-6 class locomotives were all state of the art, built with cast steel beds, fitted with Timken roller bearings on all axles, pressure and mechanical lubrication, and Baker vale gear with McGill multiroll bearings throughout. Like their predecessors, the Y-5s, the Y-6 class gained its distinctive look from the 8° forward slanted exhaust stack, which was designed to allow room for the front end throttle,a slightly off center smoke box door, inverted "Y" exhaust and crossed handrails on the pilot deck, and the massive low slung front cylinders.

Norfolk & Western's business was hauling coal from pits to port and after gaining experience with other people's Mallets for this work, Norfolk & Western in 1918 had built by their own shops at Roanoke, Virginia, a really successful coal-mover in the form of a 2-8-8-2 compound articulated steam locomotive. This Y-2 class received wholesale recognition because, later the same year, the United States Railroad Administration based their standard 2-8-8-2 on this excellent design. N&W were allocated 50 by USRA, who were then running the nation's railroads. The new locomotives, classified Y-3, came from Alco and Baldwin. In 1923 they were augmented by 30 more, designated Y-3a and a further 10 in 1927, Class Y-4. The Y-4s were the last new

Above: N&W Class Y-6b No2172 on a westbound freight at Shaffers Crossing, Roanoke, Va.

steam locomotives not to be built by N&W at Roanoke; they came from Alco's Richmond Works.

N&W moved forward, consolidating the ideas that had brought them success so far however the successive improvements made between the Y-2, Y-3, Y-3a and Y-4 classes were modest. For example, provision of feed-water heaters began with the Y-3a s, but earlier locomotives were altered to bring them into line so that operationally the group of classes could be regarded as one.

In 1930 the first of an enlarged version, Class Y-5, was produced. Over the next 22 years modern developments were successively introduced through classes Y-6, Y-6a and Y-6b and, as before, the earlier locomotives were rebuilt to bring them into line so that the fleet was kept uniformly up-to-date. The last Y-6b, completed in 1952, was the last main-line steam locomotive to be built in the US. The introduction of cast-steel locomotive beds with integrally-cast cylinders on the Y-6s was followed by rebuilding of the Y-5s to match. The old Y-5 frames were then passed on to the Y-3/Y4 group, by now relegated to local mine runs. One problem with the older Mallets was that they tended to choke themselves with the large volume of low-pressure steam because of inadequately-sized valves, ports, passageways and pipes. This was corrected in the Y-5 design and these locomotives now had the freedom to run and pull at speeds up to 50mph which can be considered an exceptional figure for a compound Mallet. Roller bearings came on the Y-6s of 1936, and with the Y6bs of 1948 was booster equipment for extra power, whereby a modicum of live steam, controlled by a reducing valve, could be introduced to the low-pressure cylinders while still running as a compound. It was quite separate from the conventional simpling valve used for starting. An interesting feature was the coupling of auxiliary tenders behind locomotives in order to reduce the number of water stops.

It says enough of the efficiency of these engines that following a line relocation they displaced in 1948, N&W's pioneer electrification in the Allegheny Mountains. In due time though, Norfolk & Western became the only railroad which saw a long-term future for steam and this position became untenable in the later 1950s. With reluctance, then, the Company was forced to follow the crowd and the last Y-6 set out on the last mine run in the month of April 1960. A Y-6a No2156 is preserved in the National Museum of Transportation at St Louis.

Class K-4 2-8-4 C&O (1943)

SPECIFICATIONS
Gauge: 4ft 8 1/2 in
Cylinders: (2) 26 x 34 in
Driving wheels: 69 in
Steam Pressure: 245 psi
Tractive effort: 69,350 lb
Weight: 460,000 lb

During World War II, the C&O turned to the 2-8-4 wheel arrangement to handle the fast freight schedule demanded by the war-time needs. The C & O had watched the development of the 2-8-4 on the Nickel Plate Road and the Pere Marquette through the "Advisory Mechanical Committee" which was common to the four railroads controlled by the Van Sweringens. It based its 2-8-4 design on the NKP and Pere Marquette "Berkshires." However it chose to name them "Kanawhas" after the Kanawha River, which paralleled its main line.

Between 1943 and 1947, the C & O purchased ninety, Class K-4, 2-8-4 Kanawhas, twenty from the Lima Locomotive Works and seventy from the American Locomotive Company. These locomotives were road numbered 2700 through 2789.

By mid 1952, the C & O had received enough diesels that it began to retire even the Kanawhas, which still had service time, and by 1957 all were retired. All but the thirteen that were donated to various cities were scrapped by May 1961.

Below: Chesapeake and Ohio Railway's "Big Mike" No2705 at the B&O Railroad Museum in Baltimore, Maryland, U.S. Built by American Locomotive Company ("Alco") in 1943, the Class K-4 2-8-4 steam locomotives were called "Kanawhas" on the C&O, also known elsewhere in the U.S. as "Berkshires."

USATC S160 Class 2-8-0 (1943)

SPECIFICATIONS
Gauge: 4ft 8 1/2 in
Cylinders: (2) 27 x 32 in
Driving wheels: 57 in
Steam pressure: 225 psi
Tractive effort: 31,490 lb
Weight: 161,000 lb
Length: 61 ft 0 in

During the 1930s, the United States Army Transportation Corps approved update of a Baldwin Locomotive Works World War I design in contingency for war transportation, to create the S159 Class. During the period of World War Two when America was neutral, the government of Franklin D. Roosevelt approved the Lend-Lease supply to the United Kingdom of the S200 Class, designed specifically to fit into the restricted British loading gauge.With America's entry to World War Two, the USATC needed a developed design from which to create a volume of locomotive power for the wrecked railways of Europe, which they could use to deploy military hardware and civilian goods. Hence the design created by Major J. W. Marsh from the Railway Branch of the Corps of Engineers learnt from both previous locomotives, designed on austerity principles and built using methods which created efficient and fast construction speed over long life, such as axlebox grease lubricators and rolled plates preferred to castings.

With cast frames and cast wheels, the front two driving axles were sprung independently from the rear two driving axles to allow for running on poor quality track. The larger tender layout was derived from the similar design for the WD Austerity 2-8-0, with the coal bunker inset above the water tank to improve visibility when running backwards.

Our example was originally numbered 5846 and left the Lima works in April 1945. Now featured as Norfolk & Western No606 at the Crewe Railroad Museum in Virginia.

800 locomotives were constructed in 1942-1946 in thirteen batches, split between ALCO, Baldwin and Lima Locomotive Works.

Above: S160 607 at the US Army Transportation Museum.

Norfolk & Western No1218 2-6-6-4 (1943)

Norfolk & Western 1218 is the sole survivor of Norfolk & Western Class A locomotives numbered 1200-1242, and the only surviving 2-6-6-4 steam locomotive in the world.

The Class A locomotives were in their day the strongest-pulling operational steam locomotives in the world. Their design is a four-cylinder, simple articulated locomotive with a 2-6-6-4 wheel arrangement. The Norfolk & Western Railway built No1218 in 1943 at its Roanoke Shops in Roanoke, Virginia, and was part of a fleet of fast freight locomotives. It was retired from regular revenue service in 1959, but Norfolk & Western's successor, the Norfolk Southern Railway, operated it in excursion service from 1987 to 1992. During No1218's excursion career, it was the most powerful operational steam locomotive in the world, with a tractive effort of 114,000 pounds which was well above the next-strongest-pulling operational steam locomotive (Union Pacific 3985, with a tractive effort of 97,350 lb). Unlike diesel-electric locomotives of similar high tractive effort (for starting heavy trains) but typical for a steam locomotive, it could easily run in excess of at 70 miles per hour.

Right: A Class Engine No1219 at Crewe,Virginia, in December 1953.

Below: Norfolk and Western Railway No1218 in railfan service in 1987.

Famed railroad photographer O. Winston Link's most famous photograph, "Hotshot Eastbound," was of one of 1218's sister engines at speed, passing a drive in theater in Ieager, West Virginia, in August 1956.

Today, 1218 is owned by the Virginia Museum of Transportation in Roanoke, Virginia, in the museum's Claytor Pavilion. It has been cosmetically restored, though not operational, since an overhaul started in 1992 was never completed. Although the undertaking would be considerable, she is very capable of being returned to operation, with the uncompleted boiler and firebox repairs being the primary scope of work remaining. No1218 is, on rare occasions, moved outside the museum grounds for special events. In 2007, Norfolk Southern pulled it (cold), to its Roanoke Shops for the shops' 125th anniversary celebration.

SPECIFICATIONS
Gauge: 4ft 8 1/2 in
Cylinders: (4) 24 x 30 in
Driving wheels: 70 in
Axle load: 72,000 lb
Steam pressure: 300 psi
Tractive effort: 114,000 lb
Weight: 573,000 lb
Length: 121 ft 9 in

No261 S-3 Class 4-8-4 (1944)

Chicago, Milwaukee, St.Paul & Pacific RR ("The Milwaukee Road") "S-3" Class 4-8-4 No.261 built by the American Locomotive Company (ALCO) in Schenectady, New York in June 1944, No261. 52 "S" Class were built:2 S1 in 1930/38built by Baldwin and the Milwaukee's own shops; 40 S2's by Baldwin in 1937-40 and 10 S3's by ALCO in 1944.. No261 operated with the railroad until retiring in 1954, and eventually donated to the National Railroad Museum in Green Bay, Wisconsin. As the new museum's first acquisition, No261 was moved to the museum site in 1958.

In 1991, the newly formed "North Star Rail" selected No261 for restoration for mainline excursions. It was selected for a variety of reasons. The engine was large enough to handle the expected trains at track speed. It featured several modern features for a steam locomotive, including axle roller bearings which are easier to maintain. It also already had its asbestos lagging removed, which is very expensive

Right: No261 at Harrison Street in Minneapolis, MN preparing for an excursion from Minneapolis Junction, September 2008.

SPECIFICATIONS
Gauge: 4 ft 8 1/2 in
Driving wheels: 74 in
Cylinders: (2) 26 x 32 in
Steam pressure: 250 psi
Tractive effort: 62,119 lb
Weight: 460 lb
Length: 109 ft 8 1/2 in

261
261

to remove for environmental and safety reasons. Finally, No261's relatively short 10 year service life meant that the engine's boiler was in better shape, meaning it would take less work to rebuild the engine.

North Star Rail and the National Railroad Museum came to an agreement in November 1991 for a ten year lease (which was later renewed ten years later). No261 was moved from Green Bay to Minneapolis to the GE shops at Humboldt Yard in September 1992. There, a full-time staff rebuilt the engine. Work progressed quickly, allowing for a hydrostatic test in June 1993, a test fireup in July, and the eventual restoration completion in September. After passing the FRA inspection on September 14, the engine deadheaded over Wisconsin Central in time for its first public excursions on September 18–19, 1993. The engine later returned to its new home at the leased Burlington Northern Minneapolis Junction.

The following year, 261 had an extensive season including excursions on the Wisconsin Central Railroad and the Twin Cities and Western Railroad. Notable events included "Chocolate City Days" excusions, campaign trains, a movie shoot painted as "Lackawanna 1661," running over CSX tracks for the famed "New River Train," and a wrap up celebrating the engine's 50th birthday.

The engine participated in the Steamtown National Historic Site's grand opening in July 1995. Over five days, No261 deadheaded from Minneapolis to Scranton, Pennsylvania. The locomotive stayed in Scranton for the next year pulling numerous excursions, including rare mileage trips, a rare snow plow run, and the engine's first steam doubleheader with Susquehanna No142. No261 returned to the Midwest after almost a year at Steamtown. On its way home, the engine made its first runs over the newly formed Burlington Northern Santa Fe Railway. It pulled a few sets of excursions in 1997 and 1998 over BNSF and TC&W tracks.

On September 29, 2012, No261 was test fired and ran under its own power once again and since then has ran on several prestigious excursions including National Train Day 2013,when No261 ran on an excursion north from Minneapolis to Duluth, where it met Soo Line No2719 for the first time returning to Minneapolis. The Fall Color excursions in 1913 to Willmar, Minnesota and Boylston, Wisconsin and again in 2014 to Duluth, Minnesota.

S-2 Class No765 2-8-4 (1944)

SPECIFICATIONS
Gauge: 4ft 8 1/2 in
Cylinders: (2) 25 x 34 in
Driving wheel: 69 in
Steam pressure: 245 psi
Tractive effort: 64,155 lb
Weight: 100 ft 8 1/2 in
Length: 440,800 lb

No765 is a steam locomotive built for the Nickel Plate Road in 1944 by the Lima Locomotive Works. The locomotive's construction was completed on September 8, 1944. Classified as an S-2 type steam locomotive it is based on a 2-8-4 wheel arrangement. It operated freight and passenger trains until retirement in 1958. Following a restoration in 1979 and after a major overhaul in 2005, No765 operates in public exhibition and passenger excursion train service. It is owned and maintained by the Fort Wayne Railroad Historical Society, Inc (FWRHS).

At the turn of the 20th Century, as freight and passenger traffic increased, railroads faced problems with the limitations of the steam technology of the day. Most railroads still used the older design 4-4-0 configuration engines that could pull heavy trains but at low speeds. The Lima Locomotive Works developed a new wheel arrangement: 2-8-4, to allow an increase in the size of the locomotive's firebox which allowed more coal combustion and subsequent heat output, improving the amount of steam developed and therefore increasing horsepower. These and other modifications

Below: Nickel Plate No765 leads an excursion during TrainFestival 2009, near Owosso, Michigan.

created the concept of locomotives that had excellent hauling power as well as speed Named for the rugged terrain of the Berkshire Mountains in which the design was proven, the 2-8-4 Berkshire-type locomotive, with two pony wheels, eight driving wheels, and four trailing wheels, became the first embodiment of the "Super-power" locomotive design that would change the course of locomotive development in the United States.

The 2-8-4 design was quickly adopted by other railroads like the New York Central, Erie Railroad, Illinois Central, Pere Marquette, Boston & Maine, and Chesapeake & Ohio, and the Nickel Plate Road.

The Nickel Plate Road was able to eventually employ 80 Berkshires on high-speed freight and passenger trains with the first order (designated S-1) supplied by Alco in 1941 based on Lima's design and the next two constructed and delivered by Lima in 1944 (S-2) and 1949 (S-3), respectively. As a group, these engines were referred to as the "Seven Hundreds. An additional number of Berkshires (S-4 class) were acquired when the Nickel Plate Road leased the Wheeling and Lake Erie Railroad in 1949. As a direct result of the Berkshire class, the railroad earned a reputation for high-speed service, which later became its motto.

Given their immense numbers up until the end of steam power in the United States, several Berkshires, like No765, would be preserved for future generations as others were sold for scrap.

Below: Nickel Plate Road No765 at Carland, Michigan.

Class 2900 ATSF 4-8-4 (1944)

The Santa Fe main line crossed the famous Raton Pass in the New Mexico with its 1 in 28 1/2 (3 1/2 percent) gradient, as well as the less impossible but still severe Cajon Pass in eastern California. East of Kansas City across the level prairies 4-6-2s and 4-6-4s sufficed until the diesels came, but for the heavily graded western lines Santa Fe in 1927 took delivery of its first 4-8-4s.

It was only by a small margin that the Northern Pacific Railroad could claim the first of the type as its own. These early 4-8-4s ,Nos. 3751 to 3764, were remarkable for having 30in diameter cylinders, the largest both in bore or volume in any passenger locomotive, apart from compounds.

This first batch burnt coal, subsequent 4-8-4s being all oil-burners. More 4-8-4s Nos.3765 to 3775 came in 1938 and a further batch was built in 1941. The final group Nos.2900 to 2929, on which the specification given here are based. Because of wartime restriction on high-tensile steel alloys, ordinary metal had to be used, which pushed up the weight

SPECIFICATIONS
Gauge: 4ft 8 1/2 in
Tractive effort: 79,960 lb
Axleload: 74,000 lb
Cylinders: (2) 28 x 32 in
Driving wheels: 80 in
Heating surface: 2,366 sq ft
Superheater: 2,366 sq ft
Steam pressure: 300 psi
Grate area: 108 sq ft
Fuel (oil): 7,000 gall
Water: 24,500 gall
Adhesive weight: 294,000 lb
Weight: 961,000 lb
Length: 120 ft 10 in

Above: No2925 is one of the final batch of Class 2900 built in 1944.

of the locomotives making them the heaviest passenger locomotives ever built. They managed this feat by a very small margin, but when those immense 16-wheel tenders were included and loaded there were no close rivals to this title. The big tenders were fitted to the last two batches; and as well as being the heaviest passenger locomotives ever built, they were also the longest.

Apart from the early diesel incursions, these 4-8-4s that totaled 65 ruled the Chicago-Los Angeles main line from Kansas City westwards. It was normal practice to roster them to go the whole distance -1,790 miles via Amarillo or 1,760 miles-via the Raton Pass. For steam locomotives these were by far the longest distances ever to be scheduled to run without change of locomotive. Speeds up in the 90-100 mph range were both permitted and achieved.This journey was not made without changing crews. Water was taken at 16 places and fuel nearly as often, in spite of the enormous tenders.

These magnificent examples of the locomotive builder's art were conventional in all main respects. One unusual feature was the "hot hat" smoke-stack extension shown on the picture above; absence of overbridges and tunnels over many miles of the Santa Fe route meant that this could be raised for long periods with beneficial effect in keeping smoke and steam clear of the cab. Another detail concerned a modification to the Walschaert's valve gear on some of the 4-8-4s. To reduce the amount of swing— and consequent inertia forces— needed on the curved links, an intermediate lever was introduced into the valve rod. This was so arranged as to increase the amount of valve travel for a given amount of link swing.

A number of AT&SF Northerns have been preserved. You can see another ATSF Class 2900, No2903, at the Illinois Railway Museum Train Shed and ATSF No2912 gradually moldering away on the Pueblo Railway Museum

Above: ATSF Class 2900 Northern (4-8-4) type No2913 is displayed in Riverview Park beside the Mississippi River in Fort Madison, IA.

The last group of 30 were built in 1943 and 1944. Wartime shortages of material resulted in ordinary metals being used for their construction. As a result, they were the heaviest Northerns ever built. They out weighed their nearest rivals by over 2000 pounds. This group was known as Class 2900 and included road numbers 2900 through 2929.

U-1-f 4-8-2 CN (1944)

SPECIFICATIONS
Gauge: 4ft 8 1/2 in
Tractive effort: 52,500 lb
Axleload: 59,500 lb
Cylinders: (2) 24 x 30 in
Driving wheels: 73 in
Heating surface: 3,584 sq ft
Superheater: 1,570 sq ft
Steam pressure: 260 psi
Grate area: 70.2 sq ft
Fuel: 40,000 lb
Water: 9,740 gall
Adhesive weight: 237,000 lb
Weight: 638,000 lb
Length: 93 ft 3 in

Canadian National Railways U-1-f class locomotives, were a class of twenty 4-8-2 or Mountain type locomotives built by Montreal Locomotive Works in 1944. They were numbered 6060–6079 by CN and nicknamed "Bullet Nose Bettys" due to their distinctive cone-shape smokebox door cover. It was in 1923, very soon after the formation of Canadian National Railways, that 4-8-2 locomotives were first introduced into passenger service there. This was the original "Ul-a" a batch consisting of 16 locos, built by the Canadian Locomotive Company. Then 1924 and 1925 brought the "Ul-b" and "Ul-c" batches of 21 and five

Above: Canadian National No6060 leaving Union Station with the last steam train to leave that station. Canadian National Railways U-1-f class locomotives, were a class of twenty 4-8-2 or Mountain type locomotives built by Montreal Locomotive Works in 1944.

Below: No6077 preserved as a static exhibit at The Northern Ontario Railroad Museum and Heritage Centre is a rail transport museum located in the community of Capreol in Greater Sudbury, Ontario, Canada.

Above: The tender of 6077 can hold 5,000 gallons of fuel oil and 11,000 gallons of water.

Above right: Here is the famous bullet nose.

from Canadian and from Baldwin respectively. The latter were for CN's Grand Trunk Western subsidiary in the USA. In 1929 and 1930 there followed five "U1-d" and 12 "U1-e" from Canadian and from the Montreal locomotive works.

Thus in seven years, fifty-nine 4-8-2s, numbered from 6000 to 6058, became available, although by now the class had become overshadowed by the 4-8-4s introduced in 1927. There were also four 4-8-2s acquired by the Central Vermont Railway, another CN subsidiary but one which did not then number or classify its locos as part of the main CN fleet. It did use the CN method of classification, though, so these 4-8-2s were also Class "U1-a." In fact they were rather different in design, having been acquired from amongst a flood of 4-8-2s which the Florida East Coast Railroad had ordered but found itself unable to pay for.

The 6000s performed excellently on the then highly competitive express trains between Montreal and Toronto; speeds up to 82mph have been noted with 700 tons or so. Later, the same engines operated well in pool service in conjunction with Canadian Pacific.

In 1944, a further twenty 4-8-2s were delivered from Montreal. They were brought up to date by having cast-steel locomotive frames, disc wheels and other improvements. Some were oil-burners and all had Vanderbilt cylindrical tenders and outside bearings on the leading bogies. Most significant was a major simplification consisting of the replacement of the boiler feed pump and feed-water heater, by a device called an exhaust steam injector. Injectors are usually tucked away tidily under the side of the cab but in this case the device was hung outside the driving wheels, the large pipe which supplied the exhaust steam adding to its conspicuousness.

With just a few exceptions, CN steam locomotives were totally utilitarian, but with these excellent engines, efforts were made to make them good looking too. Side valences, a flanged British-style smokestack, green and black livery, brass numbers and placing the dome and sand container in the same box all contributed to the clean lines except of course the bullet nose.

No29 Duluth & Northwestern 0-6-0 (1944)

SPECIFICATIONS
Gauge: 4 ft 8 1/2 in
Cylinders: (2) 21 x 28 in
Driving wheels: 51 in
Steam Pressure: 190 psi
Tractive effort: 40,000 lb
Weight: 154,500 lb

No29 is a 0-6-0 switch engine built at the Lima Locomotive Works, Lima, Ohio in 1944. It is coal fired and weights 77 tons. No29 originally worked for the Bay Terminal Railroad in Toledo, OH and then for the Duluth and Northeastern in Minnesota before arriving at Prairie Village in 1976. After 12 years and much needed work on the firebox and grates, No29 run was able to run during the 2013 Jamboree under its own power.

Work on No29's own steam driven compressor was completed early in 2014 and it will be re-installed in the spring.

Above: Prairie Village Herman & Milwaukee 0-6-0 No29 (ex-Duluth & Northeastern No29) steaming up outside the replica roundhouse at Historic Prairie Village, Madison, South Dakota.

S-2 Class Pennsylvania RR 6-8-6 (1944)

SPECIFICATIONS
Gauge: 4ft 8 1/2 in
Driving wheel: 68 in
Steam pressure: 310 psi

In a final attempt to secure the continued use of steam as a power source for railroads the Baldwin Locomotive Works produced this steam turbine locomotive for the Pennsylvania Railroad in 1944.

Designated as class S-2, only one was built, road numbered No6200.

The S-2 was the sole example of the 6-8-6 wheel arrangement with a six-wheel leading truck, eight driving wheels, and a six-wheel trailing truck. The S-2 used a direct-drive steam turbine; the turbine was geared to the center pair of axles with the outer two axles connected by side rods. The disadvantage was that the turbine could not operate at optimal speeds over the locomotive's entire speed range. The S-2 was the largest direct-drive turbine locomotive design ever built.

The locomotive was originally designed as a 4-8-4, but wartime restrictions on light steel alloys increased its weight until six-wheel leading and trailing trucks were needed. Two turbines were fitted, one for forward travel and a smaller one for reversing at speeds up to 22 mph. A large boiler with a Belpaire firebox and long combustion chamber was fitted. The turbine exhaust was piped through a set of four nozzles in the smokebox, providing an even draft for the fire and exiting through a unique quadruple stack. A Worthington-pattern feedwater heater was fitted for increased efficiency. Twin air pumps for train braking were fitted below the running boards beside the smokebox front, and a large radiator assembly at the nose cooled the compressed air. The large 16-wheel tender was similar to that used on the PRR's other large passenger locomotives, the T-1 and S-1.

The locomotive proved to be powerful and capable, with reserves of power at speed and reasonable fuel economy. The turbine drive was easy on the track and allowed more power at the rail. While economical at speed, the locomotive was highly uneconomical at lower speed. The turbine used less steam than conventional locomotives above 30 mph; below that, the locomotive used too much steam and fuel. The boiler normally operated at 310 psi but at low speed the pressure could drop as low as 85 psi. The increased fuel usage at low speeds caused the firebox to run hotter, which sometimes caused stay bolts to break.

The locomotive's problems and the advantages of the emerging Diesel locomotive ensured that this design would never be duplicated. The locomotive was withdrawn from service in 1949 and scrapped in 1953.

Above: Pennsylvania Railroad No6200 at Chicago, Illinois in July 1945.

Left: The Baldwin promotional photo of the S-2.

T-1 Class Northern 4-8-4 (1945)

SPECIFICATIONS
Gauge: 4ft 8 1/2 in
Cylinders: (2) 27 x 32 in
Driving wheel: 70 in
Steam pressure: 240 psi
Weight: 441,300 lb
Length: 45 ft 10 in

This class of locomotive was developed for fast freight services by the Reading Railroad Company together with the design engineers at the Baldwin Locomotive Works to develop a plan to convert 30 of the railroad's I-10a 2-8-0 Consolidations to new 4-8-4 Northerns. It was the idea of Revelle W. Brown , the president of the Company and supervised by the company's superintendent of motive power and rolling equipment, E. Paul Gangewere.

These 30 steam locomotives (road numbers 2100 -2129) were rebuilt in the railroad's own shops in Reading, PA, between 1945 and 1947. The conversions were made with several reclaimed parts from the I-10a Consolidations (road numbers 2020-2049) and new parts furnished from Baldwin and other suppliers.

The first 20 locomotives converted were put into freight service and were regularly used for coal traffic. The last 10 were equipped for passenger service but, except for a few troop trains, were also used for freight.

The Reading Class T-1's were handsome heavy-duty locomotives which saw steady work until diesels took over freight service in the early 1950's. As they were taken out of service, the T-1's were stored in serviceable condition at the Reading roundhouse.

By 1956, steam was dead and virtually all steam locomotives but the T-1's were gone from the Reading. Three years later the T-1's would begin a new and more celebrated career pulling train loads of railfans on excursions through the Pennsylvania country side. These outings became known as "Reading Rambles" from 1962-64.

Of the 30 T-1's built by the Reading only four survive today One such lucky locomotive No2102 was built in October 1945 by the RDG Co., Reading, PA shops using components from former class I-10sa Consolidation (2-8-0) #2044. No2102 was used on the Reading Rambles. Sold in1966 to Steam Tours, and subsequently operated under many sponsors and railroads. Finally brought home to Reading in 1985 and later sold to Andrew Muller, operating excursions into the mid 90's on both the Blue Mountain & Reading tourist line and the Reading & Northern.

Left: No2102 on a ramble.

Niagara Class 4-8-4 NYC (1945)

SPECIFICATIONS
Gauge: 4ft 8 1/2 in
Tractive effort: 61,570 lb
Axleload: 71,680 lb
Cylinders: (2) 25 1/2 x 32 in
Driving wheels: 79 in
Valve gear: Baker valve gear
Superheater: 2,770 sq ft
Length: 115 ft 5 1/2 in
Weight: 907,200 lb

The first Niagara was Class S-1a No6000 in 1945; the S-1b (6001-6025) were delivered in 1945-46. By the 1940s loads being hauled on the New York Central main line from New York to Chicago were as much as the famous J-class NYC Hudson 4-6-4s could handle. The Chief of Motive Power for the railroad, Paul W. Kiefer, decided to order some 4-8-4's which could sustain 6,000 horsepower on the run between the two cities, day after day without respite.

The American Locomotive Company ALCO proposed these locomotives, and although the design owes something to the Union Pacific 4-8-4's, of which Union Pacific 844 is the best-known, the design was actually quite new. Some steam experts have claimed the Niagara to be the ultimate locomotive, as it had the speed of an FEF (the Union Pacific's nickname for their "four eight fours" was FEF)

Above: No6001 at Harmon Sheds, New York waiting for its next duty.

and the power of Northerns with smaller driving wheels.

The NYC's last steam locomotive was Class S-2 No5500 which had poppet valves. The New York Central Railroad's Niagara Class was named after the Niagara River and Falls. The class is considered one of the most efficient type 4-8-4s ever built.

The Niagaras did not have steam domes, as did most steam locomotives, which resulted in a smooth contour along the top of the boiler. A perforated pipe collected steam instead. This was necessary because of the lower loading gauge of the New York Central (15 ft 2 in versus 16 ft 2 in for other American railroads).

These locomotives had a small water capacity 18,000 gallons; in the tender, because the New York Central was one of the few in North America which used track pans. This allowed a larger coal capacity—46 tons—so the New York to Chicago run could be done with one stop for coal. The stop was said to be at Wayneport, New York, 14 miles east of Rochester, but that would leave 603 miles to Chicago via the Cleveland lakefront.

On test these locomotives reached 6,600 hp at the cylinders, and ran 26,000 miles per month. All were scrapped in the 1950s.

Above: A fine illustration of No 600 by celebrated postcard artist Howard Fogg.

Class AG-2 2-6-6-6 (1945)

SPECIFICATIONS
Gauge: 4ft 8 1/2 in
Tractive effort: 110,200 lb
Axleload: 86,350 lb
Cylinders: (4) 22 1/2 x 33 in
Driving wheels: 67 in
Steam Pressure: 260 psi
Superheater: Yes
Weight: 1,076,000 lb
Length: 130 ft 1 in

The Virginian Railway was a small railroad built for one primary purpose, coal hauling operations. The railroad was exceptionally well designed and built and operated many large and high horsepower locomotives to move coal from the mines through the Blue Ridge Mountains to the east coast of Virginia. The railroad needed more power due to increased traffic during World War II and ordered eight 2-6-6-6s. They took delivery of them from the Lima Locomotive Works between March and June of 1945. These locomotives were designated as Class AG, part of Lima's "Superpower" range of locomotives .On the Virginian these clones of the C&O "Alleghenies" were called "Blue Ridge" type locomotives and were assigned road numbers 900 through 907. They were the heaviest reciprocating steam locomotives ever built, at 389 tons for the locomotive itself plus 215 tons for the loaded tender.

They used them for hauling coal trains well into the 1950s. All of the "Blue Ridge" locomotives were retired by 1955 and were scrapped by 1960.

L-1 Class 4-6-4 (1946)

SPECIFICATIONS
Gauge: 4ft 8 1/2 in
Cylinders: (2) 27 x 32 in
Driving wheel: 70 in
Steam pressure: 240 psi
Weight: 441,300 lb
Length: 45 ft 10 in

Starting in 1946, the Chesapeake & Ohio took five of its Class F-19 Pacific 4-6-2 locomotives and converted them into Class L-1 Hudsons. This work was done in its Huntington Shops and was completed in 1947. All but one of the new L-1s were covered with a streamlined stainless steel cowl which was painted yellow and silver. The tenders were cased in fluted stainless steel and tapered at the top so they would blend exactly with the new Budd passenger cars. The "yellowbellies" as they were called by C&O crews were numbered 490 through 494 (number 494 did not have a cowl applied). Only 490 survives today and is on display at the B&O Railroad Museum in Baltimore, MD.

No490 was originally constructed by the American Locomotive Company (ALCO) for the Chesapeake & Ohio Railway (C&O) in 1926. As one of the five F-19 4-6-2 "Pacific" locomotives built for the C&O, No490 was used on passenger trains on the mainline east of Charlottesville and west of Clifton Forge. In 1930, the No490 was assigned to the "Sportsman," the premier C&O passenger train at the time. Later, No490 was transferred to the "George Washington." No490 and other 4-6-2s operated between the Cincinnati and Washington route until 1942 when the C&O replaced the "Pacifics" with new heavy Baldwin 4-8-4. The No490 and other 4-6-2s continued to run as secondary passenger trains during World War II.

Shortly after the end of the war, the C&O decided to upgrade their passenger service. The C&O was primarily a coal-hauler

and therefore wanted to improve steam locomotive technology. They developed a new luxury liner between Washington and Cincinnati named the "Chessie." The new liner was to be powered by experimental steam-turbine-electric locomotives.

In addition, the C&O's Huntington shops rebuilt the No490 and the other 4-6-2s into 4-6-4 "Hudsons." The new locomotives had roller bearings, front-end throttle, high-speed booster, cross counterbalance, and the Franklin system of steam distribution.

Above: L-1 without streamlining.

Due to the increased automobile production and the airline expansion that occurred after the war, the luxury Chessie passenger trains never ran. The rebuilt "Hudsons" instead hauled regular passenger trains until 1953. The No490 was stored in the Huntington Roundhouse until 1968 when it was moved to the Baltimore & Ohio Railroad Museum.

QR-1 Class 4-8-4 (1946)

SPECIFICATIONS
Gauge: 4ft 8 1/2 in
Cylinders: (2) 25 x 30 in
Driving wheel: 70 in
Steam pressure: 255 psi
Tractive effort: 58,126 lb

The Ferrocarriles Nacionales de Mexico (National Railways of Mexico) bought a total of 32 Northern type, Class QR-1, locomotives in 1946. NdeM purchased 16 from the American Locomotive Company and 16 from the Baldwin Locomotive Works.

On the NdeM these locomotives were called "Niágaras" and were among the few that it bought new. Most of its motive power was obtained second-hand from U. S. railroads.

All 32 of the "Niágaras" (road numbers 3025 through 3056) had 70" drivers, 25 x 30 cylinders, a boiler pressure of 250 psi, a tractive effort of 57,000 lbs and weighed 387,000 pounds making them amongst the lightest of the Northern type.

The Niagaras had a very good reputation in service and were felt to be both nimble and graceful engines which were an ideal match between a locomotive design and the environment of its railroad as was ever achieved during the steam era.

They were described as thoroughly modern in appearance and in specifications, the 32 engines brought a useful increase to the roster of the hard working N. de M. Their design included a good compromise of features that was absolutely essential for the widely diversified areas where they ran. For over 20 years these "Niagaras" served throughout Mexico, handling heavy international freight and passenger movements between Mexico City and Nuevo Laredo as well as regular runs to Guadalajara and Aguascalientes through the difficult high altitude mountain country.

Left: Nacionales De Mexico (NdeM) QR-1 class 4-8-4 No3051 slowly backs past a track gang at Tula in 1966. Tula was the helper station situated approximately 50 miles north of Mexico City. By 1966 only the standard gauge 4-8-4s and narrow gauge 2-8-0s were still in regular service on the NdeM.

M-1 Class Steam Turbine C&O (1947)

SPECIFICATIONS
Type: Steam turbine electric
Builder: Baldwin Locomotive Works
Serial number: 73079–73081
Total produced: 3
Configuration: 2-C1+2-C1-B
Gauge: 4 ft 8 1/2 in
Driver diameter: 40 in
Length: 154 ft 3/4 in
Weight: 857,000 lb
Fuel type: Coal
Fuel capacity: 29.25 short tons
Water capacity: 25,000 gallons
Boiler pressure: 310 psi
Maximum speed: 100 miles per hour
Power output: 6,000 hp (turbine)

The Chesapeake and Ohio class M-1 was a fleet of three steam turbine locomotives built by the Baldwin Locomotive Works for the Chesapeake and Ohio Railway in 1947–1948 for service on the *Chessie* streamliner. At the time of its construction it was the longest single-unit locomotive in the world. They were road numbered Nos 500-502.

As diesel locomotives became more prevalent following World War II, the C&O was one of several railroads loath to abandon coal as a fuel source, and saw steam turbine technology as a possible alternative to diesel. C&O was reluctant to dieselize because its management felt that since the principal commodity it hauled was coal, it should retain coal-fired motive power. For that reason, there was no early experimentation with diesel-electric motive power as on many other railroads. However, seeing the obvious economics of diesels, management tried to find a middle ground solution. The result was the huge M-1 class Steam-Turbine-Electrics of 1947-48. They incorporated the efficiency of electric drive, but instead of a diesel prime mover, standard steam locomotive power and steam turbine supplied the power for the electric generator. The M-1 class steam turbines were used briefly, but the maintenance required for its standard steam generating plant could not match the lower maintenance costs of the new diesels. More money and time was spent on research of a coal-gas turbine, without success. The locomotives became known as the Sacred Cows and were scrapped in 1950.

Above and right: The M-1s were 106 feet long, making them the longest locomotives ever built for passenger service. The cab was mounted in the center, with a coal bunker ahead of it and a backwards-mounted conventional boiler behind it (the tender only carried water).

L-2a Class 4-6-4 C&O (1948)

SPECIFICATIONS
Gauge: 4ft 8 1/2 in
Tractive effort: 52,100 lb
Axleload: 73,500 lb
Cylinders: (2) 25 x 30 in
Driving wheels: 78 in
Heating surface: 4,233 sq ft
Superheater: 1,810 sq ft
Steam pressure: 255 psi
Grate area: 90 sq ft
Fuel: 60,000 lb
Water: 21,000 gall
Adhesive weight: 219,500 lb
Weight: 839,000 lb
Length: 108 ft 0 in

When in 1947 the Chesapeake & Ohio Railway went to Baldwin of Philadelphia for five 4-6-4 locomotives little did they realize that they were to be the last steam express passenger locomotives supplied for home use by any of the big US constructors. The C&O divided its routes into mountain and plains divisions and the eight-coupled engines were for the former, the six-coupled ones for the latter. There was, therefore, scope for the 4-6-4s, both north-west of the Allegheny mountains on the routes to Louisville, Cincinnatti, Chicago and Detroit, as well as south-east of them in the directions of Washington and Richmond, Virginia.

The C&O ran through the big coalfields and at that time hauled more coal than any other railroad. It was therefore unthinkable that anything but coal-burning power should be used. Amongst his plans was one for a daytime streamline service actually to be known as The Chessie—and three steam-turbine locomotives with electric drive and 16 driving wheels were built in 1947-48 to haul it on the main stem and over the mountains. Conventional steam was to haul connecting portions and provide back-up. Unfortunately these proved too unreliable and uneconomical and in two years the turbo-electrics (Class M-1, Nos.500-502) were scrapped .In the meantime the whole C&O streamline project had been scrapped, but not before some older 4-6-2s (the F-19 class) had been converted into streamlined 4-6-4s to handle the new train over part of its route. The C&O ordered these L-2a Hudsons, intending them to be streamlined. Road numbers were 310 to 314 and fortunately they were as trouble-free as the turbines had been troublesome.

On various important counts the 4-6-4s were the top six-coupled locomotives of the world—in engine weight, at 443,000lb 7 1/2 per cent above those of the nearest rival, the Santa Fe. In tractive effort, both with and without their booster in action, the latter worth 14,200lb of thrust, and adhesive weight, the figures are records. The massive qualities of C&O track are illustrated by the fact that their adhesive weight is also unmatched elsewhere. Technically the engines represented the final degree of sophistication of the American steam locomotive that came from nearly 120 years of steady development of practice and details upon the original principles. The L-2a class was developed from the eight L-2 class 4-6-4s of 1941 (Nos.300-307) and differed from them mainly in having Franklin's system of rotary-cam poppet valves instead of more conventional Baker's gear and piston valves. These locomotives also were notable for having unusually clean lines with the absence of the air pumps on the smokebox door and the headlight was cleared away and mounted above the pilot beam.

Despite the excellence of these engines by 1953 C&O's passenger service had become 100 per cent dieselized.

Selkirk Class 2-10-4 CPR (1949)

SPECIFICATIONS
Gauge: 4ft 8 1/2 in
Tractive effort: 76,905 lb
Axleload: 62,240 lb
Cylinders: (2) 25 x 32 in
Driving wheels: 63 in
Heating surface: 4,590 sq ft
Superheater: 2,055 sq ft
Steam pressure: 285 psi
Grate area: 93.5 sq ft
Fuel (oil): 4,925 gall
Water: 14,000 gall
Adhesive weight: 311,200 lb
Weight: 732,500 lb
Length: 97 ft 10 in

Ten-coupled locomotives were used in most parts of the world for freight movement; in fact, the only steam locomotives in quantity production in the world today are 2-10-2s in China. Because the length of a rigid wheelbase has to be limited, five pairs of coupled wheels implies that they are fairly small ones and this in turn means (usually) low speeds. Perhaps the ten-coupled engines with the best claim to be considered as express passenger locomotives were the 2-10-4 Selkirk class of the Canadian Pacific Railway. Not only were they streamlined but a colored passenger livery was also used for them; also, of course, they handled CPR's flag train, then called the "Dominion," across the Rockies and the adjacent Selkirks.

The overall story was very similar to that of the CPR Royal Hudson. First came the slightly more angular T-1a batch; 20 (Nos.5900 to 5919) were built in 1929. A further ten T-1b with softer and more glamorous lines were built in 1938 and, finally, another six T-lc came in 1949. No5935 was not only the last of the class but the last steam locomotive built for the company and, indeed, for any Canadian railway. The "Royal Hudson" boiler was used as the basis, but enlarged and equipped for oil-burning, since all locomotives used on the mountain division had been fired with oil since 1916.

Crossing Canada by CPR was 2,882 miles from Montreal to Vancouver this being reasonably easy going apart from a section along the north shore of Lake Superior and, more notably, the 262 miles over the mountains between Calgary and Revelstoke. Until the 1950s CPR's flag train, the "Dominion," could load up to 18 heavyweight cars weighing some 1,300 tons and to haul these up the 1 in 45 (2.2 per cent) inclines required a tough haul. There was very little difference in the timings and loadings of the various types of train. The 2-10-4s were permitted to haul loads up to about 1,000 tons on the steepest sections. Typically when hauling a capacity load up a bank of 20 miles mostly at 1 in 45, (2.2 per cent) the average speed would be 10mph. The booster would be cut in if speed fell below walking pace and cut out when the train had reached the speed of a man's run. Fuel consumption would be of the order of 37 gallons per mile up grade. In the mountains downhill speeds were limited to 25-30mph by curvature, frequently as sharp as 462ft radius. On the few straight sections of line 65mph could be achieved by these locomotives.

The 2-10-4s were able to negotiate these sharp curves by dint of widening the gauge on the curves from 4ft 8 1/2 to 4ft 9 1/2 in an exceptional amount, and by giving the leading axle nearly an inch of side-play each way as well as providing it with a pair of flange lubricators.

In 1952 diesels took over the running across the mountains and after the 2-10-4s had done a stint on freight haulage across the prairies, they were withdrawn. The last one was cut up in 1959, except for No5931 in the Heritage Park, Calgary, and No5935 at the Railway Museum at Delson, Quebec.

Above: No5935 is preserved at the Railway Museum in Delson, Quebec.

H-6 Class 2-6-6-2 (1949)

SPECIFICATIONS
Gauge: 4 ft 8½ in
Cylinders: (4) 35 x 32 in 22 x 32 in
Driving wheels: 56 in
Steam Pressure: 210 psi
Tractive effort: 98,773 lb
Weight: 434,400 lb
Length: 99 ft 8 in

No1308 was one of ten 2-6-6-2 H-6 class locomotives built by Baldwin in 1949 for the Chesapeake & Ohio Railroad. They were the last steam locomotives built by Baldwin.

No1308 is a true Mallet articulated locomotive with the rear engine rigidly attached to the frame of the locomotive. The front engine rode on a truck attached to the rear frame by a hinge, so that it could move from side to side on the tightly curved lines in C&O's West Virginia and Kentucky coal country, and that's where it worked for its seven year service life. No1308 was also a compound locomotive reusing steam from the rear set of high pressure cylinders in larger, lower pressure front cylinders. Its use in heavy mountain railroading is emphasized by its two cross compound air compressors mounted on the smokebox door to supply enough air for frequent heavy braking.

It worked on the run from Peach Creek, West Virginia, to Russell, Kentucky, with an occasional trip to Hinton, West Virginia.

No1308 was the next to the last Class 1 mainline locomotive built by Baldwin, closing out more than 100 years of production, a total of more the 70,000 locomotives. The last locomotive, its sister, No1309, is being restored to operation at the Western Maryland Scenic Railroad.

While No1308 is on the face of it a very modern locomotive, with roller bearings, mechanical lubricators, stoker, and a superheater, it was the last of a series of 2-6-6-2s that the C&O began in 1911. A similar design, the USRA 2-6-6-2 was chosen by the United States Railroad Administration as one

of its standard designs thirty years earlier during World War I.

The class was unusual for the time in that they were true Mallets. While compound locomotives are more efficient than single expansion, their extra complication led to very few United States railroads using them after the turn of the century. The C&O had a long history with Mallets and they were ideal for slow speed heavy hauling work in West Virginia.

After its last run on February 29, 1956, it was stored at Russell until the C&O gave it to the Collis P. Huntington Railroad Historical Society, Inc. Collis P. Huntington is best known as one of the Big Four who built the Central Pacific Railroad from San Francisco to Promontory, Utah, following that he spent at least ten years as a leading figure of the C&O. The town where No1308 now sits is named for him.

The locomotive was added to the National Register of Historic Places as Chesapeake and Ohio 1308 Steam Locomotive in 2003.

Opposite page top: No1309 on a double header coal train.

Above: No1308 is now on display at The Collis P. Huntington Railroad Historical Society located at Memorial Boulevard and 14th St West in Huntington, West Virginia.

Below: No1309 was one of ten 2-6-6-2 H-6 class locomotives built by Baldwin in 1949 for the Chesapeake & Ohio Railroad. It was actually the last Mallet ever constructed in the US, and the last domestic steam locomotive built by Baldwin. It cost $207,129.12. Like many locomotives built in the 1940s, this H-6 had a short service life. The last of the class was retired in 1957, only eight years after being built. No1309 is now preserved at The Baltimore & Ohio Railroad Museum is at 901 West Pratt Street, Baltimore, Maryland.

Norfolk & Western J-Class 4-8-4 (1950)

No611 is the sole survivor of Norfolk & Western's fourteen class "J" steam locomotives designed by their own mechanical engineers in 1940 and built at the railroad's Railway's East End Shops in Roanoke, Virginia between 1941 and 1950. The first batch, numbered 600 to 604, was built in 1941–42 and were delivered streamlined. The total cost for building No611 was $251,544 in 1950. For 18 years the "Js" pulled the Powhatan Arrow, Pocahontas and Cavalier through Roanoke on their daily 680-mile runs between Norfolk, Virginia and Cincinnati, Ohio. They also ran on the N&W portion of the joint N&W and Southern Railway routes, pulling the Pelican, the Birmingham Special and the Tennessean that operated between Washington, D.C. and southern cities.

Several of the "Js" ran almost 3 million miles each before retirement. Their superb performance and reliability allowed them to operate 15,000 miles per month, even on the relatively short, mountainous N&W routes.

This success delayed the day when progress, in the form of the diesel electric locomotive, inevitably would prevail.

SPECIFICATIONS
Gauge: 4ft 8 1/2 in
Cylinders: (2) 27 x 32 in
Steam pressure: 300 psi
Driving wheels: 70 in
Axle load: 72,000 lb
Tractive effort: 80,000 lb
Length: 109 ft 2 in
Weight: 494,000 lb

Above: Norfolk and Western No611 on a steam excursion in the late 1980s after a complete overhaul. She last ran in 1994.

611
NORFOLK AND WESTERN

611
611

611

On January 23, 1956, while traveling westward with the Pocahontas, No611 derailed on a wide curve near Cedar, West Virginia and almost fell into the Tug river. As a result of the extensive repairs made necessary by the accident, No611 was in good condition when the "Js" were retired in January 1959. A request by the Roanoke Chapter of the National Railway Historical Society to operate a passenger excursion later that year led the N&W to pull the 611 out of a group of "Js" destined for the scrap yards at Portsmouth, Ohio after completing the excursion between Bluefield, West Virginia and Roanoke in October, 1959, the 611 was donated to the City of Roanoke's Transportation Museum, the present owner.

In 1981, No611 was towed from the museum to the Southern Railway's Norris Yard steam shop at Birmingham, Alabama to be rebuilt. Restored to mint condition, the 611 steamed into Roanoke in August, 1982 with N&W Chairman Robert Claytor at the throttle. After renovation No611 was returned to active service as part of the railroad's steam program. The engine was used for some special excursions along with engine No1218, until the early 1990s. The engine last ran in 1994.

At the time of writing No611 is once again undergoing a new renovation to steam once again.

Below: Norfolk and Western steam locomotive N611 positioned for a special photo session at the Virginia Museum Of Transportation on July 12, 2013.

Diesel Power: 1900-Present

McKeen Car Roslyn (1905)

The McKeen Motor Car Company of Omaha, Nebraska was a builder of internal combustion-engined railroad motor cars (railcars), constructing 152 between 1905–1917. Founded by William McKeen, the Union Pacific Railroad's Superintendent of Motive Power and Machinery, the company was essentially an offshoot of the Union Pacific and the first cars were constructed by the UP before McKeen leased shop space in the UP's Omaha Shops in Omaha, Nebraska. The UP had asked him to develop a way of running small passenger trains more economically, and McKeen produced a design that was ahead of its time. Unfortunately, internal combustion engine technology was not, and the McKeen cars never found a truly reliable powerplant.

Most, although not all, McKeen cars had the distinctive "wind-splitter" pointed aerodynamic front end and rounded tail. The porthole windows were also a McKeen trademark, adopted allegedly for strength after the 7th production car.

Originally, McKeen cars used engines from the Standard Motor Works of Jersey City, New Jersey, but switched to an engine of their own design from the eighth car produced, M8 on the Union Pacific. All engines were straight-6 in configuration, of power ratings between 100 horsepower on the first car (M1) and a maximum of 300 horsepower on the most powerful later cars. The cylinders were vertical and the engine mounted transversely across the car in all cases.

Below: The then futuristic design with the pointed aerodynamic front end and port hole windows must have seemed like something out of Jules Verne novel in 1910.

SPECIFICATIONS
Type: Diesel electric railcar
Gauge: 4 ft 8 1/2 in
Propulsion: Standard Motor Works straight 6
Power: 100hp early cars 300hp later

Left: The Nevada State Railroad Museum undertook a restoration of McKeen car No22 in 1997. The restored motor car was unveiled in 2010, a century after it was originally delivered to the Virginia and Truckee.

GE 57-ton Gas-Electric Boxcab B-B (1913)

Before Diesel engines were perfected in the early 1900s, many companies chose to use the gasoline engine for rail motive power. The first GE Locomotive was a series of four-axle (B-B) boxcab gasoline-electric machines. They were closely related to the GE doodlebugs. These were a line of self-propelled passenger cars that were built in the early 1900s.

One of their first major customers was the Minneapolis, St. Paul, Rochester and Dubuque Electric Traction Company. This was better known as the Dan Patch Electric Lines after the owner's prize horse of the same name. The company was founded on the principle of not using steam power if they could avoid it, so the company asked GE to make them a series of locomotives based on their doodlebugs. GE complied, and created a number of locomotives originally claimed to be the first engines using an engine to drive a generator for traction motors. However, historians later determined that a narrow-gauge diesel-electric locomotive had been built in 1912.

The Dan Patch Electric Lines purchased the second GE demonstrator. By 1913 they owned eight diesel electric locos. When the need arose for a straight locomotive, a General Electric product was a natural answer.

SPECIFICATIONS
Type: Diesel electric locomotive
Gauge: 4 ft 8 1/2 in
AAR wheel arrangement: B-B
Fuel type: Gasoline
Prime mover: 2 x GM-16C4
Engine type: V-8
Propulsion: GN16-C4
Traction Motors: GE205D
Wheel diameter: 33in
Height: 14 ft 6 3/4 in
Width: 10 ft 5 in
Weight: 114,000l b
Length: 36 ft 4 in

Below: The Minnesota Transport Museum restored locomotive No100 to operating condition in the mid-1970's. General Electric participated in rebuilding its trucks. After restoration, it toured Minnesota on excursions with steam locomotive NP 328. Its destinations included Northfield and Duluth. In 1987 MTM acquired trackage in Stillwater and began regularly scheduled excursions. No100 is preserved at the museum.

Above: This Canadian Boxcab was built by GE in 1915 and was very similar to the Dan Patch locos.

M-300 Single Unit Railcar (1924)

SPECIFICATIONS
Type: Gasoline electric railcar
Gauge: 4ft 8 1/2 in
Propulsion: Winton 175hp (130kW) 6-cylinder four-stroke gas engine coupled to a generator driving two rose-suspended motors on the leading bogie.
Weight: 70,000 lb
Adhesive weight: 35,000 lb
Max. axleload: 17,500 lb
Length: 57 ft 4 in
Max. speed: 60 mph

The Electro-Motive Division of General Motors Corporation was to become the largest locomotive builder in the Western world. The story began here with this humble railcar and the company ended up displacing steam from the railways of the world. However when Chicago Great Western 39-seat railcar No.M-300 was delivered in 1924, the Electro-Motive Company's assets consisted of hope, and a dream and a one-room rented office in Cleveland, Ohio.

Although not strictly a diesel powered locomotive it unmistakably bore the genesis of later trains like the M-10001 trainset of 1934.

The Winton gasoline engine was basically a marine (submarine) powerplant with added electrical equipment from General Electric; carbody. construction was by the St Louis Car Company, who also assembled the unit. One or two other builders had been supplying these gas-electric cars since 1910, but Electro-Motive quickly became the major supplier, just as they were to do in the diesel locomotive field 20 years later.They became known as Doodlebugs and by 1930, when the market for these locos collapsed, some 400 of these cars had been constructed, about 80 percent of the total supplied by Electro-Motive to all US railroads.

Simplicity and standardisation was the secret of Electro-Motive's success then as now. Whilst it was not possible to dictate the physical layout of the cars supplied to customers, most of the equipment was well-tried and easily available. For example, controls were only provided at the engine end, but this did not matter because turntables were generally available for turning.

Later, more powerful twin-engine cars with trailer-hauling capability came into being. They reached in excess of 500hp a weight of 160,0001b and 75ft length and used distillate rather than gasoline.

No1 an M-100 example has been preserved on Pennsylvania's East Broad Top Railroad.

Above: A gas-propelled railcar of the Long Island Rail Road provides economical transport on the Sag Harbour Branch, 1935.

Below: The preserved and still fully operational gas electric railcar of the East Broad Top Railroad in Pennsylvania is a good example of the class.

No1000 Bo-Bo Algeir (1924)

SPECIFICATIONS
Type: Diesel-electric switching locomotive
Gauge: 4ft 8 1/2 in
Propulsion: Ingersoll-Rand 300hp 6-cylinder four-stroke diesel engine and GEC generator supplying current to four nose-suspended traction motors geared to the axles.
Weight: 120,000 lb
Max Axleload: 30,000 lb
Length: 32 ft 6 in

This 120,000 lb, 6-cylinder, 300 hp locomotive was the first commercially successful diesel-electric to operate in the US. Built in 1925, it was a joint effort of Alco, General Electric and Ingersoll Rand. Alco issued orders, built and shipped the car bodies to GE at its Erie, PA, works. Ingersoll-Rand produced the engines in Phillipsburg, NJ, and shipped them to GE, which delivered completed units back to Ingersoll-Rand.

No1000 worked as a switcher at Central of New Jersey's New York Yard. In 1928, it replaced CNJ's 0-4-0 tank engine at the railroad's Bronx Terminal yard. The change was necessary because of the city's new smoke ordinances, although the 0-4-0T was used off and on as a standby engine. No1000 was retired in 1957.

Including No1000, twenty-six of these units were produced from 1925 to 1930 in both 60 ton and 100 ton versions. The 32ft 6in, 60 ton units like CNJ No1000 had a tractive effort of 37,200 lbs and a top speed of 30 mph.

GE 60-Ton Boxcab (1928-1930)

The General Electric boxcabs were diesel-electric switcher locomotives. General Electric built the chassis and running gear, generator, motors and controls, and Ingersoll Rand provided the diesel engine. The principle of operation was the same as modern locomotives, the diesel engine driving a main generator of 600 volts DC with four traction motors, one per axle.

SPECIFICATIONS
Power type: Diesel-electric
Builder: GE Transportation Systems
Model: 60-ton
Build date: 1928-1930
Total produced: 2
AAR wheel arrangement: B-B
Prime Mover: Ingersoll Rand 300

Two models were the 60-Ton with a six-cylinder four-stroke in-line engine of 300 hp and the 100-Ton with two of the same engines. Thirteen of these units were produced between 1928 and 1930.

These locomotives were originally produced by a consortium of ALCO, GE and Ingersoll Rand. ALCO dropped out of the arrangement in 1928, after acquiring their own diesel engine manufacturer in McIntosh & Seymour and went on to start its own line of diesel switchers.

The only surviving example of these boxcabs, Foley Bros. number 110-1, a 100-ton dual-engined locomotive, could be found at the Western Pacific Railroad Museum in Portola, California. In December 2011, the Foley Brothers number 110-1 twin engine 600 hp GE-IR boxcab was moved to the California State Railroad Museum in Sacramento, California. These boxcabs were often termed "oil-electrics" to avoid the use of the German name "diesel," which became an unpopular term after the First World War.

GE 100-Ton Boxcab B-B (1928-1930)

The GE boxcabs were diesel-electric switcher locomotives. General Electric built the chassis and running gear, generator, motors and controls, and Ingersoll Rand provided the diesel engines for the models. The principle of operation was the same as modern locomotives, the diesel engine driving a main generator of 600 volts DC with four traction motors, one per axle.

SPECIFICATIONS
No. of Examples: 11
AAR Wheel Arrangement: B-B
Power Unit: Ingersoll Rand 300 (2)

Two models were produced with the same power plants. The 60- and 100-ton Boxcabs were fitted with six-cylinder four-stroke in-line engine of 300 hp. The 60-ton engine was fitted with one of these while the 100-ton engine was equipped with two of the same engines. Eleven of the 100-ton units were produced between 1928 and 1930.

These locomotives were originally produced by a consortium of ALCO, GE and Ingersoll Rand, ALCO dropped out of the arrangement in 1928, after acquiring their own diesel engine manufacturer in McIntosh & Seymour and went on to start its own line of diesel switchers.

The only surviving example of these boxcabs, Foley Bros. #110-1, is a 100-ton dual-engined locomotive. This can be found at the Western Pacific Railroad Museum in Portola, California.

These boxcabs were often termed "oil-electrics" to avoid the use of the German name "Diesel," unpopular after World War I.

No9000 Canadian National Railways (1929)

SPECIFICATIONS
Type: Main line Diesel Electric
Gauge: 4ft 8 1/2 in
Propulsion: Beardmore four-stroke V12 diesel engine and generator, originally supercharged, supplying direct current to four nose-suspended traction motors geared to the axles.
Power: 1,330 hp
Weight: 255,644 lb
Adhesive weight: 374,080 lb
Max. axleload: 63,920 lb
Overall length: 47 ft 0 in
Tractive effort: 50,000 lb
Max. speed: 75 mph

Above: Canadian National's two-unit 2-D-1 type 9000 of 1929, the first large diesel in North America, on a passenger train test run Montreal - Vancouver, later split into 2 units (Canadian National).

Canadian National Railways was among the earliest users of diesels in North America establishing a number of records. It began late in 1924 with a tour of Europe by CNR motive power officials including C.E. (Ned)Brooks, Chief of Motive Power, their last stop being at the Glasgow, Scotland plant of the Wm.Beardmore Co. builders of airship motors. These very lightweight (many comparable powered diesel engines at the time weighed over twice as much) and very advanced-design engines were already being used in London, Midland & Scottish Railway rail cars built in 1922.

In 1929 the Canadian Locomotive Company built a two-unit diesel at their Kingston Ontario shop. The units separately were road numbered 9000 and 9001.The Scottish –built 12-cylinder Beardmore engine fitted was rated at 1,330 hp. After several test runs including one to Vancouver and back, No9000 was put into regular service between Toronto and Montreal.The two units were withdrawn from service in 1939 but No9000 was armor-plated for wartime use on the Prince Rupert line.Both were scrapped in 1946.

Left: CNR 9001

Following page: No9000 has its Beardmore power plant installed at the Canadian Locomotive Company shop in Kingston,Ontario in 1929.The Scottish built unit was a lightweight ex airship engine rated at 1,330 HP.

9000
CANADIAN NATIONAL
9000
C.N.R.
C.N.R.

GE 110-Ton Switcher B-B (1930s)

SPECIFICATIONS
AAR wheel arr.: B-B
UIC classification: Bo Bo
Gauge: 4 ft 8 1/2 in (1,435 mm)
Locomotive weight: 110 short tons (98 long tons; 100 t)
Prime mover: 2 x Cummins
Engine type: 2 Diesel engines
Engines: Twin Cummins
Horsepower: rated 670 hp
Air Brakes: 14
Traction Motors: 763
Couplers: E Type
Air Compressor: Gardner Denver
Trucks: Roller bearing
Main Generators: 558

Above: A classic center cab type GE 110 ton switcher still perfectly serviceable and up for sale in Greenville, SC.

The GE 110-ton switcher is a diesel-electric locomotive model built by GE Transportation Systems. It was intended for use in light switching duties. The typical switcher is optimised for its job, being relatively low-powered but with a high starting tractive effort for getting heavy cars rolling quickly. Switchers are geared to produce high torque but are restricted to low top speeds and have small diameter driving wheels.

American switchers tend to be larger, with bogies to allow them to be used on tight radiuses. European switchers, or shunters, tend to be smaller and often have fixed axles. Heavily used switch engines wear out quickly from the abuse of constant hard contacts with cars and frequent starting and stopping.

M-10001 Union Pacific Railroad (1934)

In February 1934 the Union Pacific Railroad only failed by a technicality to be the first in America with a diesel-electric high speed train. A suitable diesel engine was not quite ready and UP's stunning train had to have a spark-plug engine using distillate fuel. This was the yellow and gray No. M-10000, consisting of three articulated streamlined light-alloy cars weighing, including power plant, 93 1/2 tons, in total hardly more than a single standard passenger car as then existing. The train was built by Pullman-Standard in late 1934 with an engine from General Motors Electro-Motive Corporation and General Electric generator, control equipment and traction motors. It was the UP's second streamliner after the pioneering M-10000, the first equipped with a diesel engine and was a much longer train (six cars) than its three-car predecessor. All cars were articulated—trucks were shared between each car. It was delivered on October 2, 1934 and was used for display, test and record-setting runs for the next two months before being returned to Pullman-Standard for an increase in its power and capacity, following which it was placed into service as the City of Portland train.

When built, the M-10001 was a fully articulated six-car train, 376 ft in length, comprising a 48 ft turret-cab power car, a Railway Post Office/baggage car, three Pullman

Right: General Motors advertisement for its new line of Winton diesel engines in 1935. Union Pacific's M-10001, shown in the ad, was the first train to be equipped with the new version. The train was delivered to the railroad in late 1934.With this close-up, the many rivets that held the aluminum train together can be seen, making it much different from the Budd Company's stainless steel, shot-welded Zephyr streamliners for the Burlington.

SPECIFICATIONS
Type: Diesel-electric high-speed passenger train
Gauge: 4ft 8 1/2 in
Propulsion: Electro-Motive V-16 two-stroke diesel engine and generator supplying current to four 250hp (187kW) nose-suspended traction motors geared to the axles of the two leading bogies.
Power: 1,200 hp
Weight: 143,260 lb
Adhesive weight: 413,280 lb
Max. axleload: 35,815 lb
Overall length: 376 ft 0 in
Max. speed: 120 mph

sleeping cars, and a rounded-tail coach/buffet/observation car. Its construction differed from the Pioneer Zephyr in that it used riveted aluminum rather more in the style of aviation construction. It was powered by a 900 hp (670 kW) V12 version of the Winton 201-A engine, driving the first two trucks of the train. Both the 900 hp and 1,200 hp (670 and 890 kW) Winton diesels were the first of that type to be installed in a production train.

The Pullman-Standard rebuilding lengthened the entire train to 455 feet and seven cars. The power car had 12 ft added to accommodate a larger, 1,200 hp (750 kW) V16 Winton diesel engine. The RPO/baggage car was lengthened by eight feet to take a steam generator for train heating, and was followed by the addition of a diner/lounge car. The rebuilt train was re-delivered on May 23, 1935, and after some test runs was dedicated as the first City of Portland on June 5 at the Portland Rose Festival, entering service between Portland, Oregon and Chicago the following day. It remained in that service until June 1939, when it was replaced in service by the M-10002 trainset, reassigned from the City of Los Angeles. In December, the diesel engine, generator, trucks, and the steam generator from the RPO/baggage car were removed and installed in a new carbody to become the third power unit on the CD-07 set for the City of Denver, along with the former M-10003's two power cars. The remainder of the train was stored until August 13, 1941 when it was sold for scrap.

Right: William E. Marcom, posing with the M-10000 at the Century of Progress Exposition in 1933.

Pioneer Zephyr Three Car Train CB&Q (1934)

Chicago, Burlington and Quincy Railroad (CB&Q) President Ralph Budd decided to find a way to encourage the public to travel by rail during the depression years of the early 1930s. In 1932 Ralph Budd met Edward G. Budd (no relation), an automotive steel pioneer who was founder and president of the Budd Company. Edward Budd was demonstrating his new carbody construction in a prototype rail motorcar built of stainless steel. They decided to use this technology to create a unique new streamliner initially named the Zephyr.

The construction included innovations such as shotwelding (a specialized type of spot welding) to join the stainless steel, and articulation to reduce its weight.

The cars were permanently articulated together using Jacobs bogies. As a result of General Motors purchase of both the Electro-Motive Company and their engine suppliers; the Winton Engine Co. GM had concentrated their efforts on making a diesel engine suitable for rail transport and it was this unit that powered the Pioneer Zephyr.

The train entered the regular revenue service on November 11, 1934, between Kansas City, Missouri; Omaha, Nebraska; and Lincoln, Nebraska. It operated this and other routes until its retirement in 1960, when it was donated to

Below: Pioneer Zep 2:CB&Q Pioneer Zephyr 9900 at Lincoln, Nebraska in July, 1959.

Chicago's Museum of Science and Industry, where it remains on public display. The train is generally regarded as the first successful streamliner on American railroads.

On May 26, 1934, it set a speed record for travel between Denver, Colorado, and Chicago, Illinois, when it made a 1,015-mile non-stop "Dawn-to-Dusk" dash in 13 hours 5 minutes at an average speed of 77 mph For one section of the run it reached a speed of 112.5 mph just short of the then US land speed record of 115 mph The historic dash inspired a 1934 film and the train's nickname, "The Silver Streak."

Below: Engine room with Electro-Motive Type 201E inline two-stroke diesel engine.

SPECIFICATIONS
Type: High-speed articulated streamlined diesel-electric train
Gauge: 4ft 8 1/2 in
Propulsion: Electro-Motive Type 201E inline two-stroke diesel engine and generator feeding two nose-suspended traction motors on the leading bogie.
Power: 600 hp
Weight: 90,360 lb
Adhesive weight: 175,000 lb
Max. axleload: 45,180 lb
Overall length: 196 ft
Max. speed: 110 mph

B&O No50 Bo-Bo (1935)

SPECIFICATIONS
Type: Passenger diesel-electric locomotive
Gauge: 4ft 8 $^1/2$ in
Propulsion: 2 Winton 201A V-12 diesels
Power: 1800 hp

Baltimore & Ohio No50 is one of five experimental 1,800 hp passenger diesel locomotives built by EMC(later to become EMD) in 1935. They were the first non-articulated diesels to work on US main line railroads. It was one of five experimental diesel locomotives designed for passenger train service. In addition to No50 the order consisted of two company-owned demonstrators, No511 and No512, and two units for the Atchison, Topeka and Santa Fe Railway, Diesel Locomotive No1 and 1a.They were the mechanical ancestors to EMD's successful E-units, with identical pairs of 900 hp Winton 201-A diesel engines, although they ran on AAR type B two-axle trucks instead of the A1A trucks of E-units. When delivered, the units were fitted with shrouding around their trucks, but this did not last long.

The boxy carbodies of all but the Zephyrs were the work of GE's Erie, Pennsylvania works, EMC having not yet developed the ability to produce their own bodywork. Like most boxcabs, they had control cabs at both ends, a feature that would only rarely be repeated in future North American locomotives, although it would become common elsewhere.

No50 hauled B&O's first diesel-powered Royal Blue service until the EA/EB units were introduced in 1937 .In 1938, No50 was transferred to the Chicago & Alton and then became No1200 under the Gulf, Mobile & Ohio Railroad. It was retired in 1958 and donated to The St. Louis Museum of Transportation.

Above: No50 Fresh from a coat of blue paint in 1972.

Below: A pale version of its former self, No50 is to be found at the St Louis Museum of Transportation.

The Flying Yankee Three-car train (1935)

SPECIFICATIONS
Type: High-speed articulated streamlined diesel-electric train
Gauge: 4ft 8 1/2 in
Propulsion: Winton 201A 8 cylinder 600hp inline two-stroke diesel engine and generator feeding two nose-suspended traction motors on the leading bogie.
Power: 600 hp
Weight: 90,360 lb
Adhesive weight: 175,000 lb
Max. axleload: 45,180 lb
Overall length: 196 ft 0 in
Max. speed: 110 mph

The Flying Yankee was a virtual clone of the Chicago, Burlington and Quincy Railroad's Pioneer Zephyr diesel-powered streamliner.The Flying Yankee was the name of the famous train that took passengers from Portland, Maine, then to Boston, Massachusetts, followed by a return to Portland and continuing to Bangor, Maine, returning through Portland to Boston and finally returning to Portland late in the day, a distance of 750 miles per day.

The diesel powered unit consisting of three articulated cars built in 1935 for the Maine Central Railroad and the Boston and Maine Railroad by the Budd Company and with mechanical and electrical equipment from Electro-Motive Corporation. The Flying Yankee was in fact the third streamliner train in North America after the Union Pacific Railroad's M-10000 and the Chicago, Burlington and Quincy Railroad's Pioneer Zephyr. Unlike the latter, the Flying Yankee dispensed with the baggage/mail space to seat 142 in its three articulated cars. The lightweight train was constructed with welded stainless steel using Budd's patented process. It was fitted with air conditioning in all cars. No dining car was provided; instead, meals were prepared in a galley and served to passengers in trays that clipped to the back of the seat in front.

Because the heavy schedule took place over six days a week; the three-car unit spent Sundays undergoing maintenance. The train proved extremely successful, attracting new travelers and earning a profit for its owners. Later on, as newer equipment replaced it on the prime route, it would be switched to other routes, bearing the names The Cheshire, The Minuteman, The Mountaineer, and The Business Man. As railroad passenger numbers declined in the 1950s the Yankee was also showing its age and thus the trainset. then running as The Minuteman, was retired on May 7, 1957.

Above: The train boosted passenger numbers on its arrival on the Portland -Boston route.

Below: The flying Yankee leaving the Budd works in 1935.

After some years on static display in 1997, the train was moved to the Claremont Concord Railroad's shops in Claremont, New Hampshire for a complete restoration once purchased by the State of New Hampshire, which is ongoing. By 2004, the major structural restoration had been completed, and detailed restoration of components was proceeding. The eventual goal is to restore the train completely to running condition. The train was moved to Lincoln, New Hampshire, on August 10, 2005, to the Hobo Railroad where the mechanical restoration is taking place.

Above: The Flying Yankee awaits restoration minus its bogies.

Below: The "Flying Yankee" trainset at Edaville heritage railroad in August 1991.

Illinois Central No121 (1936)

ILLINOIS CENTRAL'S GREEN DIAMOND GLIDES BETWEEN CHICAGO AND ST. LOUIS IN 4 HRS. 55 MINS.

SPECIFICATIONS
Type: High-speed articulated stream lined diesel-electric train
Gauge: 4ft 8 1/2 in
Propulsion: Two straight-8 600 hp Winton 201-A engines
Power: 1200 hp
Cylinders: 16
Fuel capacity: 725 gall

Left: Illinois Central was scrapped in 1950 and only this postcard gives testament to the train's sweeping good looks.

The Green Diamond was a streamlined passenger train operated by the Illinois Central Railroad between Chicago, Illinois and St. Louis, Missouri. Constructed by Pullman-Standard the train's original fixed consist included a power car, baggage/mail car, coach, coach-dinette, and kitchen-dinette-parlor-observation car. It was the last fixed-consist train built in the 1930s for a railroad in the United States and was the Illinois Central's first streamliner. Initially the service operated with Illinois Central No121, but after 1947 it operated with more conventional streamlined equipment until its discontinuance. The name honored the "green diamond" in the Illinois Central's logo as well as the Diamond Special, the Illinois Central's oldest train on the Chicago-St. Louis run.

No121's cars were numbered 121-125. The coach seated 56, while the coach-dinette seated 44 in the coach section and the dinette area had seating for 16. The parlor car had seating for 22 passengers, whilst a second generator within the power car provided electricity for the lights, and separate steam generator heated the train. The train's interior was art deco, including lot of aluminum detailing, as was popular in the period and use of a bright two-tone green paint scheme elsewhere.

Right: Illinois Central No121 in Milwaukee, Wisconsin at the Milwaukee Road depot on April 24, 1936 ready to haul The Green Diamond.

No2027/2028 Seaboard Air Lines (SAL) (1936)

SPECIFICATIONS
Type: Diesel-electric railcars
Gauge: 4ft 8 1/2 in
Propulsion: Winton 201-A
Power: 600 hp
AAR wheel arr.: B-2
Engine type: 2-stroke diesel
Generator: DC generator
Traction motors: DC traction motors
Cylinders: (8) 8 in x 10 in
Locomotive brake: Straight air
Train brakes: Air

Nos 2027 and 2028 were lightweight, streamlined Diesel-electric railcars built by the St. Louis Car Company in 1936. Two units were manufactured for the Seaboard Air Line Railroad (SAL). Electromotive Corporation supplied the 600 hp eight-cylinder Winton Diesel 201-A prime mover and electric transmission components. The units had a B-2 wheel arrangement, mounted on a pair of road trucks. The rear section was divided into two separate compartments: one was used to transport baggage and the other served as a small railway post office, or RPO (the forward door, located just behind the radiator louvers, was equipped with a mail hook).

The last usage of these railcars was in May, 1971. No2027 was destroyed in a collision with a gas tanker truck at Arcadia, Florida in 1956. No2028 was renumbered to No4900 after the Seaboard-ACL merger in 1967 and was eventually retired and scrapped after Amtrak took over national passenger service in 1971.

Above: No2028 in service with the Seaboard Air Lines in the early 1960s.

Below: The engine room housed the Electromotive Corporation 600 hp eight-cylinder Winton Diesel 201-A propulsion unit and electric transmission components.

GE 125-Ton Ford Switcher (1937)

SPECIFICATIONS
Type: Diesel electric center cab switcher
Gauge: 4ft 8 1/2 in
Propulsion unit: Ford
Power: 1200 hp

Several 125-Ton Center Cab Switcher locomotives were built for Ford by General Electric. The Ford units introduced the styling used on most GE switchers of the 1940s and 1950s, and the cabs are almost identical with that of the the 44-Ton switcher. When they were first built, these units were painted blue to match the color of the Ford blue oval badge and the lettering was chrome plated. It is said that Henry Ford asked GE to style the switchers in the same way as his 1937 Ford cars. They were decked out with chrome rails and had a large Ford logo on the side.

The 125-Ton Switcher engines had a B-B wheel arrangement. This means that there are two identical trucks. Each truck has two powered axles, a popular configuration used in high-speed, low-weight applications, such as intermodal trains, high-speed rail, and switching.

Right: Ford switcher in US Army service.

Above & below: The Ford Switcher used styling cues from the 1930s Ford Car range and were painted in Ford Blue with the company logo.

E-series A-1-A A-1-A EMD GM (1937)

The EMD E series (originally the design of the Electro-Motive Corporation before it became a division of General Motors) was developed primarily for passenger service and was important in that it instituted the general conversion of the American train to diesel operation, eventually seeing many of the most famous steam trains out. In their heyday the US had an undisputed world lead in passenger train speeds. Geared for up to 117mph the E-series were the fastest diesel locomotives in the world, and yet their construction was rugged and straightforward.The design was first developed in the 1930s as the EA, featuring a slant-nose and sweeping carbody.
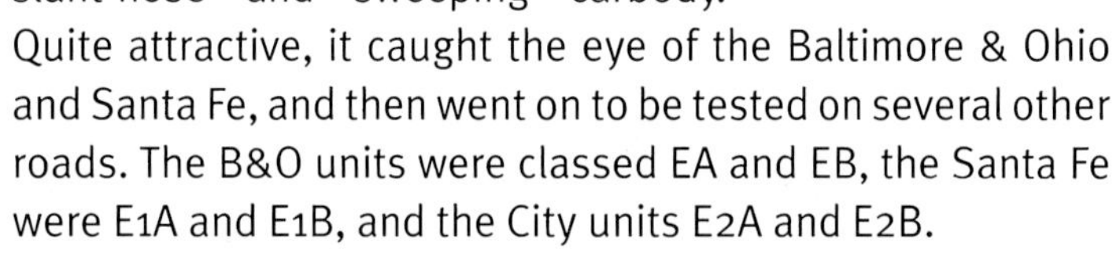
Quite attractive, it caught the eye of the Baltimore & Ohio and Santa Fe, and then went on to be tested on several other roads. The B&O units were classed EA and EB, the Santa Fe were E1A and E1B, and the City units E2A and E2B.

Above: The Southern Pacific Shasta Daylight, which traveled the West Coast between Portland, Oregon and Oakland, California between 1949 and 1966. This is train No10, going north to Portland. It appears to be one of the early 1949 runs of the train as it's equipped with EMD E-7 locomotives; the locomotives were switched to Alco models soon after its inaugural 1949 run. The train is also shown with its original colorful paint scheme; SP changed to red and black in 1959.

The E series went through a few carbody modifications before it was a decided that a more-standard, "Bulldog" nose design would be used for both passenger and freight service.

It started when EMD moved into its own purpose-built works at La Grange, Illinois in 1936, and work commenced on the first of the "E" series, known also as the "Streamline" series. Like the four earlier locomotives, they had two 900hp Winton engines, but the chassis and body were completely new. The body had its main load-bearing strength in two bridge-type girders which formed the sides. The bogies had three axles to give greater stability at high speeds, but as only four motors were needed, the center axle of each bogie was an idler, giving the wheel arrangement A1A-A1A. The units were produced in two versions, A units with a driver's cab and B units without. Progress at La Grange was rapid. At 900hp the Winton engine was reaching its limit, and an EMD engine was therefore developed. Designated 567 (the capacity of a cylinder in cubic inches), it was available in three sizes with 8, 12, and 16 cylinders, giving 600, 1,000 and l,350hp Simultaneously La Grange began to manufacture its own generators, motors and other electrical equipment. The first all-EMD locomotives were an order from Seaboard Air Line for 14A and five B units, which appeared from October 1938 onwards. They had two l,000hp engines and were operated as three 6,000hp units. These were the E-4s, E-3 and E-5s, the former comprising 18 units for the Sante Fe and the latter 16 for the Burlington.

The next series, the E-6, which appeared in the same

SPECIFICATIONS
Type: Express passenger diesel-electric locomotive; A units with driving cab, B units without
Gauge: 4ft 8 1/2 in
Propulsion: Two EMD 567A 1000hp 12-cylinder pressure-charged two-stroke Vee engines and generators, each supplying current to two nose-suspended traction motors geared to the end axles of a bogie.
A Unit Weight:
212,310 lb adhesive, 315,000 lb total
A Units Weight:
205,570 lb adhesive, 305,000 lb total
Max. axleload (A): 53;080 lb
Max. axleload (B): 51,390 lb
Length (A): 71 ft 1/4 in
Length (B): 70 ft 0 in
Max. speed: 85mph 92mph , 98mph or 117mph according to gear ratio fitted. (Figures refer to E-7 variant 1945)

Below: Union Pacific No949 from Cheyenne, Wyoming. This is an E-9 locomotive. It is in Gibbon, Nebraska providing backup for UP No8444, the steam engine pulling a short train from Omaha, NE to Cheyenne, WY in July 2012.

month in 1939 as the first freight demonstrator, was therefore a standard off-the-shelf unit, with the minimum of options. This was the start of real diesel mass production and 118 units had been built by the time the War Production Board terminated building of passenger locomotives in February 1942.

Construction of passenger locomotives was resumed in February 1945 with the first of the E-7 series. With locomotive fleets rundown by wartime traffic, the railroads were even more eager to acquire passenger diesels, and Electro-Motive Division settled down to a steady production of E-7s, averaging 10 per month for four years. During this time 428 A units and 82 B units were built so that the E-7 outnumbered the passenger diesels of all other US makers put together.

In 1953 the l,125hp 567B engine became available, and this was incorporated in the next series, the E-8. By this time most of the principal passenger services were dieselized, so the impact of the E-8 was less spectacular than that of the E-7. By the time the final version appeared, the E-9 with l,200hp 567C engines, the need for passenger diesels had almost been met, and only 144 units were sold between 1954 and 1963, compared with 457 E-8s. In the 1960s the American passenger train traffic had declined rapidly in the face of air and coach competition, and many of the later Es had short lives, being traded in against the purchase of new general-purpose locomotives. Today, several examples of Es are preserved and a handful remain operational.

Right: This photo is of Rock Island E-6, a locomotive that until just a few years ago was at the Midland Railway in Baldwin City, Kansas. The E-Unit has been sold to an out of state owner and may be renovated.

Below: E-9 panorama.

630
BN-3
BURLINGTON NORTHERN

Left: Built in 1939, this E series diesel locomotive was one of the first mass production models and provided service between New York City and Miami. It is one of the exhibits at the North Carolina Transportation Museum. Located midway between Charlotte, NC, and Greensboro, NC, the museum is located at the site of the former Spencer Railroad Shops in Spencer, NC.

EMC E-1 Atchison T&SF (1937)

SPECIFICATIONS
Type: Diesel-electric
Builder: Electro-Motive Corporation
Build date: 1937–1938
Total produced: 8 A units, 3 B units
AAR wheel arr.: A1A-A1A
Gauge: 4 ft 8 1/2 in
Propulsion: Winton 201-A, 2 off
Power output: 1,800 hp
Number(s): 2–9 (A units), 2A–4A (B units)

The EMC E-1 was an early passenger-train diesel locomotive that was built during 1937 and 1938 for the Atchison, Topeka and Santa Fe Railway for a new generation of diesel-powered streamlined trains. Many of the other railroads were making similar purchases from by Electro-Motive Corporation of La Grange, Illinois who were a division of General Motors. The E-1—along with the more-or-less simultaneous EA/EB for the Baltimore and Ohio Railroad and the E-2 for the Union Pacific Railroad, Chicago and North Western Railway and Southern Pacific Railroad—represented an important step in the evolution of the passenger diesel locomotive. While the EA, E1 and E2 were each built for a specific railroad, they were largely identical mechanically and were a step further away from the custom-built, integrated streamliner and towards mass-produced passenger diesel locomotive. The power output of all the trains was broadly similar at 1800hp from the two 900 hp Winton 201-A engines, each engine driving its own generator to power the traction motors.

Eight cab-equipped lead A units and three cabless booster B units were built. The initial three locomotives were AB pairs built to haul the Santa Fe's Super Chief diesel streamliners, while the others were built as single A units to haul shorter trains., with each engine driving its own generator to power the traction motors. The E-1 was the second model in a long line of passenger diesels of similar design known as EMD E-units.

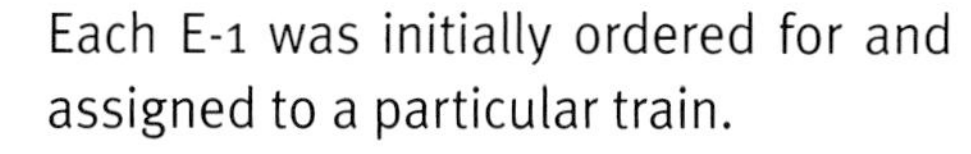

Each E-1 was initially ordered for and assigned to a particular train.

- 2 and 2A - for the original streamlined Super Chief.
- 3 and 3A - for the second streamlined Super Chief trainset.
- 4 and 4A - backup power for the Super Chief.
- 5 - for the El Capitan.
- 6 - for the El Capitan.
- 7 - for the San Diegan.
- 8 - for the Golden Gate.
- 9 - for the Golden Gate.

GE 20-and 23-Ton Boxcab (1938)

General Electric produced a total of five 20-Ton Boxcab diesel electrics which were all completed during June, 1938. Three of these Industrial Locomotives were purchased by Lehigh Portland Cement. These were engines 12447, 12448 and 12449 and were delivered during July 1938. They gave the company over half a century of service. The company subsequently became General Portland Industries. They are currently part of the Midland Railway Historical Association collection at Baldwin city, Kansas. The other two were sold to Wisconsin Steel, a subsidiary of International Harvester. The 20 Ton Box Cabs had single panel doors and fewer windows than the GE 23-Ton Boxcab engine that was introduced in the following year. It also has body-side grab irons and two-inch thin deckplates. The 23-Ton Boxcabs had multi panelled doors, extra windows, body-side grab irons which wrapped around the ends. It also had grab irons on the ends and thicker three-and-a-half inch deckplates.

The power was supplied by a Cummins engine. Cummins Inc. is a Fortune 500 corporation that designs, manufactures, and distributes engines, filtration, and power generation products. Cummins also services engines and related equipment, including fuel systems, controls, air handling, filtration, emission control and electrical power generation systems. Headquartered in Columbus, Indiana, Cummins sells in approximately a hundred-and-ninety countries and territories through a network of more than six-hundred company-owned and independent distributors and approximately six thousand dealers. Cummins reported net income of $1.64 billion on sales of $17.3 billion in 2012.

SPECIFICATIONS
Power type: Diesel-electric
Builder: GE Transportation Systems
Model: 20-Ton switcher
Build date: 1938
Prime mover: Cummins
Power output: 150 hp (112 kW)

Below: General Electric produced a total of five 20 Ton Box Cab Diesel-electrics which were all completed during June, 1938.Three of these Industrial Locomotives were purchased by Lehigh Portland Cement.

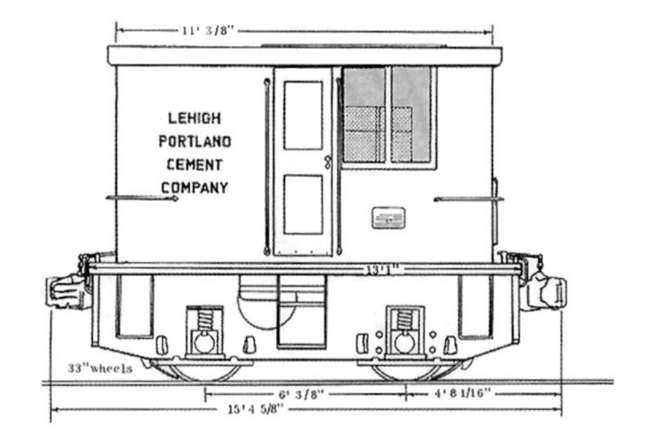

Below: The Lakeshore Railway Museum's 1939 GE 23 ton boxcab locomotive being pulled by the its GE 25 ton workhorse.

EMD FT Bo-Bo (1939)

SPECIFICATIONS
Type: Diesel-electric locomotive
Model: FTA (cab unit), FTB (regular booster), and FTSB (short booster)
Build date: November 1939 – November 1945
Total produced: 555 A units, 541 B units
Gauge: 4 ft 8 1/2 in
Power output: 2,700 hp

Left: Santa Fe 124, an EMD FT built in 1944, in an A-B-A combination at Denver, Colorado on August 14, 1957. It was retired in 1965.

The EMD FT was a 1,350-horsepower diesel-electric locomotive produced between November 1939, and November 1945, by General Motors' Electro-Motive Division (the F stood for 1400 horsepower (rounded from 1350) and the T for twin, as it came standard in a two-unit set). All told 555 cab-equipped A units were built, along with 541 cabless booster B units, for a grand total of 1,096 units. The locomotives were all sold to customers in the United States. It was the first model in EMD's very successful F-unit series of cab unit freight diesels, and was the locomotive that convinced many U.S. railroads that the diesel-electric freight locomotive was the future. Many rail historians consider the FT one of the most important locomotive models of all time.

Multiple EMD FT units survive today. They include the lead A-unit from demonstrator No. 103 displayed at the Museum of Transportation in St. Louis, Missouri. It is paired with one of the two original FT B-units from the EMD 103 demonstrator set. (B Unit is on loan from the Virginia Museum of Transportation). Both units are cosmetically restored and painted in the original GM demonstrator paint.

An FT A unit, FSBC 2203-A on display in Mexico, which was originally built for the Northern Pacific Railway

Three B-units from the Southern Railway are preserved. No960604 is at the Southeastern Railway Museum in Duluth, Georgia, and No960602 is in Conway, South Carolina.

Below left: Part of General Motors' original FT demonstrator trainset No103 was featured at Electro-Motive's La Grange, Illinois plant during an open house ceremony in September of 1989.

Below: A consist of GN EMD FTs

F-Series BoBo EMD GM (1939)

F-Series Diesel Electric locomotives were introduced in 1939 by GM's Electromotive Division and manufactured at the La Grange plant in Illinois and at the GMDD plant in London, Ontario, Canada.

The term F-Series refers to the model numbers given to each successive type, all of which began with F. The F originally meant Fourteen, as in 1,400 horsepower, not as might be assumed F as in Freight. All the same the F-units were originally designed for freight service, although many hauled passenger trains. Almost all F-units were B-B locomotives; they ran on two Blomberg B two-axle trucks with all axles powered.

Structurally, the locomotive was a carbody unit, with the body as the

Above: A Metro-North F10 that was formerly an F3 (originally GM&O and later an Illinois Central No880A) in Bridgeport, Connecticut, operates for Metro-North Railroad in 2005. This locomotive also ran for the MBTA.

Below: Western Pacific Railroad No913, an EMD F7 locomotive on display at the California State Railroad Museum in Sacramento.

main load-bearing structure, designed like a bridge truss and covered with cosmetic panels. The so-called bulldog nose was a distinguishing feature of the locomotive's appearance, and made the F-series instantly recognizable.

The F series used a 16 cylinder version of the 567 series diesel engine, introduced in 1939. The 567 was designed specifically for railroad locomotives, a supercharged 2

Above: No274 , an F-7, on the Oregon Coast Scenic Railroad.

Right: BN F-3 No9762, ex-NP No6502, leading the North Coast Hiawatha into Yakima, Washington in August 1971.

GULF MOBILE & OHIO
43
43
Rock Island

stroke 45 degree V type with 567 cu in displacement per cylinder, for a total of 9,072 cu in An ongoing improvement program saw the FT's 1,350 hp up-rated to 1,800 hp in the FL9. A D.C. generator powered four traction motors, two on each truck. The Blomberg B truck first used in the FT became the EMD standard, being used through 1995. EMC/EMD has built all of its major components since

The F-units were the most successful "first generation" road Diesel locomotives in North America, and were largely responsible for superseding steam locomotives in road freight service.

The series was continuously improved during its 30 years service starting from the original FT introduced in 1939 with the 1,350 hp 567 engine and Blomberg B trucks, instantly proving itself as a successful design, with 1096 being produced during WWII.

SPECIFICATIONS
Type: All-purpose diesel-electric locomotive, A units with cab, B units without.
Gauge: 4ft 8 1/2 in
Propulsion: One EMD 5,67B 1,500hp 16-cylinder pressure-charged two-stroke Vee engine and generator supplying current to four nose-suspended traction motors geared to the axles.
Weight: 230,000 lb
Max. axleload: 57,500 lb
Length (A): 50 ft 8 in
Length (B): 50 ft 0 in
Tractive effort: 57,500 lb
Max. speed: Between 50mph and 120mph according to which of eight possible gear ratios fitted.
*Dimensions refer to the F-3 variant of 1946

The F-2 of 1946 retained the 1350hp power output with a total of 104 produced of A & B units.

The F-3 of 1946 had a different roof arrangement, and slightly different dimensions, than the FT. The 567B engine was uprated to 1,500 hp. From 1946 to 1949 over 1800 were produced.

In 1949 the F-7 was launched which was undoubtedly the best selling F-series with 3849 being produced from 1949-53.A version extended by 4 feet, the FP7, was introduced at the same time to allow for extra heating and air conditioning required for passenger duties.

The F-9 of 1953 had a 1,750 hp 567C engine. A louver arrangement over the vents changed their appearance from the F3.255 F-9s were produced between 1953 and 1960, plus 90 of the 4 foot longer FP-9.

Only one F model did not have Bloomberg B trucks, the FL9 of which 60 were produced between 1956-60 had a lightweight Flexicoil B in front and a standard passenger A-1-A at the rear.This model also had an uprated 1800hp power plant.

Above: Ex-New Haven EMD FL-9 diesel electric locomotive No2006 at Danbury Railway Museum, Danbury, Connecticut, USA.

Right: D&RGW F9 No5771, 2009. Note the carbody filter grille ahead of the front porthole, the only reliable distinguishing feature of an F9.

Below: No's 902 & 903 are two of the six EMD FP-7 units ordered by the Reading Company in 1950 ,road numbered 900-905). Two more were ordered in 1952 No's 906 and 907), and all eight units worked into the 1960s.

Above: An EMD FL9 locomotive at Danbury Railway Museum in New York Central's passenger locomotive colors.

Above: Western Pacific No805-A is an FP-7 was mainly used to pull passenger trains, specifically the California which was operated jointly by the Western Pacific, Denver and Rio Grande Western, and the Chicago, Burlington and Quincy Railroads.

Above: F-series F-3 in the distinctive colors of the Jersey Central Lines.

Below: KCS No2 F-Unit in active service in 2015. This F-unit pulls the Kansas City Southern Railway business train around the railroad and in December hauls the Christmas train to towns along its service area.

NW2B-R EMD GM (1939)

SPECIFICATIONS
Type: Diesel-electric switching locomotive
Gauge: 4 ft 8 1/2 in
Propulsion: Type 567 12-cylinder two-stroke 1,000hp diesel engine and generator supplying current to four dc traction motors geared to the axles
Weight: 250,160 lb
Max. axleload: 62,540 lb
Tractive effort: 62,500 lb
Length: 44ft 5 in

As with General Motors' Electro-Motive Division takeover of main-line rail movement from steam power with their E and F series locomotives, they did not forget humbler switching operations in yards and depots. The N series were produced and sold a total of 50 between 1937 and 1939. The differences between the model numbers were connected with variations in the electrical equipment. Series production began in earnest in 1939 with the l,000hp NW2 model, the subject of the details given above, which was the first EMD locomotive to use their own electrical equipment. This model also had a 12-cylinder version of EMD's famous standard 567 series engine. Over 1,100 examples were sold between 1939 and 1949 and many are still hard at work.

There were a few NW3s and NW5s built in 1942 and 1947 respectively which were road-switcher versions of the NW2 extended to house steam generator equipment, used to pre-heat or pre-cool passenger trains before they began their journeys. These machines were equipped with EMD's "road" trucks and larger fuel tanks.

Since all EMD switchers by now had welded frames, SW was taken in 1949 to mean "switcher" and the NW2 line was modified to become successively the SW7 and then the SW9 model. The 12-cylinder "567" engine continued to be used, but uprated to give 20 percent more power. In 1954, the designation was again changed; from now on the model number was to indicate the horsepower, hence the SW1200.

Top: Maryland and Pennsylvania locomotive No81, an EMD NW-2 at the Railroad Museum of Pennsylvania, Strasburg, PA .

Above: Midland Railway EMD NW-2 No524, at Baldwin City, Kansas.

Left: Erie Railroad's No436 is an SW-9 model.

In 1966 a completely new series of switchers incorporating the later 645 series was introduced Around l,oooohp could now be offered using an eight-cylinder engine and production of the SW1000 and SW1001 models (for freight clearances) continues to this day, while roads needing more powerful switchers can buy the "Multi-Purpose" light road-switcher, model MP15, with a 12-cylinder l,500hp engine. Switching was the first steam citadel to fall and, while all the other diesel manufacturing firms offered similar locomotives, a high proportion of the railroads in the United States, Canada and Mexico, as well as many users in general industry, had these units on the roster.

Above: NW2 switcher 1.

Left: Burlington Northern Railroad 440, an EMD SW1000 working the yard at Eola, Illinois, in September 1992.

GE 65-Ton Switcher (1940)

SPECIFICATIONS
Power type: Diesel-electric
Builder: GE Transportation Systems
Model: 65-ton
AAR wheel arrangement: B-B
Prime Mover: Ingersoll Rand 300

The General Electric Switchers were diesel-electric switcher locomotives. The GE 65-ton switcher is a diesel-electric locomotive with a B-B wheel arrangement, with models producing 400-550 hp. The engine is an upgraded GE 44-ton with a heavier frame and a more powerful diesel engine. The first commercially successful diesel-electric locomotives to be produced in the United States incorporated a diesel engine manufactured by Ingersoll-Rand. It had a ten-inch bore and a twelve-inch stroke. It had an idle speed of 250-275rpm depending upon its auxiliary load. It had a full speed of 550rpm and was rated at 300 brake horsepower.

This diesel engine was developed by Ingersoll-Rand for use in locomotives. This diesel engine really bears no resemblance in any way to common diesel engines for locomotive service that appeared later. For example, the use of exposed push rods was common at the time (even in submarine diesels) and so it was carried over here; no doubt, the problems of dirt and dust in locomotive service weren't yet well appreciated. Note also the use of totally exposed cylinder units; these were a common thing in early, large gasoline, distillate and oil (otherwise known as Diesel) engines at the time as well but later engines of all builders enclosed the cylinders fully in the engine frame.

Below: No12 is a 65 ton center-cab diesel-electric switcher built by General Electric in 1943. The Northern Pacific Railroad museum uses it to haul passenger trains along a very short section of track across the old freight yard.

GE 44-Ton Center Cab (1940-1956)

SPECIFICATIONS
Power type: Diesel-electric
AAR wheel arrangement: B-B
UIC classification: Bo Bo
Gauge: 4 ft 8 1/2 inches
Locomotive weight: 44 short tons
Prime mover: Caterpillar D17000 (2 off) except: Hercules DFXD (2 off) 9 locomotives; Buda 6DH1742 (2 off) 10 locomotives; Caterpillar D342 (2 off) 4 locomotives.
Engine RPM Range: D17000: 1,000 (max), 6DH1742: 1,050 (max), DFXD: 1,600 (max), D342: 1,200 (max)
Engine type: D17000: V8 diesel, All others: 6-cyl diesel
Aspiration: Normally aspirated
Traction motors: Four
Cylinders: D17000: 8, All others: 6
Cylinder size: D17000: 5.75 in x 8 in, 6DH1742: 6.5 in x 8.375 in, DFXD: 5.5 in x 6 in, D342: 5.75 inx 8 in
Power output: 360 to 400 hp

Between 1940 and 1956, General Electric Transportation Systems built some small diesel locomotives. The GE 44-ton switcher is a 4-axle diesel-electric locomotive. It was designed for industrial and light switching duties, often replacing steam locomotives that had previously been assigned these chores. This locomotive's specific 44-short ton weight was directly related to one of the efficiencies the new diesel locomotives offered compared to their steam counterparts: reduced labor intensity. In the 1940s, the steam to diesel transition was in its infancy in North America, and railroad unions were trying to protect the locomotive fireman jobs that were redundant with diesel units. One measure taken to this end was the 1937 so-called "90,000 Pound Rule." This meant that locomotives weighing 90,000 pounds or 45 short tons or more required a fireman in addition to an engineer. Industrial and military railroads had no such stipulation. The 44-ton locomotive was born to skirt this requirement. Other manufacturers also built 44-ton switchers of center-cab configuration. 276 examples of this locomotive were built for U. S. railroads and industrial concerns, ten were exported to Canada, ten were exported to Cuba, one was exported to the Dominican Republic, five were exported to France, three were exported to India, six were exported to Mexico, five were exported to Saudi Arabia, one was exported to Sweden, two were exported to Trinidad, ten were exported to Uruguay, and fifty-seven were built for the U. S. Military. Many remain, in service and in museums.

Below: Visalia Electric Railway No502 GE 44-Ton diesel electric yard switcher.

GE 45-Ton Switcher B-B/B'B' (1940-1956)

SPECIFICATIONS
Power type: Diesel-electric
Gauge: 4 ft 8 1/2 inches
Total weight: 43 to 50 short tons
Prime mover: Two Cummins 6-cylinder HBI-600
Maximum speed: 20 mph
Power output: 2 x 150 hp
Engine RPM range: 1,800 rpm (max)
Cylinders: 6 in each of two engines

The GE 45-ton switcher is a four-axle diesel locomotive built by GE between 1940 and 1956. The locomotive was equipped with two 150 horsepower Cummins diesel engines, each driving a generator which, in turn, drove one of the two traction motors, one per truck. In early models, the second axle on each truck was driven with side rods. Later models had chain drives inside the trucks that served the same purpose. A traditional train air brake was optional, but all came with two compressors (one per engine) and a straight-air independent (locomotive) brake. The cabs were spacious for the size of the locomotive, and both the engineer's seat and the fireman's seat were raised two feet on platforms (under which was the brake equipment, if applicable), so as to afford better views during switching. The GE 45-ton was extremely versatile and many variants existed. It has a high weight to power ratio, and has excellent traction, rated to be able to pull twenty loaded freight cars on level track. They were built with a short wheelbase for use in industrial plants, yards, and other places where clearances were tight. Although intended as switchers, they sometimes served mainline duties, although nearly all had an imposed speed limit of 20 mph due to the double reduction gearing of their traction motors. Current owners and operators include the Old Colony and Newport Scenic Railway operates two of these locomotives on a regular revenue basis. The Lake Superior And Mississippi railroad of Duluth, Minnesota also owns and operates one. Nearby Lake Superior Railroad Museum also has a former Minnesota Power 45-ton, which is now used for switching rolling stock around the museum. Hudson Bay Mining and Smelting operate two of these for slag operations in the smelter. Michigan State University's on-campus T. B. Simon Power Plant uses a 45-ton painted in school colors to position coal.

Above: A GE 45-Ton Switcher located a the Texas Transportation Museum.

Left: A GE 45-Ton Switcher, located at the Cass Scenic Railroad in West Virginia.

GE 50-Ton Boxcab (1940s)

In the early 1940s, General Electric built some small diesel locomotives as part of the war effort. Among the models offered was the 50-Ton Boxcab. These early versions were all constructed for the U.S. Navy and featured side rods on the trucks. One of the surviving examples is now owned by North Sunbury Bulk Transfer of Sunbury, Pennsylvania. This engine also includes some unusual swing couplers. The pictured example is owned by NYC Transit (serial number 38943). The Association of American Railroads wheel arrangement is B-B. This means that there are two identical trucks. Each truck has two powered axles. This is a popular configuration used in high-speed, low-weight applications, such as intermodal trains, high-speed rail, as well as switching. Examples include the general purpose units. High speed freight trains, with guaranteed schedules often use B-B locomotives of 3,800 HP (950 HP per axle), but this application, too, has largely been replaced by higher-powered, 4,500 HP C-C locomotives (750 HP per axle). An American colloquialism for B-B is four axle.

Above: This GE 50 ton boxcab has been modified to run on the New York Subway.

GE 25-Ton Boxcab (1941)

A long series of locomotives were produced by GE Transportation Systems. They were all built at Erie, Pennsylvania. Most (except the electrics, the switchers, the AC6000CW, and the Evolution series) were powered by various versions of GE's own FDL diesel prime mover, based on a Cooper Bessemer design and manufactured at Grove City, Pennsylvania. The 25-Ton Boxcar version was built between 1941 and 1974. A total of 510 engines of this type were produced. Powered by a Cummins engine.

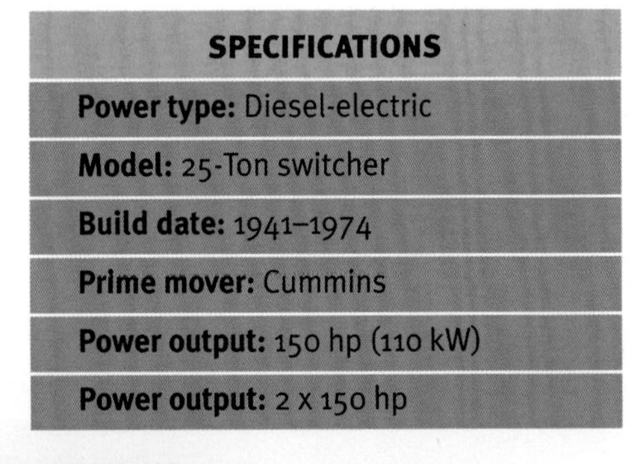

SPECIFICATIONS
Power type: Diesel-electric
Model: 25-Ton switcher
Build date: 1941–1974
Prime mover: Cummins
Power output: 150 hp (110 kW)
Power output: 2 x 150 hp

Alco RSD-1 6-Axle Road Switcher Co-Co (1942)

The Alco RSD-1 was a diesel-electric locomotive built by American Locomotive Company .This model was a road switcher type rated at 1,000 horsepower and rode on three-axle trucks, having a C-C wheel arrangement. It was often used in much the same manner as its four-axle counterpart, the Alco RS-1, though the six-motor design allowed better tractive effort at lower speeds, as well as a lower weight-per-axle. It was developed to meet the wartime Soviet demands of a locomotive with a lower axle load. On the other hand, due to the traction generator and appurtenant control apparatus being sized for four axles and yet having two additional powered axles, it had poorer performance at higher speeds. 150 were produced between 1942 and 1946.

There were three different specifications issued that covered the RSD-1 model; E1645 and E1646 were for wartime production for the US Army, while E1647 was a post-war order for the Mexican National Railways (FDeM).

SPECIFICATIONS
Type: Diesel-electric road switcher
Gauge: 4 ft 8 1/2 in
Length: 55 ft 5 3/4 in
Width: 10 ft 0 in
Height: 14 ft 5 in
Weight: 247,500 lb
Fuel capacity: 1,000 gal
Propulsion: Alco 539T
Engine type: Straight-6 Four-stroke diesel
Aspiration: Turbocharger
Displacement: 9,572 cu in
Generator: DC generator
Traction motors: DC traction motors
Cylinders: 12 1/2 in x 13 in
Transmission: Electric
Maximum speed: 65 mph
Power output: 1,000 hp
Tractive effort: 40,425 lb
Locomotive brake: Straight air
Train brakes: Air

Centipede 2-Do-Do-2 PRR (1945)

Above: Intended for express passenger trains Seaboard Air Line's No.4500 hauls freight. The original design was referred to as a 4-8-8-4.

Baldwin Locomotive Works, like some other famous steam locomotive builders, was slow to recognize that the main-line or road diesel posed a genuine threat to the future of steam. In 1939 the company launched a range of diesel switchers designed by a new diesel group which was almost independent of the firm's steam designers. One of this group, a Swiss named Max Essl, took out some patents for road diesels which were to be powered by small configuration engines mounted transversely- the number of engines being varied to suit the power output required. It was with a demonstration locomotive built to one of Essl's patents that Baldwin eventually entered the road diesel field.

The unit was of the 2-Do-Do-2 layout, although the maker designated it 4-8-8-4 in steam terminology. There were two massive frames, each with four rigid motored axles and an end truck to lead the unit into curves. The frames were hinged together in the manner of a Mallet steam locomotive. Compared with contemporary conventional diesels it had a lower axleload and took more kindly to rough track. It was intended to have eight engines giving a total power of 6,000hp, in token of which it was numbered 6000. The unit was designed to reach 120mph, and to haul 14 80-ton cars at 100mph .The concentration of so much power in one unit was attractive because, amongst other reasons, there was uncertainty

SPECIFICATIONS
Type: Diesel-electric locomotive for express passenger trains.
Gauge: 4 ft 8 1/2 in
Propulsion: Two Baldwin Model 608SC eight-cylinder 1,500hp four-stroke diesel engines and Westing-house generators supplying current to eight nose-suspended traction motors geared to the main axles.
Weight: 409,000 lb adhesive, 593,700 lb total
Max. axleload: 51,200 lb
Tractive effort: 102,250 lb
Overall length: 91 ft 6 in
Max. speed: 93 mph

as to whether the labor unions would demand an engine crew on each unit of a multiple-unit. Wartime restrictions delayed completion of the locomotive, but eventually in 1943 it was able to make some successful trial runs fitted with four engines.

By this time the firm's accountants had totaled the cost of the unit and decided that they could not hope to compete pricewise with a conventional double-bogie diesel. Furthermore, agreement by the unions that only the leading unit of a train need have a crew reduced the interest in large diesels, and No.6000 was laid aside, never having become more than 3000. Baldwin then turned to production of more conventional road diesels.

In 1945 Seaboard Air Line was troubled with EMD diesels damaging the track on curves at high speed, and this prompted Baldwin to rescue the big locomotive from the scrap line and rebuild it with two l,500hp engines of a new design; it was designated 4-8-8-4 "1500/2 DEI." Seaboard bought the locomotive in December 1945 and was sufficiently impressed to order 13 more. In the meantime two other roads, which were already buying Baldwin A1A-A1A diesels, also placed orders for the 4-8-8-4. Pennsylvania Railroad was attracted by the high power, and ordered 12 double units to work back-to-back. Nacionales de Mexico was interested both in the high power and in the ability to run fast on indifferent track; they ordered a double unit, but later canceled. BLW completed this pair as a demonstrator, but it won them no more orders; the conventional layout prevailed.

Above: Baldwin Centipede diesel at the PRR Altoona station, in Altoona, Pennsylvania - September 10, 1951.

Delivery of these orders was delayed by strikes at BLW's Eddystone plant and at suppliers' works, so delivery was slow by BLW standards; 53 units were delivered between March 1947 and July 1948.

Despite the intention that the locomotives should work express passenger trains, they all moved fairly quickly to freight work. At their best they were good, but all BLW road diesels suffered from the firm's late entry into this field, and much of the detail design was troublesome, particularly pipework. The engines also gave trouble. The NdeM locomotives were sent back to the maker in 1953 for rebuilding, and in their improved condition some of them remained in service until 1971. Their US counterparts spent much of their time on helper (banking) duties, where their lack of dynamic (rheostatic) braking and inability to work in multiple with other types was little of a disadvantage, and their unusually high starting tractive effort could be exploited. The Seaboard locomotives were withdrawn by 1961 and all the PRR units were withdrawn in 1962. By this time conventional diesels had reached 3,000hp.

Although technically unsatisfactory there was no doubt that the big Baldwin diesels were impressive, if only for the densely-packed wheels, which earned them their unofficial nickname of "Centipede."

Alco PA Series A1A-A1A (1946)

SPECIFICATIONS
Type: Diesel-electric express passenger locomotive; A units with cab, B units without.
Gauge: 4 ft 8 1/2 in
Propulsion: One Alco 244 2000hp 16-cylinder turbocharged four-stroke V-16 engine and gearbox, supplying four nose-suspended traction motors geared to the end axles of the bogies.
Weight: 204,000 lb adhesive, 306,000 lb total
Max. axleload: 51,0200 lb
Tractive effort: 51,000 lb
Max. speed: 80mph, 90mph, 100mph or 117mph according to gear ratio fitted.

The American Locomotive Company was mainly a builder of steam locomotives until the end of World War II, but it had already achieved considerable success with diesel switchers, and in 1940 had produced a 2,000hp twin-engine passenger locomotive, of which 78 were built before construction ceased during the war. In the following year Alco produced a l,500hp road-switcher, but the railroads were not yet accustomed to the idea of a locomotive which could combine two functions. All these locomotives had engines made by specialist manufacturers, but in 1944 Alco produced its own engine, designated the 244, the last two digits indicating the year in which it first ran. It was a turbocharged V engine made in two versions, one with 12-cylinders producing l,500hp and the other 16 cylinders producing 2000 hp. In early

Right: Nickel Plate Road No190 at the Oregon Rail Heritage Center, where it is undergoing restoration.

190
190
190
190
NICKEL
PLATE
ROAD

1944, development started on the new design, and by November 1945, the first engines were beginning to undergo tests. This unusually short testing sequence was brought about by the decision of Alco's senior management that the engine and an associated line of road locomotives had to be introduced no later than the end of 1946. In September 1946, the first production units, an A-B-A set of PA1s in Santa Fe colors were released from the factory, and sent to New York's Waldorf-Astoria Hotel, which had a private railroad siding, for exhibition before being launched into road service. In practice the Alco V-16 diesel propulsion unit proved to be the undoing of the PA: The engine had been rushed into production, and proved to be unreliable. The PA locomotives failed to capture a marketplace dominated by General Motors Electro-Motive Division and their E-units. However fans deemed the PA one of the most beautiful diesels and an "Honorary Steam Locomotive," as noted by Professor George W. Hilton in a book review in September, 1968 *Trains Magazine*. When accelerating, until the turbocharger came up to speed thick clouds of black smoke would pour from the exhaust stacks, due to turbo lag. The PA locomotives failed to capture a marketplace dominated by General Motors Electro-Motive Division and their E-units. The original Santa Fe three unit set #51L, 51A and 51B was re-engined in August 1954 with EMD 16-567C engines rated at 1,750 hp However fitting new engines to the PAs was economically unfeasible and the remaining Santa Fe units retained their 244 engines. The later Alco 251-series engine, a vastly improved propulsion unit was not available in time for the company to recover the loss of reputation caused by the unreliability of the 244. By the time the improved 251 engine was accepted into widespread use, General Electric (which ended the partnership with Alco in 1953) had fielded their competitive entries into the diesel-electric locomotive market. General Electric eventually supplanted Alco as a manufacturer of locomotives and the company's loss of market share led to its demise in 1969.

Below: Nickel Plate PA/PB unit hurries an eastbound train through Dunkirk, NY , in the early days of 1952.

Baldwin DRS-4-4-1500 Bo-Bo (1947)

SPECIFICATIONS
Type: Diesel-electric road switcher
Gauge: 4 ft 8 1/2 in
Length: 57 ft 10 3/4 in
Height: 14 ft
Weight: 240,000lb
Propulsion: Baldwin 608SC
Engine type: Four-stroke diesel 8 cyl
Aspiration: Turbocharged
Displacement: 15,832 cu in
Generator: DC generator
Traction motors: DC traction motors
Cylinders: 12 3/4 in x 15 1/2 in
Transmission: Electric
Maximum speed: 65 mph
Power output: 1,500 hp
Tractive effort: 61,510 lb
Locomotive brake: Straight air
Train brakes: Air

The Baldwin DRS-4-4-1500 was a diesel-electric locomotive of the road switcher type rated at 1,500-horsepower that rode on two-axle trucks, having a B-B wheel arrangement. It was manufactured from 1947 until it was replaced in 1950 by the 1,600-horsepower AS-16.

Nine railroads bought 35 locomotives, with five railroads later buying the successor model.

Above: LV No200 the only Baldwin DRS road switcher on the Lehigh Valley Diesel roster pushes a caboose past three RS-11's at Sayre , PA. in May 1967.

Left: Lehigh Vally No200 was the only Baldwin Road Switcher on the Lehigh Valley diesel roster. The brick station in Sayre was built in 1881 and is preserved today by the Sayre Historical Society

Fairbanks- Morse H-15-44 Bo-Bo (1947)

SPECIFICATIONS
Type: Diesel-electric road switcher
Gauge: 4 ft 8 1/2 in
Length: 51 ft 0 in
Weight: 250,000lb
Propulsion: FM 38D-8
Engine type: Two-stroke diesel 8 (Opposed piston)
Aspiration: Roots blower
Displacement: 8,295 cu in
Cylinders: 8.125 in x 10 in
Transmission: DC generator
Maximum speed: 65 mph
Power output: 1,500 hp
Tractive effort: 42,125 lb
Locomotive brake: Straight air
Train brakes: Air

Above: Fairbanks-Morse diesel No. 200, the AC&Y's only H15-44, poses for her builder's photo in Beloit, Wisconsin in 1949.

The FM H-15-44 was a road switcher manufactured by Fairbanks-Morse from September 1947 to June 1950. The locomotive was powered by a 1,500-horsepower eight-cylinder opposed piston engine as its propulsion unit, and was configured in a B-B wheel arrangement mounted atop a pair of two-axle AAR Type-B road trucks with all axles powered. The H-15-44 featured an offset cab design that provided space for an optional steam generator in the short hood, making the model versatile enough to work in passenger service as well as freight duty.

Raymond Loewy heavily influenced the look of the unit, which emphasized sloping lines and accented such features as the radiator shutters and headlight mounting, as is found on CNJR No1501 and KCS No40. The cab-side window assembly incorporated "half moon"-shaped inoperable panes which resulted in an overall oblong shape. The platform was shared with F-M's 2,000-horsepower end cab road switcher, the FM H-20-44, as was the carbody to some extent. The platform and carbody was also utilized by the H-15-44's successor, the FM H-16-44.

Only 35 units were built for American railroads and none is thought to exist today.

Left: Seen in Central Railroad of New Jersey service at Communipaw, Jersey City, is an H-15-44 model road switcher of which 35 were manufactured by Fairbanks-Morse from September 1947 to June 1950. The Bo-Bo type locomotive was powered by a 1,500-hp 2-stroke eight-cylinder opposed piston prime mover fitted with a Roots blower. The H-15-44 featured an offset cab design that provided space for an optional steam generator in the short hood, making the model versatile enough to work in passenger service as well as freight duty.

Alco RS-1 (1948)

SPECIFICATIONS
Gauge: 4 ft 8 1/2 in
Wheel diameter: 40 in
Wheelbase: 40 ft 5 in
Length: 55 ft 5 3/4 in
Locomotive weight: 247,500 lb
Fuel capacity: 1,000 US gal
Prime mover: ALCO 539T
Engine type: Four stroke diesel
Aspiration: Turbocharger
Displacement: 1,595 cu in per cylinder 9,572 cu in total
Generator: GE GT-553-C DC generator
Traction motors: (4) GE 731 DC traction motors
Cylinders: (6) 12 1/2 in x 13 in
Maximum speed: 65 mph
Power output: 1,000 hp
Tractive effort: 40,425 lb

The ALCO RS-1 was a 4-axle diesel-electric locomotive built by Alco-GE between 1941 and 1953 and the American Locomotive Company from 1953 to 1960. The Montreal Locomotive Works built three RS-1s in 1954. This model has the distinction of having the longest production run of any diesel locomotive for the North American market. The RS-1 was in production for 19 years from the first unit Rock Island #748 in March 1941 to the last unit National of Mexico #5663 in March 1960.

The carbody configuration of the RS-1 pioneered the road switcher type of diesel locomotive. Most locomotives built since have followed this basic design. In 1940, the Rock Island Railroad approached ALCO about building a locomotive for both road and switching service.

The first thirteen production locomotives were requisitioned by the US Army, the five railroads affected had to wait while replacements were manufactured. The requisitioned RS-1s were remanufactured by ALCO into six axle RSD-1s for use on the Trans Iranian Railroad to supply the Soviet Union during World War Two.

Below: No467 is an Alco built RS-1 switcher, one of four delivered to the Long Island Railroad in 1950 road numbered 466-469). After leaving the LI, it went through a number of owners, including Indiana Hi-Rail. It is currently owned by a private individual and may be restored at a future date.

GE 70-Ton Switcher (1947)

SPECIFICATIONS
Power type: Diesel-electric
Model: 70-ton switcher
Total produced: 238
AAR wheel arr.: B-B
UIC classification: Bo Bo
Gauge: 4 ft 8 1/2 inches
Locomotive weight: 70 short tons
Prime mover: Cooper-Bessemer FWL-6T
Power output: 500–660 hp

Right: The GE 70 ton End Cab type No16 of the Rahway Valley RR.

The GE 70-ton switcher is a four-axle diesel locomotive built by General Electric between about 1942 and 1955. It is classified as a B-B type locomotive. The first series of 70 tonners were a group of seven center-cab locomotives built for the New York Central Railroad in November 1942. These units differ from the later end-cab versions. The Modesto and Empire Traction Company used nine of these reliable 70-ton locomotives on its railroad along with two former-Southern Pacific EMD SW1500s. The company has since retired and sold all of these locomotives except number 600. The GE locomotives are also used by the Santa Maria Valley Railroad.

The Belfast and Moosehead Lake Railroad still has locomotives 50, 51 and 53. Down East Scenic Railroad's 54 is also in service. The Oregon Pacific Railroad owns the former Southern Pacific 5100. The locomotive is painted in the historic Southern Pacific scheme and is currently out-of-service.

Right: Former Southern Pacific 5100, a GE 70-ton switcher, on display at the Oregon Rail Heritage Center, in Portland, Oregon.

5100
SOUTHERN PACIFIC 5100
DF 200 139
Please Keep Off
Please Keep Off

BL-2 Bo-Bo (1949)

Shortly before General Motors Electric-Motive Division swept the board with their GP-7 road-switcher, they had put on the market a similar product but very differently packaged. This was the BL-2 which never reached 60 units sold, whilst GP-7 sales were counted in thousands. One reason perhaps was that BL stood for "Branch Line," and branch line was a naughty word in the railroad industry at the time, being synonymous with accounts written in red ink.

These controversial units had the same 567B engine and running gear as the contemporary best-selling F-3 cab unit. The shape of the bodywork concealed the trusses upon which the unit's strength depended, while at the same time it allowed good visibility in both directions from the cab. A steam boiler for heating and cooling passenger trains was an optional extra, but it found few takers. Incidentally, there was a BL-1 which was a single demonstrator unit completed in February 1948, but this was no different from the BL-2 production series.

Because the machinery was the same as the F-series units, and because the BLs could be run in multiple with other units, there were no difficulties in keeping them running or finding them work. So most lived out normal lives, being in due time traded-in for the latest model in true automobile fashion. Many of their parts could then be re-purposed for resale. Several BL-2s ended up on the 463-mile Bangor & Aroostook Railroad in northeastern Maine, close to the Canadian border. A railroad that once ran its romantically-named "Potatoland Special" complete with sleeping and

Right: At North Maine Junction in 1980, a 30-year-old diesel-electric displays some smart new paintwork.

Below: Bangor & Aroostook veteran Electro-Motive BL-2 Branch Line pattern diesel-electric locomotive at Oakfield, Maine in July 1980.

buffet cars along the length of its main line. These first-generation diesel locomotives operated on BAR until they were museum pieces. The economic downturn of the 1980s coupled with the departure of heavy industry from northern Maine forced the railroad to seek a buyer and end operations in 2003.

SPECIFICATIONS
Type: Diesel-electric locomotive originally intended for passenger and freight traffic on branch lines but now used for freight traffic generally.
Gauge: 4 ft 8 1/2 in
Propulsion: Type 567B 16-cylinder 1,500hp two-stroke diesel engine and generator supplying current to four traction motors geared to the axles.
Weight: 217,600 lb
Max. axleload: 54,000 lb
Tractive effort: 54,420 lb
Overall length: 57 ft 8 in
Max. speed: 70 mph

RF-16 Shark Nose Bo-Bo (1949)

SPECIFICATIONS
Type: Diesel-electric freight locomotive.
Gauge: 4 ft 8 $^1/_2$ in
Propulsion: Baldwin Model 608A eight-cylinder 1,600hp four-stroke diesel engine
Weight: 248,000 lb (adhesive and total)
Max. axleload: 62,000 lb
Tractive effort: 73,750 lb
Length: 54 ft 11 in
Max. speed: 70 mph

The Baldwin Locomotive Works, in conjunction with the Westinghouse Company, kept trying for a number of years in the diesel-electric age to regain the position it held for so long in steam days, that of being the world's largest producer of locomotives.

As we have seen, the earliest offerings in this direction were futuristic designs such as the "Centipede" diesels and were not wholly successful. But by 1949 satisfactory units were on the market and found takers amongst several of the larger eastern railroads. The earliest of these were EMD look-alikes, except for the larger front windows which earned them the nick-name "baby-face."

When production got under way, the appearance was changed to what was described as the "shark-nose" outline. As with the General Motors range, there were passenger A1 A-A1A and freight Bo-Bo units offered in both cab and booster forms. The shape of the front end was based on that of the Pennsylvania Railroad's T-l class 4-4-4-4 steam locomotives. In fact, PRR was the first and only customer for the A1 A-A1A' shark-nose' passenger sets, taking 18 cab and 9 booster units in 1948. These were known respectively as models DR-6-4-20 and DR-6-4-20B, meaning that they were Diesel Road Units with 6 axles, 4 of which were driven, 2,000hp and B if Booster. To give this power, two 6-cylinder engines were used.

The freight units were at first classified DR-4-4-15 and DR-4-4-15B and 105 were sold between 1947 and 1950. The last 68 of these (34 cabs, 34 boosters) were shark-nose style and went to the Pennsylvania Railroad. The model RF-16 (Road Freight, 1600hp) appeared in 1950, and 160 (including 50 boosters) were sold between then and 1953 to the New York Central, Pennsylvania, Baltimore & Ohio and Monongahela, Delaware and Hudson railroads. At last Baldwin had a rugged, reliable and saleable product, but it was to no avail, for there were no further takers. It says enough that locomotive production at what had now become the Baldwin-Lima-Hamilton Corporation ceased in

Below: A Pennsylvania RR loco formed from three "Shark-nose" units at Cresson, Pa, May 1949.

Above: Baldwin "Shark-nose" unit as operated by the Delaware & Hudson Railway, 1977.

1956 after 120 years.

Baldwin used the De La Vergne marine diesel engine for locomotives. This differed greatly from the General Motors model 567 engines, being four-stroke instead of two-stroke. It also revolved at only three-quarters of the speed and had 121/2 x 151/2in cylinders instead of 81/2x10in Baldwin diesels unlike other manufacturers units would not would work in multiple with any except their own kind. This lack of compatibility in the changing face of railroad operations would play yet another part in Baldwin's downfall.

Alco RS-2 Bo-Bo (1950)

SPECIFICATIONS
Type: Diesel-electric 4-axle switcher
Total produced: 378
Gauge: 4 ft 8 1/2 in
Wheel diameter: 40 in
Minimum curve: 57°
Wheelbase: 39 ft 4 in
Length: 56 ft
Width: 10 ft
Height: 14 ft 5 in
Weight: 249,600 lb
Fuel capacity: 800 gal
Propulsion: Alco 244-B
Engine type: Four stroke diesel V12
Aspiration: Turbocharger
Generator: GE 5GT-564B-1
Traction motors: (4) GE 752-A
Cylinders: 9 in x 10 1/2 in
Power output: 1,500 hp @ 1,000 rpm
(later models 1,600 hp)
Tractive effort: 62,500 lb

The Alco RS2 was a further development of the road switcher concept. It had more horsepower than the RS1, and was better suited for heavy road service. Externally, the RS2 bodywork was more rounded, mechanically the new 244 engine was introduced. It was manufactured by American Locomotive Company from October 1946 to May 1950, and 378 were produced — 369 by the American Locomotive Company, and 9 by Montreal Locomotive Works in Canada. Eight of the Alco RS-2s were exported to Canada. The RS-2 has a single, 12 cylinder, model 244 engine, developing 1,500 horsepower. Thirty-one locomotives built by Alco between February and May 1950 with the 12 cylinder 244C 1,600 horsepower engine.

Alco built the RS-2 to compete with EMD, Fairbanks-Morse, and Baldwin Locomotive Works. In 1947, Fairbanks-Morse introduced the 1,500 hp H-15-44. Also in that year, Baldwin introduced the 1,500 hp DRS-4-4-1500. In a competitive market Alco, Fairbanks-Morse, and Baldwin, each increased the power of an existing locomotive line from 1,500 to 1,600 hp and added more improvements to keep ahead of the others.

EMD, however, kept its competing GP-7 at 1,500 hp In 1954, EMD introduced the GP-9. It was rated at 1,750 hp .

EMD produced 2,734 GP-7s. ALCO produced 378 RS-2s, and 1,370 RS-3s. Fairbanks-Morse produced 30 H-15-44s, and 296 H-16-44s. Baldwin produced 32 DRS-4-4-1500s, and 127 AS-16s.

Very few RS-2s survive today. An RS-2 is in active service on the Texas State Railroad (rebuilt as an RS-2-CAT). It is the former Union Railroad No608.

Above: One of few surviving RS-2s Texas State Railroad engine No7 at Rusk terminal - Alco built in 1947.

C-Liners A1A-A1A CNR (1950)

SPECIFICATIONS
Type: Diesel-electric express passenger locomotive
Gauge: 4 ft 8 1/2 in
Propulsion: Two-stroke Fairbanks-Morse opposed piston diesel
Cylinders: 8, 10 or 12
Bore/Stroke: 8.125 in x 10 in
Power output: 1600, 2000, 2400 hp according to no of pistons
Maximum speed: 90mph

The Consolidated line, or C-line, was a series of diesel-electric railway locomotive designs produced by Fairbanks-Morse and its Canadian licensee, the Canadian Locomotive Company. Individual locomotives in this series were commonly referred to as "C-liners." A combined total of 165 units (123 cab-equipped lead A units and 42 cabless booster B units) were produced by F-M and the CLC between 1950 and 1955.

In December 1945 Fairbanks-Morse produced its first streamlined, cab/carbody dual service diesel locomotive as direct competition to such models as the Alco PA and EMD E-unit. Assembly of the 2,000 horsepower unit, which was mounted on an A1A-A1A wheel set, was subcontracted out to General Electric due to lack of space at F-M's Wisconsin plant. GE built the locomotives at its Erie, Pennsylvania facility, thereby giving rise to the name "Erie-built." F-M retained the services of renowned industrial designer Raymond Loewy to create a visually impressive carbody for the Erie-built version. The line was only moderately successful, as a total of 82 cab and 28 booster units was sold through 1949, when production was ended. The Erie-built's successor was to be manufactured in Beloit and designed from the ground up; the result of this effort was the Consolidated line, which debuted in January 1950.

C-liners took many of their design cues from the Erie-builts, and appeared in the F-M catalog with a variety of options. All of the designs were based on a common 56 ft 3 in carbody, but the customer could choose cab or booster units equipped with 1,600 hp ,2,000 hp or 2,400 hp opposed piston engine prime movers. Each option was also offered in both passenger and freight configurations. By 1952, orders had dried up in the United States, with a total production run of only 99 units. The units proved relatively more popular in Canada, particularly with the CPR, and orders continued there until 1955. Several variants were only ever produced by the Canadian Locomotive Company, and Canadian roads accepted a total of 66 units.

Below: The Canadian Pacific Railway also accepted C-Liners in service.

Alco RSD-4 Co-Co (1951)

SPECIFICATIONS
Type: Diesel-electric road switcher
Gauge: 4 ft 8 1/2 in
Trucks: Alco trimount
Wheel diameter: 40 in
Minimum curve: 21°
Wheelbase: 42 ft 3 in
Length: 56 ft 6 in
Width: 10 ft 1 7/8 in
Height: 14 ft 5 1/4 in
Locomotive weight: 278,860 lb
Fuel capacity: 800 gal
Propulsion: Alco 244
Engine RPM range: 1000 (max)
Engine type: Four stroke diesel V12
Aspiration: Turbocharger
Displacement: 8,016 cu in
Generator: GE 5GT-581A1
Traction motors: (6) GE 5GE752-C1
Cylinders: 9 in x 10.5 in
Power output: 1,600 hp
Tractive effort: 69,700 lb

Left: RSD-4 at West Helper, Utah.

The Alco RSD-4 was one of 36 diesel-electric locomotives -Specification E1663- of the road switcher type rated at 1,600 horsepower that rode on three-axle trucks, having an Co-Co wheel arrangement.

Used in much the same manner as its four-axle counterpart, the Alco RS-3, though the six-motor design allowed better tractive effort at lower speeds. Due to the inadequate capacity of the main generator, this model was later superseded in production by the Alco RSD-5

The only Alco RSD-4 that has survived is Kennecott Copper Corporation No201. As of 2007, it resides in fully restored condition at the Northwest Railway Museum (formerly known as the Puget Sound & Snoqualmie Valley Railway) in Snoqualmie, Washington, wearing a coat of bright orange paint.

Right: Kennecott Copper Corporation No201, ALCO 1951 RSD-4.

Below: The back of Locomotive 201, allegedly the last Alco RSD-4 in the world preserved at the Northwest Railway Museum, Snoqualmie, WA.

EMD SD Series: 1952-present

The General Motors SD (Special Duty) series of locomotives are six-axled locomotives were built by EMD in America and GMDD in Canada. While it was originally marketed as a light-footprint unit for running and switching on lines with low axle loading, railroads soon took advantage of the additional tractive effort they offered due to their ability to be more heavily loaded and the additional traction motors. By the 1970s SD-series units were outselling GP-series units by a fair margin. They were offered alongside the GP-series of 4-axle units and the SC/SW-series of switcher locomotives, but are now the only freight units cataloged by EMDD for use in North America.

SD-7 Co-Co (1952)

SPECIFICATIONS
Type: Diesel electric
Gauge: 4 ft 8 1/2 in
Trucks: EMD Flexicoil C
Wheel diameter: 40 in
Minimum curve: 23° (250 ft radius)
Wheelbase: 48 ft 7 in
Length: 61 ft 2 3/4 in
Width: 10 ft 8 in
Height: 15 ft 4 1/2 in
Locomotive weight: 309,000 lb
Fuel capacity: 1,200 gal
Propulsion: EMD 567B
Engine: RPM range 800
Engine type: Two-stroke diesel
Aspiration: Roots-type supercharger
Displacement: 9,072 cu in
Generator: D-12-C
Traction motors: (6) D-27-B
Cylinders: V16
Power output: 1,500 hp
Tractive effort: 77250 lb

The SD-7 was the first model in EMD's SD series of locomotives, a lengthened B-B GP7 with a C-C truck arrangement. The two extra axles and traction motors are useful in heavy, low speed freight service. SD series locomotives are still being produced today.

Built as Demonstrator No990 in 1951, this was the first of EM D's Special Duties series units, later models of which are still being produced. One hundred and eighty-eight SD-7s were built between 1952 and 1953, all for US railroads. No990 demonstrated on several railroads before being bought by the SP in October 1952 and renumbered No5308. It underwent a series of number changes before being rebuilt as an SD-7R in 1980 when it became No1518. The Union Pacific took control of the SP in 1996 and retired No1518 the following year. It was kept at the Cheyenne, WY, roundhouse until donated to the museum in 2003.

Above: Nevada Northern's SD-7 No401 has arrived at Cobre, Nevada, the north end of the line to pick up interchange from SP in 1964.

Below: SD-7 No1518 at the Illinois Railroad Museum.

SD-9 Co-Co (1952)

No0204 is one of one hundred and fifty 1,750 hp, six-axled SD-9s ordered by the Southern Pacific. Five hundred and fifteen SD-9s were built by EMD between 1954 and 1959, four hundred and seventy-one for US railroads, the rest for export.

SPECIFICATIONS
Type: Diesel-electric
Gauge: 4 ft 8 1/2 in
Fuel capacity: 1,200 or 2,400 gall
Propulsion: EMD 567C
Cylinders: V16
Power output: 1,750 hp

Above: SD-9 No0204 in the machine shop at the Nevada Northern Railway Museum.

Below: No0203 one of two SD-9s of the Algers,Winslow & Western Railroad at Yankietown, IN in 2008.

This model is, externally, similar to its predecessor, the SD-7, but this model, internally, features the improved and much more maintainable EMD 567C 16-cylinder engine generating 1,750 horsepower. The principal identification feature is the classification lights on the ends of the locomotive, above the number board which are on a small pod, canted outward. The last phase of construction had a carbody similar to the SD-18 and SD-24, and used two 48-inch cooling fans instead of four 36-inch cooling fans.

No0204 was built in 1956 and delivered as No5468, but was renumbered No3942 in 1965. In 1977, it was rebuilt and renumbered again as No4426.

The unit was retired in July 1995 and sold to Progress Rail Service, a dealer in used locomotives. The following year it was purchased by the Nevada Northern Railway, and it now runs excursions and provides engineer rentals.

SD-24 Co-Co (1958)

This immaculately liveried EMD SD-24 CBQ No504 in was built in 1959. One of sixteen delivered that year to the CB&Q it was also one of sixty-eight built for various US railroads with a high short hood. One hundred and seventy-nine SD-24 A units and forty-five cableless B units were built by EMD between 1958 and 1963, all for US railroads. Weighing 390,000 lbs and 60ft 8in long, they were equipped with a 567D3 16 cylinder propulsion unit powering six GM-D47 traction motors, one on each axle, producing 2,400 hp. The unit delivered 97,500 lbs starting tractive effort and 72,300 lbs continuous tractive effort at 9.3 mph with a top speed of 65 mph. No504 is one of only two SD-24s to have survived.

SPECIFICATIONS
Type: Diesel-electric
Gauge: 4 ft 8 1/2 in
Fuel capacity: 1,200 or 2,400 gall
Propulsion: EMD 567C
Cylinders: V16
Power output: 1,750 hp

Below: You can see the other survivor, WC SD-24 No2402, at the National Railroad Museum.

SD-35 Co-Co (1964)

SPECIFICATIONS
Type: Diesel-electric
Gauge: 4 ft 8 1/2 in
Length: 60 ft 8 1/2 in
Fuel capacity: 3,000 gal
Propulsion: EMD 567D3A
Cylinders: V16
Power output: 2,500 hp

The EMD SD-35 is one of 360 6-axle diesel-electric locomotive built by General Motors Electro-Motive Division between June 1964 and January 1966. Power was provided by an EMD 567D3A 16-cylinder engine which generated 2,500 brake horsepower A 3,000- gallon fuel tank was used on this unit. This locomotive model shared a common frame with the EMD SD-28 , giving it an overall length of 60 feet 8 1/2 inches. 360 examples of this locomotive model were built for American railroads.

The SD-28, built between July 1965 and September 1965 was basically a non-turbocharged version of the EMD SD-35. Only 6 examples of this locomotive model were built for American railroads

Power was provided by an EMD 567D1 16-cylinder engine which generated 1,800 horsepower.

Below: Southern Pacific Railroad SD-35 No3106 at West Colton Yard, CA

Below left: SD-35 No7402 is now preserved at the B&O Museum, Baltimore, Maryland.

SD-40 Co-Co (1966)

SPECIFICATIONS
Type: Diesel electric
Gauge: 4 ft 8 $^{1}/_{2}$ in
Driver diameter: 40 in
Length: 65 ft 8 in
Width: 10 ft 0 in
Height: 15 ft 5 $^{1}/_{4}$ in
Weight: 360,000 lb
Fuel capacity: 3,200 gall
Lubricant capacity: 243 gall
Coolant capacity: 295 gal
Sandbox capacity: 56 cu ft
Proplusion: EMD 16-645-E3
Aspiration: turbocharged
Alternator main: AR-10
Auxiliary: D14
Traction motors: D-77
Cylinders: V16
Maximum speed: 83 mph
Power output: 3,000 hp
Locomotive brake: Independent air
Optional: dynamic brakes
Train brakes: Air, schedule 26-L

The EMD SD-40 was built between January 1966 and August 1972. Like its predecessor the SD35, the SD40 is a high-horsepower, six-motor freight locomotive.

In 1966, EMD updated its locomotive catalog with entirely new models, all powered by the new 645 diesel engine. These included six-axle models SD-38, SDP-40 and SD-45, in addition to the SD-40. All shared standardized components, including the frame, cab, generator, trucks, traction motors, and air brakes. The primary difference was the power output: SD-38-2,000 hp from a non-turbocharged V16, SD-40- 3,000 hp from a turbocharged V16, and SD-45-3,600 hp from a turbocharged V20.

Above: An EMD SD-40 at Winnipeg, Manitoba in 2014.

Below: On the left, NS EMD SD-40-2 No3370, still bearing its Conrail livery, hauls a mixed freight east with NS EMD SD-40-2 No6166 and EMD GP-38-2 No5298. NS EMD SD-40-2 No3235 is standing on the adjacent track.

SD-45 Co-Co (1965)

A total of 1260 SD-45, 6 axle diesel locomotives were built for American railroads by EMD from 1965 to 1971. As built, they were powered by an EMD 645E3 twenty cylinder engine generating 3,600 hp. which had several teething problems. Reliability was not as high as anticipated; the twenty-cylinder unit could break its own relatively long crankshaft and there were difficulties initially in setting and maintaining the firing sequence of the 20 cylinders. Though it produced 600 horsepower more than the 16-645E3 in the SD-40, some railroads felt it wasn't worth it, even after EMD redesigned the block to reduce crankshaft flexing, thereby producing the 645F crankcase and crankshaft. But, the redesigned block and crankshaft formed the basis of the exceptionally reliable 710G engine, which is the cornerstone of EMD's current range.

No3607 was bought by the Erie Lackawanna in 1967. It is one of thirty-five SD-45s bought by the railroad between 1967 and 1968. 65' 9 1/2" long and weighing 368,000 lbs, it delivered continuous tractive effort of 82,100 lbs at 11 mph with a top speed of 65 mph. After EL's bankruptcy in 1972, Conrail acquired the locomotive and donated it to the museum in 1986.

SPECIFICATIONS
Type: Diesel-electric
Total produced: 1,260
Gauge: 4 ft 8 1/2 in
Propulsion: EMD 645E3
Cylinders: V20
Power output: 3,600 hp

Many SD-45s have survived and you can see the first production unit built by EMD in 1965, now GN No400, at the Lake Superior Railroad Museum.

SD-40-2 Co-Co (1972)

SPECIFICATIONS
Type: Diesel-electric
Gauge: 4 ft 8 1/2 in
Driver diameter: 40 in
Wheelbase: 43 ft 6 in between bolsters
Length: 68 ft 10 in
Width: 10 ft 3 1/8 in
Height: 15 ft 7 1/8 in
Weight: 368,000 lb
Fuel capacity: 3,200–4,000 gall
Proplusion: EMD 16-645-E3
Aspiration: turbocharged
Cylinders: V16
Power output: 3,000 hp

The EMD SD-40-2 is a 3,000 horsepower diesel-electric locomotive built by EMD from 1972 to 1989.3982 units were produced.The SD40-2 was introduced in January 1972 as part of EMD's Dash 2 series, competing against the GE U30C and the ALCO Century 630. Although higher-horsepower locomotives were available, including EMD's own SD-45-2, the reliability and versatility of the 3,000-horsepower SD-40-2 made it the best-selling model in EMD's history and the standard of the industry for several decades after its introduction. The SD-40-2 was an improvement over the SD-40, with modular electronic control systems similar to those of the experimental DDA40X.Peak production of the SD-40-2 was in the mid-1970s however its sales began to diminish after 1981 due to the oil crisis, increased competition from GE's Dash-7 series and the introduction of the EMD SD-50, which was available concurrently to late SD-40-2 production. The last SD-40-2 delivered to a United States railroad was built in July 1984. As of 2013, nearly all still remain in service.

Above: Montana Rail Link XDM SD-40-2 diesel locomotive No250 at Everett, Washington, USA, January 1994.

Below left: CSXT No8029 is waiting for another train at the siding at Tunnel Hill, Georgia, on the Western and Atlantic Sub

Below: Union Pacific Railroad No3045, EMD SD40-2.

SD-50 (1980)

SPECIFICATIONS
Type: Diesel-electric
Gauge: 4 ft 8 1/2 in
Length: 71 ft 2 in
Propulsion: EMD 16-645F3B
Alternator main: AR11
Power output: 3,500 hp, increased to 3,600 hp November 1984

Four hundred and twenty seven EMD SD-50 3,500-horsepower diesel-electric locomotives were built by General Motors Electro-Motive Division. The type was introduced in May 1981 as part of EMD's "50 Series," but prototype SD-50s locomotives were built from 1980; production ceased in January 1986. The SD-50 was in many respects a transitional model between EMD's Dash 2 series which was produced throughout the 1970s and the microprocessor-equipped SD-60 and SD-70 locomotives. The SD-50 used an updated version of the V16 645 used in the SD-40-2, uprated to 3,500 hp and later 3,600 hp—at 950 rpm from 3,000 hp at 900 rpm. This proved to be a step too far; the 50 series models were plagued by engine and electrical system problems which harmed both sales and the reputation of EMD.

SD-60 (1984)

The EMD SD60 is a 3,800 horsepower 6-axle diesel-electric locomotive built by General Motors Electro-Motive Division. Intended for heavy-duty drag freight or medium-speed freight service. It was introduced in 1984, and production ran until 1995. The SD-50's electrical reliability was poor and, similarly, the 3,500 horsepower 16-645F engine had poor mechanical reliability, both believed to be largely due to the 950 maximum rpm of the 645F propulsion unit. It was time to develop a replacement for the venerable 645 engine which, in its earlier 16-645E form, had proved to be exceptionally reliable. EMD therefore quickly commenced development of the SD-60 series, which would eliminate the weaknesses of the SD-50. The lessons learned in developing the 645F crankcase and crankshaft (for the earlier 20-645F, and the then-current 16-645F) were incorporated in the replacement, the 710G. The SD-60 proved to be more reliable and fuel-efficient than the SD-50, with 1,140 produced but it was not a resounding success in terms of regaining the market share that was lost due to the electrical and mechanical issues that plagued the earlier SD-50.

SPECIFICATIONS
Type: Diesel-electric
Gauge: 4 ft 8 1/2 in
Propulsion: EMD 16-710G3A
Alternator: AR-11
Traction motors: D-87
Power output: 3,800 hp

SD-70 (1992-present)

SPECIFICATIONS
Type: Diesel-electric
Gauge: 4 ft 8 1/2 in
Proplusion: EMD 16-710-G3B
Cylinders: V16
Tractive effort: 113,100 lb
Top Speed: 70 mph

No2712 is one of GE's SD-70 series, over four thousand of which have been built since 1992. Both the original SD-70 and the later SD-70M develop 4,000 hp, powered by a 16 cylinder 710G3B propulsion unit using six GM D-100 DC traction motors. The SD-70 uses the smaller standard cab, common on older 60 Series locomotives, instead of the larger, more modern "comfort" cab. This makes it hard to distinguish from the nearly-identical SD-60, the only difference being the use of the HTCR radial truck instead of the HT-C truck mounted under the SD-60. Another difference is the frame on the SD70 is approximately 1-2 inches higher than the frame on the SD-60. This model is equipped with direct current traction motors, which simplifies the locomotive's electrical system by obviating the need for computer-controlled inverters (as are required for alternating current power).

Total produced
SD-70: 122
SD-70M: 1,646
SD-70I: 26
SD-70MAC: 1,109
SD-70ACe: 1,034
SD-70ACe/lc: 64
SD-70ACS: 25
SD-70M-2: 331

Above: Billiton Iron Ore EMD SD-70ACe No4352 Lightning, at the head of a loaded train at Boodarie, near Port Hedland, Western Australia. The train was awaiting clearance to continue towards the Finucane Island loop, at the northern end of the Goldsworthy railway, to unload.

GTE 8500 UPR No18 (1952)

No18 is one of only two Union Pacific gas-turbine locomotives to survive complete with its B unit and auxiliary tender which is based on a steam locomotive design, it holds 24,000 gallons of oil in addition to the 2,500 gallons provided by the B unit.

Unlike the Big Boys, Centennials and other smaller locomotives donated to museums and cities, both turbines were not directly donated by the UP. In 1975, they were in a dead line at Intercontinental Engineering's facility in Riverside, MO, ready to be cut up for scrap and parts for other uses.

UP's order for thirty turbines followed testing of a smaller prototype, No50 built by Alco-GE, and delivery of twenty-five smaller units (Nos51-75). The huge, two-unit locomotives Nos1-30, were delivered during 1958 to 1961. All were equipped with fuel tenders from the start, as well as multiple unit connections so one or more trailing diesels could be controlled from the turbine's cab. Some were eventually upgraded to 10,000 hp.

Once in operation, they quickly earned the nickname "Big Blows" because of the deafening noise they made. The UP calculated a single turbine unit could haul seven hundred and thirty-four freight cars (a train over seven miles long) at

Below: Publicity photo of a first generation Union Pacific GTEL locomotive and a circa 1923 electric auto in Fremont, Nebraska. The auto was owned by a local woman and the locomotive was on its way west to haul freight between Wyoming and Utah.

SPECIFICATIONS
Gauge: 4 ft 8 1/2 in
Trucks: 4
Length: 83 ft 6.5 in (Prototype)
Weight: 500,000 lb (Prototype)
849,212 lb for 3rd. generation
Fuel type: Bunker C: heavy fuel oil (UP 57 used compressed propane fuel from May 1953 to January 1954)
Propulsion: GE 5-Frame Gas Turbine 3rd. Generation.
Engine type: Cummins 250 hp donkey engine 1st and 2nd generation. Cooper-Bessemer 850 hp 3rd. Generation.
Traction motors: GE 752E1 1st and 2nd Generation, GE 752E3 3rd Generation.
Maximum speed: 65 mph
Power output: 4,500 hp 1st & 2nd Generation
8,500 hp 3rd Generation
Tractive effort: 212,312 lb 3rd Generation
Safety systems: Twin Leslie Tyfon A-200 air horns 1st & 2nd generation, Leslie S-5T-RF air horn
Total produced: 56

a steady 12 mph.

The "Big Blows" started running from Ogden, UT, to Los Angeles, CA, but, after four months with temperatures nudging 115°F that didn't allow the turbines to perform at their best, they were pulled back to work from Ogden, UT, to Council Bluffs, IA. There may also have been complaints from LA residents about the turbines' noise.

The Union Pacific rostered the largest fleet of turbine freight locomotives in the world: at one time, they claimed turbines hauled 10% of their entire freight. They were initially quite cost-effective, despite poor fuel economy, because they used residual fuel oils, "Bunker-C oils," the same fuel used in oil burning steam locomotives.

However, as other uses were found for these heavier petroleum by-products, particularly plastics, the turbine units became too expensive to operate.

No18 A is over 80ft long, and the combined A-B unit is 165ft 11in long and weighs 849,248 lbs. The A unit housed the cab, an 850 hp diesel engine and other control equipment. The B unit housed the electric generators and 8,500 hp GTEL turbine that powered the twelve GE 752E4 traction motors.

All the turbines had been retired by 1970. The trucks and traction motors from the other units were used to build EMD's U-50C diesel-electrics. Several of the tenders were also retained and converted to hold water for UP's operating steam locomotives, including UP FEF-3 No844 and UP Challenger No3985.

Trainmaster H-24-66 Fairbanks & Morse (1953)

Fairbanks Morse of Beloit, Wisconsin, was an engineering firm which had for a long time supplied general equipment to railroads, such as water stand pipes. In the 1930s the firm began specializing in opposed-piston diesel engines, with two pistons in each cylinder, and two crankshafts connected by gearing. The engines were fitted to carbody or cab units, with a choice of engines rated at l,600hp,2,000hp or 2,400hp supplying four traction motors. Twenty-two units with the 2,400hp engine were sold in 1952-53, but the market was changing rapidly, and carbody designs were giving way to the more versatile hood-type road switcher. Fairbanks Morse acted quickly, and in 1953 produced a 2,400hp hood unit designated "H-24-66" (Hood, 2,400hp, 6 motors, 6 axles). It was mounted on two three-motor three-axle bogies of new design. Compared with its competitors everything about it was big—the dynamic brake power, the train heating boiler (if fitted), the fuel supplies, the tractive effort. Although EMD offered a twin-engined 2,400hp carbody unit, the FM engine was the largest then on the market. Marketed as the "Trainmaster," it was unveiled at a Railroad Manufacturers' Supply Association Fair at Atlantic City, where it stole the show. This publicity, combined with the impression made by the four demonstrator units, soon brought orders.

SPECIFICATIONS
Type: General-purpose diesel-electric locomotive
Gauge: 4 ft 8 1/2 in
Propulsion: One Fairbanks Morse 38D-122,400hp 12-cylinder turbocharged opposed-piston diesel engine and generator supplying six nose-suspended traction motors geared to the axles.
Weight: 375,000 lb
Max. axleload: 62,500 lb
Overall length: 66 ft 0 in
Tractive effort: 112,500 lb
Max. speed: 65 mph, 70 mph or 80 mph according to gear ratio fitted.

Above: Canadian Pacific Railway No8909, a Canadian Locomotive Company H-24-66 locomotive.

The peak year for Trainmaster production followed all too quickly in 1954, when 32 units were built, but orders then slowed down. Railroads encountered problems with the opposed-piston engine and with the electrical systems. One of the characteristics of EMD service had always been the prompt and thorough attention which was given to faults in the field, but customers found that the smaller FM company could not give such good service. During 1954-55 the firm was still dealing with engine problems, including pistons and bearings but then trouble within the family–owned company leaked out and a result of this trouble was that Illinois Central decided against placing an order for 50 to 60 units which it had contemplated, an order which could have changed the whole outlook for the model. In the event a total of 105 were sold to eight US railroads, and a further 22 were built in Canada. Major users were the Norfolk & Western, which acquired 33 Trainmasters as a result of mergers, and Southern Pacific, which used them extensively on commuter routes.

Alco S-6 Switcher Bo-Bo (1955)

The Alco S-6 (spec. DL 430) was a diesel-electric locomotive of the switcher type constructed by Alco of Schenectady, New York; a total of 126 locomotives were built between May 1955 and December 1960. The S-6 was a development of the earlier S-5; instead of the 800 hp Alco 251 engine on the previous locomotive, the S-6 used a 251A or 251B engine rated at 900 horsepower The locomotive rode on two-axle AAR trucks, giving a B-B wheel arrangement. A cow-calf version was produced exclusively for Oliver Iron Mining Company, who bought two: this was the SB-8/SSB-9 (DL-441). The freight train used to smash the DeLorean DMC-12 time machine in Back to the Future Part III was pulled by two Alco S-6 locomotives in Ventura County Railroad livery.

SPECIFICATIONS
Gauge: 4 ft 8 1/2 in
Length: 45 ft 5 in
Width: 10 ft
Height: 14 ft 8 in
Weight: 230,000 lb
Fuel capacity: 635 gal
Lubricant capacity: 140 gal
Coolant capacity: 110 gal
Sandbox capacity: 26 cu ft
Propulsion: Alco 251A,B
Engine RPM range: 375 – 1000 rpm
Engine type: Four-stroke diesel straight 6
Displacement: 668 cu in per cylinder, 4,008 cu in total
Generator: GE GT533
Traction motors: GE 752 (4 off)
Cylinders: 9 in x 10 1/2 in
Maximum speed: 60 mph
Power output: 900 hp

GM Aero Train B-1 (1955)

During the 1950s with greater car ownership, growth of the nation's interstate highway system and air services, railway passenger numbers began to decline. In response, railroads sought ways to attract passengers, including introducing stylish trains that combined speed and added luxury. One such effort was General Motors' "Aerotrain," unveiled in 1955. Designation LWT12.

The trainset is essentially a combination of GM Truck & Coach Division 40-seat intercity highway bus bodies re-styled and adapted to run on rails and hauled by a diesel engine. The GM "style" is evident in the windshield design, smooth front end lines with recessed headlights, and the fin-like wrap-around just above the pilot. Like all of GM's body designs of this mid-century era, this train was first brought to life in GM's Styling Section. Chuck Jordan was in charge of designing the Aerotrain as Chief Designer of Special Projects.

The first railroad to test one of the new trains in February 1956 was the PRR, but the lightweight coaches, single-axle trucks and a suspension system designed for buses, made for a very uncomfortable ride at high speed. The PRR was not convinced and returned the train to GM after a year of use.

GM took the trains on a tour of the country trying an effort to drum up interest, but the same problems arose and no-one was willing to buy.

SPECIFICATIONS
Type: Diesel-electric
Propulsion: EMD 12-567C
Cylinders: 12
Power output: 1200 hp
Locomotive brake: straight air
Train brakes: air

Finally, in 1957, they were sold to the Rock Island to pull the "Talgo Jet Rocket" train between Chicago and Peoria, where lower speeds applied but, after only nine years, in 1966 both trains were retired. Only three LWT12 units were built.Second and third units, EMD serial numbers 21463 and 21464, became integrated in the Aerotrain. Two of the three LWT12 locomotives are now in the care of museums.

Below: The futuristic lines of the GM Aerotrain failed to attract enough attention for it to be a success.

Alco RS-3 Bo-Bo (1956)

SPECIFICATIONS
Type: Diesel-electric road switcher
Gauge: 4 ft 8 1/2 in
Trucks: AAR type B
Wheel diameter: 40 in
Minimum curve: 21°
Wheelbase: 39 ft 4 in
Length: 56 ft 6 in
Width: 10 ft 1 5/8 in
Height: 14 ft 5 1/8 in
Weight: 247,100 lb
Proplusion: Alco 244-D
Engine type: Four stroke diesel V12
Aspiration: Turbocharger
Generator: GE GT-581
Traction motors: (4) GE 752
Cylinders: 9 in x 10 1/2 in
Power output: 1,600 hp
Tractive effort: 61,775 lb

The Alco RS-3 is a 1,600 hp B-B road switcher diesel-electric locomotive. It was manufactured by American Locomotive Company and Montreal Locomotive Works (MLW) from May 1950 to August 1956, and 1,418 were produced — 1,265 for American railroads, 98 for Canadian railroads, 48 for Brazilian and 7 for Mexican railroads. It has a single, 12 cylinder, Model 244 engine. It was development of the Alco RS-2. Alco built the RS-3 to compete with EMD, Fairbanks-Morse, and Baldwin Locomotive Works. In 1950, Fairbanks-Morse introduced the 1,600 hp H-16-44. Also in 1950, Baldwin introduced the 1,600 hp AS-16. In the case of Alco, Fairbanks-Morse, and Baldwin, each company increased the power of an existing locomotive line from 1,500 to 1,600 hp and added more improvements to create new locomotive lines .

Right: No1508 of the Chesaning Central & Owosso Railroad that was a short-lived tourist line, running from Chesaning to Owosso in the summer and fall of 1999.

Alco RSD-7 Co-Co (1956)

SPECIFICATIONS
Type: Diesel-electric road switcher
Total produced: 29
Gauge: 4 ft 8 1/2 in
Proplusion: Alco 244 V16
Cylinders: 16
Power output: DL-600: 2,250 hp DL-600A: 2,400 hp

The Alco RSD-7 was a diesel-electric locomotive of the road switcher type built at Schenectady, New York between January 1954 and April 1956. Two versions were built, with the same RSD-7 model designation but different specifications and power ratings, although both used the Alco 244 engine in V16 configuration. Specification DL-600, of which only two were built, developed 2,250 hp and used the 244G engine. The revised specification DL-600A, numbering 27 locomotives, was rated at 2,400 hp and used the 244H engine. The RSD-7 was superseded by the Alco 251-engined Alco RSD-15. The RSD-7 was the last ALCO diesel built with a 244 engine.

Both rode on a pair of three-axle trucks with all three axles on each truck powered; this is a C-C wheel arrangement. These trucks have an unequal axle spacing due to traction motor positioning; the outer two axles on each truck are closer together than the inner two. The RSD-7 used the GE 752 traction motor. The six-motor design allowed better tractive effort at lower speeds.

Alco RSD-15 Co-Co (1956)

The Alco RSD-15 was a diesel-electric locomotive of the road switcher type built by the American Locomotive Company of Schenectady, New York between August 1956 and June 1960, during which time 75 locomotives were produced. It was designated Model DL600B. The RSD-15 was powered by an Alco 251 16-cylinder four-cycle V-type propulsion unit rated at 2,400 horsepower it superseded the almost identical Alco 244-engined RSD-7, and was catalogued alongside the similar but smaller 1,800 hp RSD-12, powered by a 12-cylinder 251-model V-type diesel engine.

The locomotive rode on a pair of three-axle *Trimount* trucks, in an AAR C-C wheel arrangement, with all axles powered by General Electric model 752 traction motors. These trucks have an asymmetrical axle spacing due to the positioning of the traction motors. The six-motor design allowed higher tractive effort at lower speeds than an otherwise similar four-motor design.

The RSD-15 could be ordered with either a high or low short hood; railfans dubbed the low short hood version "Alligators," on account of their unusually long low noses.

SPECIFICATIONS
Type: Diesel-electric road switcher
Gauge: 4 ft 8 1/2 in
Trucks: AAR type B
Wheel diameter: 40 in
Minimum curve: 21°
Wheelbase: 39 ft 4 in
Length: 56 ft 6 in
Width: 10 ft 1 5/8 in
Height: 14 ft 5 1/8 in
Weight: 247,100 lb
Proplusion: Alco 244-D
Engine type: Four stroke diesel V12
Aspiration: Turbocharger
Generator: GE GT-581
Traction motors: (4) GE 752
Cylinders: 9 in x 10 1/2 in
Power output: 1,600 hp
Tractive effort: 61,775 lb

Above: Alco RSD-15 built in 1960 as ATSF 843. Renumbered ATSF 9843. Sold to Squaw Creek Coal Co as 9843. After several owners was purchased by an individual and is now on loan to the Arkansas Railroad Museum.

Below: Green Bay and Western Railroad 2407, an ALCO RSD-15, at the Illinois Railway Museum

No6767 CN (1959)

SPECIFICATIONS
Type: Diesel-electric passenger locomotive
Gauge: 4 ft 8 1/2 in
Proplusion: EMD 251b
Cylinders: 162
Power output: 1800 hp
Top speed: 92 mph

Canadian National No6767 was one of 27 locomotives built by the Montréal Locomotive Works in 1959.They were works numbers FRA-4 and designated CN class MPA18b.Road numbered 6767-6793, they were in service on the CN until 1975 when they were retired. They were sold to Via Rail in 1978. No6767 was sold to the Cuyahoga Valley Scenic Railway becoming CVSR 6767.

Above: No6767 in service with the CN at Spadina in the early 1970s.

Below: CVSR engine No6767 sits in the Fitzwater yard at the CVSR maintenance facility .This engine was first built in 1959 by the Montreal Locomotive Works for the CN R and was mostly used for high speed passenger service between Montreal & Toronto, Canada .

EMD GP series Bo-Bo 1959-present

EMD responded to the post-war boom in diesel sales by offering a range of models based on three main series. First, the E series of A1A+A1A express passenger locomotives, secondly the F series of Bo-Bo locomotives for freight work, but with optional gear ratios covering passenger work to all but the highest speeds, and thirdly a number of switchers and transfer locomotives for work within and between switching yards.

SPECIFICATIONS
Type: Diesel-electric road switcher locomotive
Gauge: 4 ft 8 1/2 in
Propulsion : One EMD 567D2 2,000 horsepower turbo-charged two-stroke V-16 engine and generator
Weight: 244,000 pounds to 260,000 pounds according to fittings
Max. Axleload: 61,000 pounds to 65,000 pounds according to fittings
Overall length (GP-20 variant 1959): 56 ft
Tractive effort: 61,000 pounds to 65,000 pounds according to weight
Max. speed: 65mph, 71mph, 77mph, 83mph, or 89mph according to gear ratio fitted

There was an important difference between the switchers and the other models. In the switchers the structural strength was in the under-frame, on which rested the engine, generator and other equipment. The casing or hood was purely protective and had no structural strength. The E and F series, on the other hand, had load-bearing bodies, or car bodies. These provided an engine-room in which maintenance work could be carried out whilst the train was in motion, and which were more satisfactory aesthetically than a hood.

With these models EMD captured about seventy percent of the North American market. Its ability to do so stemmed from a combination of quality of performance and reliability in the locomotive, low maintenance costs, which were helped by the large number of parts which were common to the different types, and competitive prices made possible by assembly line methods of manufacture. Full benefit of assembly line methods could only be achieved by limiting the number of variants offered to customers, and this, in turn, helped EMD's competitors to pick on omissions from, or weaknesses in, the EMD range to hold on to a share of the market. To achieve market leadership some changes were made in the range, of which the most important originated in customer enquiries received before WWII for a locomotive which was primarily a switcher, but which could also haul branch line trains, local freights and even local passenger trains. To meet this need a small number of locomotives were built with switcher bodies, elongated to house a steam generator, and mounted on trucks of the F series; these were road switchers. Construction was resumed after the war, still on a small scale, and with the design adapted to meet individual customer's requirements.

By 1948, EMD's competitors, particularly Alco, were achieving success with a general purpose hood unit for branch line work. For this application, ability to gain access to the working parts was more important than protection for technicians to work on the equipment on the road, and the hoods also gave the enginemen a much wider field of view. In 1948, therefore, EMD offered a branch line diesel, designated BL, incorporating the 1500 horsepower 567B engine, and other equipment including traction motors from the F series. These were accommodated in a small semi-streamlined casing. The main advantage compared with a car body was the improved view from the cab. This model proved to be too expensive to produce.

EMD subsequently designed a true hood unit for general purpose duties, designated GP.

Richard Dilworth, EMD's chief engineer, said that his aim was to produce a locomotive that was so ugly that railroads would be glad to send it to the remotest corners of the system (where a market for diesels to replace steam still existed), and to make it so simple that the price would be materially below standard freight locomotives.

Above: GSWR No702 GP-9 EMD Locomotive of Georgia Southwestern Railroad, Ex SOU Southern Railway Built 1955 at Valdosta Railway Rail Yard, Clyattville, Lowndes County, GA.

Although the GP was offered as a radically new design, many parts were common to the contemporary F-7 series. The power plant was the classic 567 engine, which like all EMD engines was a two-stroke V-16 design; this was simpler than a four-stroke but slightly less efficient. Much development work was devoted over the years to improving the efficiency of the EMD engines to meet.

Over the years, much development work was devoted to improving the efficiency of the EMD engines to meet the competition of four stroke engines. The trucks were of the Blomberg type, a fairly simple design with swing-link bolsters, which were introduced in the FT series of 1939 and are still, with changes in the springing system,

Left: Georgia Railroad No1026, EMD GP-7 is on permanent display in Duluth, Georgia.

Left: The EMD GP-7 is a four-axle (Bo-Bo) road switcher diesel-electric locomotive built by General Motors Electro-Motive Division and General Motors Diesel between October 1949 and May 1954. Power was provided by an EMD 567B V16-cylinder engine which generated 1,500 horsepower The GP-7 was offered both with and without control cabs. Engines built without control cabs were called a GP-7B. Five GP-7Bs were built between March and April 1953. The GP-7 was the first EMD road locomotive to use a hood unit design instead of a carbody design. This proved to be more efficient than the cab unit design as the hood unit cost less, had easier and cheaper maintenance, and had much better front and rear visibility for switching.

standard in EMD Bo-Bo models to this day. The cab afforded a good view in both directions and the hood gave easy access to the equipment. Despite Dilworth's intentions, EMD's stylists produced a pleasing outline. Electrical equipment was simplified from the F series, but nevertheless it gave the driver tighter control over the tractive effort at starting. It also had a more comprehensive overall control to suit the wide range of speeds envisaged.

Of the 2,734 GP-7s built, 2,620 were for American railroads (including five GP-7B units built for the Atchison, Topeka and Santa Fe Railway), 112 were built for Canadian railroads, and two were built for the Mexican railroads. This was the first model in EMD's GP General Purpose series of locomotives.

The C series engine was introduced in 1954. This was the next development of the 567 engine. It had 1,750 hp and was introduced into the range as the GP-9. The engine differed in detail from the GP-7, mainly to bring still further reductions in maintenance. By this time the hood unit was

Below: An Aberdeen and Rockfish Railroad locomotive photographed at the company's yard at Aberdeen, North Carolina in 2008.

514
N.Y.C.&STL
NICKEL PLATE ROAD
514
514

CALIFORNIA NORTHERN
113

Left: The GP-9 is a four-axle diesel-electric locomotive built by General Motors' Electro-Motive Division in the United States, and General Motors Diesel in Canada between January 1954 and August 1963. American production ended in December 1959, although an additional thirteen units were built in Canada, including the last two in August 1963. Power was provided by an EMD 567C sixteen cylinder engine, which generated 1,750 horsepower. This locomotive type was offered both with and without control cabs. Locomotives built without control cabs were nominated GP-9B locomotives. All GP-9B locomotives were built in the United States between February 1954 and December 1959.

Right: EMD produced a turbo-charged version of the 567 engine, 567D2, which was fitted to the GP-18. 390 units of this locomotive were sold over thirteen years.

widely accepted, and sales of the GP at 4,157 established another record. The GP was now America's best selling diesel locomotive.

The EMD GP-15-1 was built by General Motors Electro-Motive Division between June 1976 and March 1982. Intended to provide an alternative to the rebuilding programs that many railroads were applying to their early road switchers, it is generally employed as a yard switcher or light road switcher. This locomotive is powered by a 12-cylinder EMD 645E engine, which generates 1,500 horsepower. The GP-15-1 uses a 50ft 9inframe, and has a wheelbase of 29 ft 9 in. Over couplers it has a length of 54ft 9 in. A total of 310 units were built for American railroads. A number of GP-15-1s remain in service today for yard work and light road duty.

The GP-15T was built between October 1982 and April 1983. It was a very close cousin to the GP-15-1, but used a turbocharger in order to generate more power from a smaller engine. Power was provided by an eight-cylinder diesel engine that generated 1,500 horsepower the same as the GP15-1, but with four fewer cylinders.

Twenty-eight examples of this locomotive model were built for American railroads. The Chessie System received the majority of them as C&O 1500-1524 (twenty-five units), while the rest went to the Apalachicola Northern in Florida as AN 720-722 (three units).

The GP-15-AC model differs from the EMD GP15-1 due to Missouri Pacific specifying new AR10 AC alternators instead of rebuilt D32 DC generators. The only external difference between the GP15AC and the GP15-1 is a straight side sill (shared with the EMD GP15T) not related to the transmission difference.

So far the EMD engines had been pressure-charged by a Roots blower driven mechanically from the engine, but with its competitors offering engines of higher power. EMD now produced a turbo-charged version of the 567 engine, 567D2,

Left: The EMD GP-15-1 was built by General Motors Electro-Motive Division between June 1976 and March 1982. These locomotives were generally employed as a yard switcher or light road switcher.

Right: EMD equipped their GP-20 locomotives with a turbo-charged version of the 567 engine.

Above: The EMD GP-39DC was a four-axle diesel-electric locomotive built by General Motors Electro-Motive Division built from June 1970. Its power was provided by a turbocharged, 2,300 horsepower, twelve-cylinder engine.

giving 2,000 horsepower. For customers for whom the extra power did not justify the expense of the turbo-blower, the 567D1 at l,800 horsepower was available. Both of these models had a higher compression than their predecessors, which, combined with improvements in the fuel injectors, gave a fuel saving of five percent.

These engines were incorporated in the GP-20 and GP-18 series, respectively. By this time, American railroads were fully dieselized, and this, combined with a decline in industrial activity, reduced the demand for diesels. EMD therefore launched its Locomotive Replacement Plan. The company claimed that three GP-20s could do the work of four F-3s, so it offered terms under which a road traded in four F-3s against the purchase of three GP-20s, parts being reused where possible. It was claimed that the cost of the transaction could be recovered in three to four years, and the railroad then had three almost new units in place of four older ones with much higher maintenance costs. Despite this, only 260 GP-20s and 390 GP-18s were sold over thirteen years.

A further phase of GP development with the 567 engine came in 1961 with the 567D3 of 2,250 horsepower in the GP-30. For this model it was claimed that maintenance was reduced by sixty percent compared with earlier types. The GP-30 was succeeded by the GP-35 of 2,500 horsepower. With trade reviving, and many more early diesels in need of replacement, these models achieved sales of 2,281.

An EMD GP39DC is a four-axle diesel-electric locomotive built by General Motors Electro-Motive Division in June 1970. Power was provided by a turbocharged twelve-cylinder EMD engine that generated 2,300 horsepower. This locomotive model was basically a GP39 with DC main generator instead of an alternator. Two examples of this locomotive model were built for Kennecott Copper Company as engines 1 and 2. Later, these were sold to the Copper Basin Railway as 401 and 402.

The EMD GP39X is a 2,600 horsepower diesel-electric locomotive built by the General Motors Electro-Motive Division. All six units built were constructed for the Southern Railway with Southern's characteristic high short hood. They were updated by EMD at Norfolk Southern's (Southern's

Above: Six units of the EMD GP39X were built. It was a 2,600 horsepower diesel-electric locomotive. All six were built were constructed for the Southern Railway. This became the Norfolk Southern line.

Left: EMD's GP-39-2 locomotives were built between 1974 and 1984. It was a four-axle diesel locomotive and 239 examples were built for the American railroads.

successor) request to EMD GP49 standards in 1982. Norfolk Southern retired these engines in 2001, and they now are owned by Tri-Rail, after being upgraded with head end power and a lowered short hood.

The EMD GP39-2 is a four-axle diesel locomotive that was built by General Motors' Electro-Motive Division between 1974 and 1984. 239 examples of this locomotive were built for the American railroads. Part of the EMD Dash 2 line, the GP39-2 was an upgraded GP39. The power for this locomotive was provided by a turbocharged twelve-cylinder EMD 645E3 diesel engine, which could produce 2,300 horsepower (1,720 kW).

Unlike the original GP39, which sold only twenty-three examples as railroads preferred the reliable un-turbocharged GP38, the GP39-2 was reasonably successful. This success can be ascribed to its better fuel economy relative to the GP38-2 which became of more interest in the energy-crisis of the 1970s, and to its better performance at altitude.

The GP40 is a four-axle diesel-electric road-switcher locomotive that was built by General Motors Electro-Motive Division between November 1965 and December 1971. It has an EMD 645E3 sixteen-cylinder engine generating 3,000 horsepower.

The GP40 is 3 ft longer than its EMD 567D3A-engined predecessor, the GP35, and distinguished visually by its three 48-inch radiator fans at the rear of the long hood, while the GP35 has two large fans and a smaller one in between. It was built on a 55 ft frame; the GP35 was built on a 52 ft frame - as were the GP7, 9, 18, and 30 engines. The difference in length can be seen in the GP40's ten handrail stanchions compared to the GP35's nine.

1,187 GP40's were built for twenty-eight American railroads, sixteen were built for one Canadian carrier, Canadian National; and eighteen were built for two Mexican carriers, Ferrocarril Chihuahua al Pacifico and Ferrocarriles Nacional de Mexico. Sixty units were built with high-short-hoods and dual control stands for Norfolk & Western Railway. Two passenger versions, the GP40P and GP40TC, were also built, but on longer frames to accommodate steam generators and HEP equipment.

On January 1, 1972 the GP40 was discontinued and replaced by the GP40-2, which has a modular electrical system and a few minor exterior changes.

Thirteen GP40Ps were built in October 1968 for the Central Railroad of New Jersey (CNJ) and were paid for by the New Jersey Department of Transportation. The CNJ put the units in service on the Raritan Valley Line and the North Jersey Coast Line (New York & Long Branch).

The CNJ was folded into Conrail in 1976, and in 1983, New Jersey Transit began operating passenger rail service in the state. Shortly after, the steam generator, which had occupied the flat end of the locomotive's long hood, was replaced with a diesel HEP generator, and the units were reclassified as GP40PH. They would later be rebuilt as GP40PH-2 units in 1991-92

Top: The GP-40 was built by General Motors Electro-Motive Division between November 1965 and December 1971.

Above: Thirteen GP-40Ps run on the Raritan Valley and North Jersey Coast lines.

The GMD GP40TC engine was built by General Motors Diesel (GMD), for GO Transit in Toronto, Ontario (The TC stood for Toronto Commuter). A total of eight units were manufactured between 1966-1968. They were built on an enlarged frame to accommodate a HEP genset. GO Transit numbered these units 600-607. GO Transit sold the fleet to Amtrak in 1988; Amtrak numbered them 192-197. For Amtrak service the 575-volt HEP engine/generator set was replaced with one for 480-volt HEP. The units were based in Chicago and used on short-haul trains. The Norfolk Southern Railway rebuilt all eight at its Juniata Shops in Altoona, Pennsylvania.

These are now classed as GP38H-3 and work maintenance-of-way trains or standby power for Downeaster trains. Amtrak renumbered them 520-527 to make room for the new GE Genesis locomotives in the 1990s.

An EMD GP40X is a four-axle diesel-electric locomotive built by General Motors Electro-Motive Division between 1977 and 1978. Power for this unit was provided by a turbocharged sixteen-cylinder EMD 645F (derivative of the EMD 645 series engine) which could produce 3,500 horsepower (2,610 kW). Twenty-three examples of this locomotive were built for North American railroads. This unit

Above: GO Transit No504 near Port Credit, Ontario in 1977.

Below: GP-40X UP No957 was originally UP No9003. This picture shows the optional HT-B trucks

was a pre-production version meant to test technologies later incorporated into EMD's 50-series locomotives GP50 and SD50. Ten GP40X were delivered with an experimental HT-B truck design that became an option (but never used) on the production GP50.

The designation GP40X was also given to an experimental locomotive built on an EMD GP35 frame in May 1965. Only one example of this locomotive was ever produced, the EMD 433A, a 3,000 horsepower (2,240 kW) test prototype that was the first four-axle locomotive to be powered by the new 645-series prime mover. The 433A served as the precursor to the EMD GP40. The 433A was purchased by the Illinois Central Railroad, and became the IC 3075. The 1965 EMD 433A has very little in common with the 1977 GP40X other than flared radiators and a 645-series prime mover.

The EMD GP40-2 is a four-axle diesel road switcher locomotive built by General Motors Electro-Motive Division as part of its Dash 2 line between April 1972 and December 1986. The locomotive's power was provided by an EMD 645E3 sixteen-cylinder engine which generated 3,000 horsepower (2.24 MW). Standard GP40-2 production totalled 861 units, with 817 built for the American railroads, and forty-four were built for Mexican roads. In addition, three GP40P-2s, passenger versions of the GP40-2, were built for Southern Pacific in 1974, and 279 GP40-2L(W) and GP40-2(W) units, equipped with wide-nosed cabs, were built by General Motors Diesel (GMD), for Canadian National and GO

Left: The GP40-2 is a road switcher locomotive was built between April 1972 and December 1986.

Below left: The GP-49 was introduced in 1979. The series included the GP-SD49.

Transit between 1974 and 1976. Of the CN units, 233 were built with a taller and lighter frame to allow for a larger fuel tank. These units were officially classified GP40-2L but are commonly referred to as GP40-2l (W). The balance of CN's fleet, thirty-five units, and the eleven unit GO Transit fleet, used standard frames and smaller fuel tanks; they are often referred to as GP40-2(W) but are classified as GP40-2. Total production of the GP40-2 and its variations totalled 1,143 units.

Although the GP40-2 was a sales success, it sold fewer units than the earlier GP40 and the contemporary GP38-2 and SD40-2 models. The popularity of high-horsepower four-axle diesels began to decline with the GP40-2, with six-axle models gaining in popularity for their superior low-speed lugging performance. Like the SD40-2 the GP40-2 has a reputation for reliability, and many of these engines are still in use. Changes such as the modular electronics system improved reliability over the GP40. Their high power-per-axle rating suited them to high-speed service rather than low-speed drag freights.

With the usual 62:15 gearing (65-70 mph maximum) EMD rated the GP40-2 at 55,400 lb continuous tractive effort; for compatibility with other units, most had the PF21 module that reduced horsepower below twenty-three mph, bringing the minimum continuous speed down to eleven mph.

The GP40-2 car body retains the high "spartan" lines of other EMD locomotives of the same era, with a beveled nose and an angular, slant-roof cab. There are three large radiator fans at the rear of the hood and a single fan in the middle for the dynamic brakes (if equipped). The radiator intakes are smaller than those of the later GP50, and the walkways lack the end "porches" of the 6-axle SD40-2. The GP40-2 can be distinguished from the earlier GP40 by the oval-shaped water-level sight glass at the right rear of the long hood; bolted (rather than hinged) battery boxes ahead of the cab; lengthened walkway blower duct; and various minor cosmetic differences in the front air intake and rear hood doors. A number of GP40-2s also came with the new Blomberg M-type trucks, with single-clasp brakes, rubber pads replacing the central leaf springs and a shock strut over each axle.

An EMD GP49 is a four-axle diesel locomotive built by General Motors Electro-Motive Division. Power was provided by an EMD 645F3B twelve-cylinder engine which generated 2,800 horsepower (2.09 MW). The GP49 was marketed as one of four models in the 50 series that was introduced in 1979. This series includes GP/SD49 and GP/SD50. Both the GP and SD50 were relatively popular with a total of 278 GP50s and 427 SD50s built. The SD49 was advertised but never built and a total of nine GP49s were built.

Alaska Railroad is the only company that ordered it in two orders; the first was ARR 2801-2804 under order number 837049-1-4, built in September 1983 and the second was ARR 2805-2809 under order number 847035-1-5, built in May 1985. Six GP39Xs were built in November 1980 for the Southern Railway under order Number 786284-1-6 and upgraded to GP49s shortly thereafter.

An EMD GP50 is a four-axle diesel-electric locomotive built by General Motors Electro-Motive Division (EMD). It is powered by a sixteen-cylinder EMD 645F3B diesel engine, which can produce between 3,500 and 3,600 horsepower (2,610 and 2,685 kW). 278 examples of this locomotive were built by EMD between 1980 and 1985. BN 3110-3162 were all delivered with five cab seats, the final five of these having the cab lengthened 23 ft longer than the standard EMD cab. The GP50 retains the same overall length of 59 ft two inches

Above: The GP-50 is powered by a sixteen-cylinder diesel engine that produces 3,600 horsepower.

Right: The GP-59 was powered by a twelve-cylinder diesel engine producing 3,000 horsepower.

(18.03 meters) as the EMD GP38, EMD GP38-2, EMD GP39, EMD GP39-2, EMD GP40 and EMD GP40-2.

In 2014–2015 the Norfolk Southern Railway rebuilt twenty-five GP50 engines into low-emission GP33ECO locomotives.

The EMD GP59 is a four-axle diesel locomotive model built by General Motors Electro-Motive Division between 1985 and 1989. Power was provided by a twelve-cylinder EMD 710G3A diesel engine, which could produce 3,000 horsepower (2,200 kW). This locomotive shared the same common frame with the EMD GP60, giving it an overall length of 59 ft nine inches (18.21 m). It featured a 3,700-US-gallon fuel tank. Thirty-six examples of this locomotive were built including three demonstrators. Norfolk Southern placed the only order for the GP59 and also acquired the three demonstrators which featured an aerodynamic cab.

By adding a full cowl body, a comfort cab, and an HEP generator, the GP59 became the EMD F59PH.

In 2011 Norfolk Southern began a program to upgrade their fleet of GP59s, the only GP59s still operating. The first one was not released until March 2013 as NS 4650 GP59E. The GP59E features a new EM2000 microprocessor, an all new electrical cabinet with SmartStart auto start/stop, a rebuilt 12-710G3C-BC prime-mover with EMDEC EUI system, NS-designed split cooling, and the NS Admiral cab equipped with cab signals, LSL, and CCB26 electronic brake valve. The GP59E is set up to operate with NS class RP-M4C road slugs.

The EMD MP15DC was a 1,500 hp (1,100 kW) switcher-type diesel locomotive model that was produced by General Motors' Electro-Motive Division between March 1974 and January 1983. 351 examples of the type were built.

The Union Pacific Railroad is now perhaps the largest current user of the MP15DC, having 102 of the engine type in service. None were originally

Left: 102 MP-15-DCs run on the Union Pacific Railroad. All were acquired by merger or takeover.

owned by Union Pacific; instead, they were acquired by merger or takeover, or bought on the second-hand locomotive market. The vast majority (62) came from the Missouri Pacific Railroad, while locomotives were also acquired from the Chicago and North Western Railway (14) and Southern Pacific Railroad (9). A further fifteen were acquired from the Pittsburgh and Lake Erie Railroad, while a further two have been leased from Helm. The Alaska Railroad had four MP15DCs that were used as yard switching engines, numbered 1551-1554.

An MP15AC variant, with an AC drive, was also offered by the division. Between August 1975 and August 1984 246 MP15ACs were built, including twenty-five for export to Mexico, and a further four engines of this type were built in Canada. The MP15DC replaced the SW1500 in EMD's catalog, and is superficially very similar to the predecessor model, using the same engine (a V12 EMD 645-series powerplant) in a similar design of hood and bodywork. The primary difference is the MP15's standard Blomberg B trucks.

In the early 1970s, railroads were starting to convert to AC power, the six largest buyers, Milwaukee (64), Southern Pacific (58), Seabord (40), Nacionales de México (25), Long Island (23), and Louisville & Nashville (10) all purchased AC road locomotives. Thirty-six further units were sold to eight other customers.

Several former MP15AC Milwaukee Road units are now owned by the Soo Line Railroad (an American operating subsidiary of the Canadian Pacific Railway). Engines painted in the Canadian "Golden Beaver" scheme are often called "Bandits." Six former Milwaukee units returned to home rails in 2008, serving the growing regional Wisconsin & Southern Railroad WSOR in Milwaukee, Madison, and Horicon. In addition, Union Pacific has bought many examples on the used locomotive market. The New York & Atlantic Railway, which carries freight on Long Island, uses four former Long Island Railroad MP15ACs to haul freight along with other ex LIRR locomotives.

An EMD MP15T is a four-axle diesel switcher locomotive that was built by General Motors Electro-Motive Division between October 1984 and November 1987. Instead of a non-turbocharged twelve-cylinder EMD 645 engine it uses a turbocharged eight-cylinder engine. The external appearance of the engine remains similar to other MP15 models.

Forty-two of these locomotives were built for the Seaboard System Railroad, 1200-1241 (later merged into CSX where they kept their numbers) and one unit was constructed for Dow Chemical Company, this was engine number 957. In 2010 Progress Rail Services, a Caterpillar Company, purchased MP15T 1220 from CSX making CSX's total count forty.

Above: 246 MP-15-ACs were built between August 1975 and August 1984.

Left: MP-15Ts are four-axle diesel switchers by General Motors between 1984 and 1987.

Above: CSX No1216 in 2004 below the imposing Cargill Superior Elevator, 874 Ohio Street, Buffalo, New York

EMD FL-9 Bo-A1A (1960)

New Haven trains used both the Grand Central terminal (of the New York Central RR) and the Pennsylvania Station. New York pollution statutes did not allow diesel burning trains into the city. Both routes were equipped with conductor rails (of different patterns) supplying low-voltage direct current. This corresponded closely to the current produced in the generator of a diesel-electric locomotive and it was suggested that a standard General Motors FP-9 passenger cab unit could be modified easily to work as an electric locomotive when required. The AC electrification could then be dismantled, yet trains could continue to run without breaking the law. In fact axleload restrictions led to one quite substantial change—substitution of a three-axle trailing truck for the standard two-axle one; hence a unique wheel arrangement. The end product was designated FL9 and. 60 were supplied between 1956 and 1960. The most obvious evidence of their unique arrangements were the two-position retractable collecting shoes mounted on the trucks, to cater for New York Central's under-contact conductor rail and Long Island RR's top-contact one. Otherwise the presence of additional low-voltage control gear inside the body was the principal technical difference between an FP-9 and an FL-9.

SPECIFICATIONS
Type: Electro-diesel passenger locomotive.
Gauge: 4ft 81/2in
Propulsion: General Motors 1,750hp Type 567C V-16 two-stroke diesel engine and generator—or alternatively outside third-rail—feeding current to four nose-suspended traction motors geared to both axles of the leading truck and the outer axles of the trailing one.
Weight: 231,937 lb adhesive, 286,614 lb total.
Max. axleload: 57,984 lb
Overall length: 59 ft 0 in
Tractive effort: 58,000 lb
Max. speed: 70 mph

In the event the New Haven changed its mind over dispensing with the electrification, but the FL9s still found employment, surviving long enough to be taken over by the National Railroad Passenger Corporation (Amtrak) in the 1970s.

Top: The New Haven's unique electro-diesel Class FL9 still doing good work as Amtrak No488, seen at Rensselaer, NY, February 1983.

Alco Century Series (1963)

The C-430 is part of Alco's "Century" series, which included six and eight axle units. The C-430 was the last of the four axles units. The preceding models were the C-415, C-420, C-424 and C-425. The Century line was produced by Alco, the Montreal Locomotive Works, and A. E. Goodwin Ltd under license in Australia. Production began in 1963 and ended in 1972. MLW and Goodwin continued to build Century locomotives after Alco ended locomotive production and shut down in early 1969. A total of 841 Century locomotives, in eleven variants, were produced over the ten years of production.

Only sixteen of the C-430s were built between 1966 and 1968, all for US railroads. The New York Central purchased ten, the Reading Line two, and the last three were Alco demonstrators.

SPECIFICATIONS
Type: Diesel-electric passenger locomotive
Gauge: 4 ft 8 1/2 in
Proplusion: EMD 251b
Cylinders: 162
Power output: 1800 hp
Top speed: 92 mph

Below: Morristown & Eire Railway No18 (a C424 built by Alco in 1964), at the Morristown shop complex, 2010.

Below: No315 is the only Alco C-430 switcher bought by the Green Bay & Western. It was delivered in 1968. After suffering a broken crankshaft in 1986, it was retired. In 1987, the GB&W donated it to the museum.

Alco RSD-12 Co-Co (1963)

SPECIFICATIONS
Type: Diesel-electric passenger locomotive
Gauge: 4 ft 8 1/2 in
Proplusion: EMD 251b
Cylinders: 162
Power output: 1800 hp
Top speed: 92 mph

The Alco RSD-12 was a diesel-electric locomotive of the road switcher type rated at 1,800 horsepower that rode on three-axle trucks, having an C-C wheel arrangement.

Used in much the same manner as its four-axle counterpart, the Alco RS-11, though the six-motor design allowed better tractive effort at lower speeds. The RSD-12 was one of the final six-axle designs produced in the Road Switcher (RS) series. The locomotive sold relatively well for Alco considering most six-axle models were lucky to see more than 50 sales. The RSD-12 was essentially an extension of the RS-11, which was Alco's attempt to compete with EMD's four-axle GP-9. Alco was late to the game as EMD's SD-9 (the six axle version of its GP-9) had already been in production for two years.

A number of railroads bought significant numbers of the 171 total production, including the Southern Pacific, 21, Pennsylvanian RR, 25 and the Chesapeake & Ohio, 10.

Alco C-630 Co-Co (1965)

The Century Series locomotives were a line of locomotives produced by Alco, the Montreal Locomotive Works, and A. E. Goodwin Ltd under license in Australia.

The C-630 was launched in 1965, prompted the impending launches of locomotives by both GE and EMD of equivalent size, and was the first production locomotive to use AC

SPECIFICATIONS
Type: Diesel-electric road-switcher locomotive
Gauge: 4 ft 8 1/2 in
Propulsion: One 16-cylinder four-stroke turbocharged 3,000hp Alco 251E V-16 engine and alternator, supplying three-phase current through rectifiers to six nose-suspended traction motors each geared to one axle.
Weight: 312,000 lb
Max. axleload: 52,000 lb but could be increased to 61,0001bif desired.
Length: 69 ft 9 in
Tractive effort: 103,000 lb
Max. speed: 80 mph according to gear ratio

technology, as the complexity of DC generators were too large and complex to be used at such high powers. It was powered by a 3,000-horsepower, 16-cylinder, Model 251E engine. It was launched in 1965, and remained in production until 1967, with 77 units produced. Production of the Century Series began in 1963 and ended in 1972. MLW and Goodwin continued to build Century locomotives after Alco ended locomotive production and shut down in early 1969. A total of 841 locomotives, in eleven variants, were produced over the ten years of production.

No6342 became No6792 when Conrail took over Penn Central in 1976. Delta Bulk Terminal then bought the locomotive from Conrail, renumbered it No1001 and based it in Stockton, CA. It was bought by Genesee Valley Transportation in 2001 for its Delaware-Lackawanna subsidiary.

Left: No3642 is a later model in the Alco Century series. It was one of fifteen ordered by PRR road numbered 6330-6344) but delivered to Penn Central in 1968, the latter having just formed from the PRR and NYC merger.

GE U30B B-B (1966)

SPECIFICATIONS
Power type: Diesel-electric
Build date: May 1966 – March 1975
Total produced: 296
AAR wheel arr.: B-B
Gauge: 4 ft 8 1/2 in
Prime mover: GE FDL-16
Power output: 3,000 horsepower

General Electric's U30B was a further development of the U28B diesel-electric locomotive, with a 3,000 horsepower, sixteen-cylinder prime mover. It remained in production for over eight years. The U30B competed with the EMD GP40 and the ALCO Century 430, but was not as successful at the GE U30C. The engine's original owners included the Atlantic Coast Line Railroad, the Illinois Central Railroad, the Louisville & Nashville Railroad, the New York Central Railroad, and the Western Pacific railroad.

GE U36B B-B (1969)

SPECIFICATIONS
Power type: Diesel-electric
Build date: January 1969 – December 1974
Total produced: 125
AAR wheel arr.: B-B
Gauge: 4 ft 8 1/2 in
Prime mover: GE FDL-16
Power output: 3,600 horsepower

The GE U36B was a diesel-electric locomotive produced by General Electric beginning in 1969. The U36B was GE's 3,600 horsepower answer to the power race with EMD. Only three railroads bought this locomotive. Major purchasers of U36Bs were Auto-Train Corporation and Seaboard Coast Line, which hosted most Auto-Train Corp. traffic. In 1976 four U36Bs that were built by GE for Auto-Train Corporation, which had purchased thirteen earlier, were instead sold and delivered to Conrail due to Auto Train's financial difficulties. These units, built to Seaboard Coast Line specifications, were originally fitted with Blomberg trucks; when Conrail purchased them, the railroad asked GE to replace the trucks with AAR Type B instead.

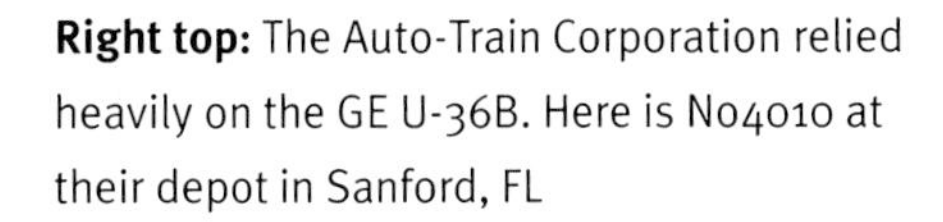

Right top: The Auto-Train Corporation relied heavily on the GE U-36B. Here is No4010 at their depot in Sanford, FL

Right middle: CSX Transportation GE U36B at Waycross, GA.

Right: Transkentucky Transportation GE U-36 B at Paris, KT

GE U23C C-C (1968)

SPECIFICATIONS
Power type: Diesel-electric
Build date: March 1968 – 1976
Total produced: 223
AAR wheel arr.: C-C
Gauge: 4 ft 8 1/2 in
Prime mover: GE FDL-12
Cylinders: V12

The GE U23C, developed towards the end of the 1960s, was a late model of the line manufactured by General Electric for use in yard, transfer, and heavy drag service. It featured a variation of GE's FDL prime mover and was meant to compete with similar models being offered by the Electro-Motive Division (EMD), notably the SD38 and SD39. While final U23C sales appeared to be rather successful most purchases were by Rede Ferroviária Federal S.A. (Brazil);as few were actually ordered by US Class I systems (less than 100). The specialized nature of the U23C was not of particular interest to many by the late 1960s, especially considering that other models being cataloged at the time could perform the very same tasks, notably EMD's phenomenally successful SD40 series released in 1966. Today, there is one U23C known to be preserved, Lake Superior & Ishpeming No2300 (in its original colors and number) located at the Arkansas Railroad Museum. The 2300 horsepower GE U23C diesel-electric locomotive model was first offered by GE in 1968, and featured a V-type 12 cylinder version of the standard GE FDL diesel motor. Designed as a competitor to EMD's

SD38 and SD39 series, it was intended for duties where speed was not a priority. Other than six tall hood doors matching six power assemblies per side, there are very few

Below: Burlington & Northern liveried GE U 23C.

features which distinguish the U23C from the U30C. The U30C has eight tall hood doors per side, a function of the V16 within. A total of seventy-three units were built at Erie including twenty for export to Brazil. An additional 150 units were built by GE de Brazil from 1972-1976, some with kits supplied by GE.

Distribution

Atchison, Topeka and Santa Fe Railway; 20 road numbered 7500-7519
Chicago, Burlington and Quincy Railroad; 9 road numbered 460-468
Lake Superior and Ishpeming Railroad 5; road numbered 2300–2304
Penn Central; 19 road numbered 6700-6718
Rede Ferroviária Federal S.A. (Brazil); 170 road numbered 3801-3880, 3901-3990.

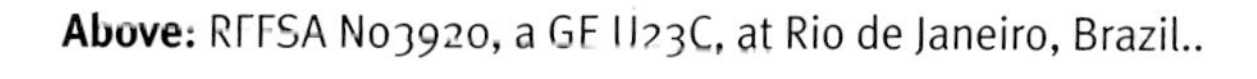
Above: RFFSA No3920, a GE U23C, at Rio de Janeiro, Brazil..

EMD DDA-40X Union Pacific (1969)

The DDA4oX was a 6,600 hp D-D diesel-electric locomotive built by the General Motors EMD division of La Grange, Illinois for the Union Pacific Railroad. Nicknamed "Centennial" and "Big Jack," the DDA4oX uses two diesel engines (each 3,300 hp) and is the most powerful single-unit diesel locomotive ever built thereby producing a single locomotive which obviated the need to have multiple header trains. In the past over the Transcontinental divide up to 6 or even 8 locomotives were used to haul trains up Sherman Hill. The DDAA 40X is also the longest single-unit diesel locomotive ever built. In 1969 Union Pacific was retiring the later gas turbine-electric locomotives. Union Pacific had ordered EMD DD35s and DD35As to replace the turbines, and the DDA4oX was a further development. Forty seven were built between June 1969 and September 1971, except the first one delivered in April in time to participate in the celebrations of the centennial anniversary of the completion of the Transcontinental Railroad driving the "Gold Spike Limited" and arriving in Salt Lake City, Utah, on the morning of May 10, 1969. The units were road numbered from 6900 to 6946, with No6936 still in service.

The frames were fabricated by an outside contractor, the John Mohr Company of Chicago, since the frame length exceeded the abilities of EMD's plant. Using more than one propulsion unit in a single locomotive was not new; the E-series were popular dual-engine locomotives, and Baldwin had produced (but not sold) a locomotive with four diesel engines.

The "X" in the designation stood for experimental, as the DDA4oX locomotives were used as the testbeds for technology that would go into future EMD products. The modular electronic control systems later used on EMD's Dash-2 line of locomotives were first used on the DDA4oX. The locomotive was the first to be able to load-test itself using its dynamic braking resistors as an electrical load so that external equipment was not required. The DDA4oX used the wide-nosed cab from the FP45 cowl units. This design was superficially similar to the Canadian comfort cab introduced by Canadian National soon afterwards in 1973, but it lacked the structural reinforcements introduced in the CN design that were carried over to future wide-nosed cabs.

SPECIFICATIONS

Type: Diesel-electric

Gauge: 4 ft 8 1/2 in

Wheelbase:
(Between truck centers) 65 ft

Truck wheelbase: 17 ft 1 1/2 in

Length: 98 ft 5 in

Width: 10 ft 4 in

Height cab roof: 14 ft 11 3/8 in

Height overall: 16 ft 4 in

Weight: 521,980 lb

Fuel capacity: 8,230 gall

Propulsion: 2 x EMD 645E3A

Engine type: V16 diesel

Aspiration: turbocharged

Cylinders: 16

Power output: 6,600 hp

Max Speed: 90 mph

Safety systems: Leslie Supertyfon model S5TRRO or S3LR horns, US&S Type "EL" Cab Signals: (No6936 equipped with US&S MicroCab ATC & CCS)

GE U23B Diesel-Electric B-B (1968)

SPECIFICATIONS
Total produced: 481
AAR wheel arr.: B-B
Gauge: 4 ft 8 1/2 in
Length: 60 ft 2 in
Propulsion: GE 7FDL-12
Cylinders: 12
Power output: 2,250 hp

GE Transportation introduced the GE U23B Diesel-electric locomotive in 1968. It was conceived as a medium horsepower roadswitcher and featured a twelve cylinder FDL engine. It was one of the most successful models of GE's Universal Series. In 1977 the model was replaced by the B23-7. 481 units were built, including sixteen that were exported to Peru. Not many U23Bs still exist, but a few shortline and regional railroads such as the Georgia Central Railway still have examples in everyday service. The last U23B, CR 2798 was built in 1977. Many railroads bought U23Bs, including the Atchison, Topeka and Santa Fe Railway, the Delaware and Hudson Railway, and the Louisville and Nashville Railroad.

Below: Western Pacific U-23B 2264 outside Portola's engine house, one of 15 GE diesels on the roster.

GE U34CH Diesel-Electric C-C (1970)

SPECIFICATIONS
Gauge: 4 ft 8 1/2 in
Power output: 3,600 hp
Wheel Arr.: C-C
Propulsion: GE FDL-16
Power Output: 3600 hp

The U34CH is a 3,600 hp passenger diesel locomotive was first built by General Electric between 1970 and 1973. In total, 33 units were built. 32 of these were built for the New Jersey Department of Transportation and operated by the Erie Lackawanna Railway and, later, Conrail. The last unit was a rebuild of a GE U30C for the New York MTA. The U34CH was the first GE locomotive to use steel crowned pistons to develop 3,600 hp and was the first commuter locomotive built with shaft driven HEP. To show their NJ DOT ownership, the units were painted in a dark blue and silver paint scheme with NJ DOT logo. These engines were often referred to as "Bluebirds." At the weekends, the engines were sometimes used on freight duty by the Erie Lackawanna freight service, returning to commuter service in the week.

Below: U -34CH No4172 on the "Farewell to the U34CH" excursion at Hillsdale, New Jersey, August 27, 1994 Copyright John Eric Durant.

GE U-30C No5383 Co-Co (1974)

SPECIFICATIONS
Type: Diesel-electric
Total produced: 596
Gauge: 4 ft 8 1/2 in
Propulsion: GE FDL-16
Power output: 3,000 hp
Top speed: 70 mph

Burlington Northern was the largest owner of U-30C units, eventually rostering one hundred and eighty-one of them.

No5383 was part of an order of fifty delivered to the railroad in 1974. It went into storage in April 1994 and was donated to the Illinois RR museum in September that year.

GE built five hundred and ninety-six U-30Cs from 1966 to 1976. They were bought by a range of customers in the US, from mining to general freight. One served as power for the Dept of Transportation's subway car test tracks in Pueblo, CO. The 67' 3" long units weigh 363,000 lbs. With a 3,000 hp FDL16 4-cycle propulsion unit powering six GE 752 traction motors, they delivered 92,500 lbs continuous tractive effort at 10.7 mph with a top speed of 70 mph.

EMD F-40PH Bo Bo (1976)

SPECIFICATIONS
Type: Diesel-electric
Gauge: 4 ft 8 1/2 in
Driver diameter: 40 in
Wheelbase: 43 ft 6 in between bolsters
Length: 68 ft 10 in
Width: 10 ft 3 1/8 in
Height: 15 ft 7 1/8 in
Weight: 368,000 lb
Fuel capacity: 3,200–4,000 gall
Proplusion: EMD 16-645-E3
Aspiration: turbocharged
Cylinders: V16
Power output: 3,000 hp

The last of the EMD passenger carbody diesels was built at the end of 1963, and with passenger traffic declining rapidly, the need for special passenger locomotives seemed to have disappeared. Both EMD and its competitors offered a train-heating steam generator as an optional extra on certain hood units, and this met the needs of the railroads which required replacements for aging E or F series units.

In 1968, with the railroads' enthusiasm for high-power diesels at its climax, the Atchison, Topeka & Santa Fe Railway proposed to buy from EMD some 20-cylinder 3,600hp Co-Co locomotives geared for high speed to operate its premier passenger services. The railroad asked that the locomotives should be given a more acceptable appearance for passenger work, and that the body should have less air resistance at speed than a normal hood unit. The outcome was the "cowl," a casing shaped like an angular version of the old carbody, but differing from it in that the casing

4910
4910
WEST OF HUDSON
F40PH-3C
DANGER
HIGH VOLTAGE
OVERHEAD
4910
North Railroad
DANGER
480 VOLTS
DANGER
480 VOLTS

Above: MBTA F-40 PH No1053 at Boston South Station

does not carry any load. The cowl extends ahead of the cab, giving the front of the cab more protection against the weather than a normal hood.

The model was designated FP-45, and was very similar in its equipment to the SD-45 road switcher. Another variant had a shorter frame resulting from the omission of the steam generator; it was designated F-45.

In 1971 the National Railroad Passenger Corporation (Amtrak) took over most of the non-commuter passenger services in the US, and in 1973 took delivery of its first new locomotives to replace the old E and F series. By this time, demand for engines above 3,000hp had declined, so the Amtrak units were similar to the FP-45 but with a 16-cylinder 3,000hp engine. A total of 150 were delivered in 1973-74. They were equipped with two steam generators mounted on skids, which could easily be replaced by two diesel-alternators when steam-heated stock was replaced by electrically-heated vehicles. In view of the similarity to the SD-40s, these locomotives were classified SD-P40.

In the meantime, for shorter-distance routes on which the coaches were already electrically-heated, Amtrak had ordered a four-axle 3,000hp locomotive, with an alternator for supplying three-phase current at 60Hz for train services driven by gearing from the engine crankshaft. This model is designated F-40PH, and deliveries began in March 1976.. As the well-tried Blomberg truck fitted to the F-40PH had given no cause for criticism, Amtrak decided that the SDP-40 locomotives should be rebuilt as F40PHs." The frame could be shortened by 16ft, as the steam generator was no longer needed. The F-40PHs built new had a 500 kw alternator, which drew a maximum of 710hp from the engine, but for the transcontinental Superliner trains an 800 kw alternator and larger fuel tanks were needed, so that the rebuilt F-40PHs are 4ft longer than the others.

In the fact the rebuilding was nominal, for it cost nearly 70 percent of the price of a new locomotive, and in effect the SDP-40s were scrapped when only four to five years old.

As Amtrak's F-40PH fleet was replaced by newer GE Genesis-series locomotives, Amtrak converted a number of the retired units—generally with mechanical problems limiting their value into non-power control unit cab cars. Colloquially known as "cabbages" (a portmanteau of "cab" and "baggage"), these units had their prime movers and traction motors removed and a large roll-up door installed in the side (allowing the former engine compartment to be used for baggage). NPCUs also differ from normal F-40PHs by their lack of grills and rooftop fans. The units were renumbered into Amtrak's car-series numbers by adding "90" before the former locomotive number; the original F-40PH number 200 became NPCU number 90200.

GE C30-7 Diesel-Electric C-C (1976-1986)

SPECIFICATIONS
Power type: Diesel-electric
Build date: 1976 – 1986
Total produced: 1,137 (50 were C30-7A variants)
AAR wheel arr.: C-C
Gauge: 4 ft 8 1/2 in (1,435 mm) standard gauge 1,520 mm (4 ft 11 27/32 in), Estonia
Engine type: V16 (V12 on C30-7A variants)
Cylinders: 16 (12 on C30-7A variants)
Power output: 3,000 hp (2,200 kW)

The General Electric C30-7 is a six-axle diesel-electric locomotive built between 1976 and 1986. The engine was later updated as the U30C, equipped with a sixteen-cylinder 3,000 horsepower (2,200 kW) FDL-series diesel engine. 1,137 were built for North American railroads. Fifty examples of another variant, the GE C30-7A were purchased by Conrail in mid-1984. Externally similar to the GE C30-7 model, the engine had six tall hood doors per side and had a twelve cylinder (rather than sixteen cylinder) prime mover. Both engines produced 3,000 horsepower (2,200 kW) but the C30-7A's smaller engine used less fuel. The C30-7A units were built between May and June 1984. Conrail later sold twelve of these engines to Chicago Freight Car Leasing Australia. These engines entered service in Australia in 2003.

Above: BNSF No5008 at Clay, CO

Below left: A C-30-7 of the NDEM No11107 at Houston, TX.

Below: Southern Pacific No 7784 at Rosenberg, Texas in 2001

GE B23-7 Diesel-Electric B-B (1977-1984)

SPECIFICATIONS
Power type: Diesel-electric
Build date: September 1977 – December 1984
Total produced: U.S.: 412 units; Mexico: 125 units
AAR wheel arr.: B-B
Gauge: 4 ft 8 1/2 in (1,435 mm)
Length: 62 ft 2 in (18.95 m)
Prime mover: GE FDL-12
Cylinders: V12
Power output: 2,250 hp (1,680 kW)

General Electric's B23-7 is a diesel-electric locomotive model that was first offered for sale in late 1977. The train featured a smaller twelve cylinder version of the company's FDL engine. It was the successor to GE's U23B, which was produced from early 1968 to mid 1977. But the FDL was exactly two feet longer than its predecessor. It competed with the very successful EMD GP38-2. General Electric also produced a variant, the BQ23-7, no.5130-5139, for the Seaboard Coast Line. A total of 537 B23-7 engines were built for nine American customers and two Mexican customers.

A B23-7A is a 12 cylinder B23-7 with horsepower boosted to 250 per cylinder or 3,000 horsepower. In 1980 the Missouri Pacific ordered three B23-7A engines and tested them system-wide. The result was the GE model B30-7A, B30-7 with a twelve cylinder FDL prime mover. They were not renumbered into the B30-7A series on the MP because they lacked Sentry Wheel Slip and had different engine governors.

Thirteen B23-7s were built by GE of Brazil in December 1979 for the United South Eastern Railways and the National Railways of Mexico. Seventeen B23-7s were built from GE kits in Mexico as Ferrocarriles Nacionales de México engine numbers 10047-10052 and no.12001-12011. Other B23-7s were bought by the Atchison, Topeka and Santa Fe Railway, Conrrail, the Louisville and Nashville Railroad, Ferrocaril Unidos del Sureste, the Missouri Pacific Railroad, the Providence and Worcester, the Seaboard Coast Line Railroad, the Southern Pacific Railroad Texas Utilities, and the Southern Railway. Southern's fifty-four units had their standard high-short-hoods.

These engines frequently are rebuilt as Control Car Remote Control Locomotive (CCRCL) due to their low value on the used locomotive market.

The B32-7 was also nicknamed the B-boat.

Above: BNSF 4258 switching the intermodal yards at Commerce, California, February 15, 2005.

GE B30-7 Diesel-Electric B-B (1977-1983)

SPECIFICATIONS
AAR wheel arr.: B-B
Gauge: 4 ft 8 1/2 in (1,435 mm)
Prime mover: GE FDL-16
Cylinders: 16 or 12 cylinders (B30-7A variants)
Power output: 3,000 hp (2,200 kW)

The first generation of the General Electric Transportation Systems B30-7 diesel-electric locomotive model was offered in 1977. The engine featured a sixteen-cylinder motor and was 61 feet 2 inches long. Later versions of the engine used a twelve cylinder FDL rated at 3,000 horsepower (2,200 kilowatts), these were known as the B30-7A, B30-7A1 and a cabless B30-7A.

GE B30-7A Diesel-Electric B-B (1977-1983)

SPECIFICATIONS
AAR wheel arr.: B-B
Gauge: 4 ft 8 1/2 in (1,435 mm)
Prime mover: GE FDL
Cylinders: 12 cylinders (B30-7A variants)
Power output: 3,000 hp (2,200 kW)

B30-7As were built only for the Missouri Pacific Railroad and are externally identical to the 16 cylinder version B30-7. B30-7A1s were built only for the Southern Railway.

Cabless versions of the B30-7A were built for the Burlington Northern Railroad. Shortline Railroad Providence and the Worcester Railroad acquired five ex-BN B30-7A cabless units. These were re-classified as B30-7AB units, numbered 3004 to 3008 in 2001. National Railway Equipment acquired these locomotives in 2015. The 279 engines of this type were ordered by the Burlington Northern Railroad, the Chesapeake and Ohio Railway, the Missouri Pacific Railroad, the St. Louis – San Francisco Railway, the St. Louis Southwestern Railway, the Seaboard Coast Line Railroad, Southern Pacific Railroad, and the Southern Railway.

GE B36-7 Diesel-Electric B-B (1980-1985)

SPECIFICATIONS
Power type: Diesel-electric
Total produced: 230
AAR wheel arr.: B-B
Gauge: 4 ft 8 1/2 in (1,435 mm)
Prime mover: GE 7FDL16
Aspiration: Turbocharged
Power output: 3,600 hp (2,700 kW)
Locale: North America South America

Above: CSXT 5842, a General Electric B36-7 locomotive, waits for the next job in Taft Yard, FL

GE Transportation systems B36-7 is a four-axle diesel-electric locomotive built between 1980 and 1985. 222 examples of this locomotive were built for North American railroads and eight further units were built for a Columbian coal mining operation. The units were designed as successors to GE's U36B engines. Of the 230 locomotives built, 180 of them were built for two Eastern railroads - Seaboard System Railroad (which became part of CSX Transportation in 1986) and Conrail.

These four-axle locomotives were powerful ones, creating 3,600 hp (2,700 kW). They were designed for fast and priority service, moving intermodal and container trains. Most of Seaboard's 120 units are still in service as of 2006. Conrail's units were all retired in 2000 and 2001. One notable exception among Conrail's units was CR 5045, which was destroyed in the infamous wreck of the *Colonial* at Chase, Maryland, on January 4, 1987. Another of the Cotton Belt Railroad's B36-7s was damaged in a wreck within a year of delivery and rebuilt as a B unit. This engine did not receive a special model designation to indicate its cabless status. Railfans sometimes term it a B36-7(B).

CSX was the last Class 1 railroad to roster B36-7s and GE Dash 7s in general. Though originally intended for high speed service, they spent much of their later life working on local trains in the Southeast and in the last years worked low priority MOW trains. The original owners of these engines were the Atchison, Topeka and Santa Fe Railway, Conrail, the Cerrejon Coal Project (In Colombia), Seaboard System, Southern Railway, Southern Pacific Transportation Company and the St. Louis Southwestern Railway.

GE P32-8 WH B-B (1990)

SPECIFICATIONS
Mode: Dash 8-32BWH
Total produced: 20
AAR wheel arr.: B-B
Gauge: 4 ft 8 1/2 in
Prime mover: GE 7FDL-12
Power output: 3,200 horsepower

The GE P32-8WH is also known as the Dash 8-32BWH and B32-8WH, is a passenger train locomotive used by Amtrak. It is based on GE's Dash 8 series of freight train locomotives. Twenty of these locomotives were delivered to Amtrak in 1991, numbered 500 through 519. They were nicknamed the "Pepsi Cans" by many railfans, due to being delivered in a wide-striped red, white, and blue livery. They have since been repainted in more recent Amtrak liveries. The Dash 8-32BWH operates in a diesel-electric configuration that uses DC to power the traction motors, producing 3,200 horsepower. When providing head end power to the train, the engine is speed locked to 900 rpm. Power output to the traction motors is 2,700 horsepower when running in HEP mode. Today, the Dash 8-32BWH has been relegated to yard switching (mainly in Los Angeles, Oakland, Chicago, Miami, and the Auto Train terminals) and transfer service, displaced by the newer and more powerful GE Genesis, but the Dash 8s occasionally substitute for the Genesis units if necessary and the Los Angeles based units see frequent use on the Coast Starlight and Pacific Surfliner.

GE Genesis Series Bo-Bo (1992-2001)

In the late 1980s, Amtrak was looking for a new diesel locomotive; one with an extended range and being able to withstand the tight clearances of the Northeast, as well as being more fuel-efficient, lighter, and faster than the existing EMD F40PH. Two manufacturers, EMD and GE, responded. GE was ultimately awarded the contract to build an initial 64 diesel units and 10 dual-mode locomotives. The final design was finally revealed in 1993, as the "Genesis." Initially classed as AMD-103 (Amtrak Monocoque Diesel-103mph), these would later be re-designated as the P40DC. GE built 44 P40s for Amtrak. Today, some have been sold to

Above: Metro-North dual-mode Genesis locomotive in paint scheme introduced in 2007, seen here at Poughkeepsie station.

SPECIFICATIONS	
Type:	Diesel-electric
Gauge:	4 ft 8 1/2 in
Trucks:	Krupp-MaK high-speed bolsterless
Wheel diameter:	40 in
Wheelbase:	43 ft 2 1/2 in (between truck centers)
Length:	69 ft 0 in
Width:	10 ft 0 in
Height:	14 ft 4 in
Axle load:	72,000 lb max
Locomotive weight:	P40DC, P42DC: 268,240 lb, P32AC-DM: 274,400 lb
Fuel capacity:	2,200 gal
Propulsion:	GE 7FDL16 (P40DC, P42DC), GE 7FDL12 (P32AC-DM)
Engine RPM range:	200-1050 (600-900 while supplying HEP)
Engine type:	45° V16, four stroke cycle (P40DC, P42DC), 45° V12, four stroke cycle (P32AC-DM)
Aspiration:	Turbocharged
Displacement:	10,690 cu in (7FDL16), 8,020 cu in (7FDL12)
Alternator:	GMG195 (P40DC, P42DC), GMG195A1 (P32AC-DM)
Traction motors:	GE 752AH (DC), GE GEB15 (AC)
Cylinders:	16 (P40DC, P42DC) 12 (P32AC-DM)
Cylinder size:	668 cu in
Maximum speed:	103 mph (original P40DC), 110 mph (P42DC, P32AC-DM, upgraded P40DC), 60 mph (in electric mode P32AC-DM only)
Power output:	4,250 hp (DC), 3,200 hp (AC)
Tractive effort:	P40DC, P42DC: 63,000 lb Starting 38,000 lb, Continuous @38 mph P32AC-DM: 62,000 lb Starting @ Stall to 14 mph, 25,500 lb Continuous @64 mph
Adhesion:	4.25 (DC) 4.4258 (AC)
Train brakes:	Electropneumatic

commuter railroads such as NJ Transit and ConnDOT, while many have undergone refurbishment and reactivated by Amtrak (after being mothballed and stored in 2005.)

The second variant came 2 years later, as the AMD-110 dual-mode locomotive. While similar to the original P40DC on the outside, this locomotive had AC motors, capability to operate as a diesel locomotive and as an electric locomotive, and was geared for a maximum speed of 110mph. These were also later reclassed as the P32AC-DM. A major part of the unique design of these diesel locomotives is their height. Designed to be lower than the EMD F40PH, the Genesis series is 14in shorter. This makes the Genesis series the only Amtrak locomotives that can operate on all their lines in their system (even on electrified or third-rail lines like in the Northeast Corridor). Hence, they have also become the primary locomotive used for Amtrak.

To create the low-profile, streamlined shape; GE designed the GENESIS series of monocoque construction. Its aerodynamics give it 22% more fuel efficiency, better crew safety (being classified as a "safety-cab" like with modern freight diesels) and it also produces 25% more power than the F40PH, which had a heavy hand in Amtrak's decision to replace the latter with this engine.

Left: Amtrak No806 pulling Amtrak's Empire Builder service at East Glacier, MT, westbound.

GE P40DC Diesel-Electric B-B (1992)

SPECIFICATIONS
Power type: Diesel-electric
Build date: 1992–2001
Wheel arr.: B-B, Bo'Bo'
Gauge: 4 ft 8 1/2 in
Trucks: Krupp-MaK high-speed bolsterless
Wheel diameter: 40 in
Length: 69 ft
Width: 10 ft
Fuel capacity: 2,200 US gal
Prime mover: GE 7FDL12

Above: Former Amtrak P40 DC No808 reborn as NJT No808 at Kearny, NJ in 2007.

General Electric produced three models of Genesis units, which are still in operation today: P40DC, P42DC, and P32AC-DM.

The P40DC (also known as Genesis Series 1 or the Dash 8-40BP was the first model in the Genesis series. The locomotive operates in a diesel-electric configuration that uses DC to power the traction motors. This produces 4,000 horsepower output at 1047 rpm. The P40DC is geared for a maximum speed of 103 miles per hour (166 km/h). The P40DC was succeeded in 1996 by the P42DC. Both the P40DC and P42DC allowed Amtrak to operate heavy long-distance trains with fewer locomotives. Two P40DCs could do the same work as three F40PH engines. 321 of the P40DC locomotives were built. The model went out of production in 2001.

Left: New Jersey Transit GE P40DC No4800 (ex-Amtrak No812) pulls Train 5439 into the Dunellen Railroad Station, en route to Raritan on the Raritan Valley Line.

Above: Amtrak P-40DC No828 in New Orleans.

Below: A GE P-40DC diesel electric locomotive at Albuquerque NM.

Right: P40DC No840 pushing a Shore Line East train in Stamford, Connecticut.

GE P42DC Diesel-Electric B-B (1992-2001)

SPECIFICATIONS
Total produced: 321
AAR wheel arr.: B-B
UIC classification: Bo'Bo'
Gauge: 4 ft 8 1/2 in
Trucks: Krupp-MaK high-speed bolsterless
Wheel diameter: 40 in
Wheelbase: 43 ft 2 1/2 in
Length: 69 ft
Width: 10 ft
Propulsion: GE 7FDL16

The P42DC of GE Transportation Systems Genesis Series 1 is the successor to the P40DC. It has an engine output of 4,250 horsepower. The P42DC has a maximum speed of 110 mph (177 km/h), though Via Rail Canada only permits its engines to travel at a maximum speed of 100 mph (161 km/h). P42DCs are used primarily on most of Amtrak's long-haul and high-speed rail service outside the Northeast Corridor, as well as a service with speeds up to 160 km/h (99 mph) on Via Rail's Quebec City-Windsor rail corridor when it replaced LRC locomotives in 2001. Several locomotives of this series were painted in special anniversary liveries. Between 1992 and 2001, 321 units were built for Amtrak, Metro-North, and Via Rail. The locomotives were designed in response to a specification published by Amtrak. The series is unique among current North American diesel-electric locomotives due to their low height. This height restriction allowed the locomotive to travel easily through low-profile tunnels in the Northeast Corridor.

Above: Amtrak No156 in Phase I llvery at Washington Union Station

Left: Amtrak P-42DC No66 in Phase II heritage livery.

GE P32AC-DM Diesel-Electric B-B (1992-2001)

SPECIFICATIONS
Type: Diesel electric
Power: 3200 hp
Propulsion: GE752AM
Traction motors: GEB15AC
Tractive effort: 62,000 lb
Max speed: 110 mph

The P32AC-DM, or Genesis Series II was developed for both Amtrak and Metro-North. They can operate on power generated either by the on-board diesel prime mover or, for a short period of time (approximately ten minutes) use the power from a third rail electrification system. The engine is rated at 3,200 hp.The Dual Mode P32AC-DM is unique not only because of its third-rail capability, but also because it is equipped with GE's GEB15 AC (alternating current) traction motors, rather than DC (direct current) motors as used in the other subtypes. The model is confined to services operating from New York City, where diesel emissions through its two fully enclosed main terminal stations are prohibited. P32AC-DMs are seen only on Amtrak's Empire Corridor between Penn Station and Buffalo, the Ethan Allen Express, Lake Shore Limited (New York section), Adirondack, and Maple Leaf services, and Metro-North's locomotive-hauled commuter trains to and from Grand Central Terminal. Metro-North's units have an escape hatch in the nose. It has a top speed of 110 miles per hour.

Below and right: A P32AC-DM locomotive heading south to Cold Spring station on the Hudson Line.(No206)

Bottom: Amtrak P32AC-DM No716 departing Toronto Union Station.

GE AC4400 CW C-C (1993)

SPECIFICATIONS
Total produced: 321
AAR wheel arr.: B-B
UIC classification: Bo'Bo'
Gauge: 4 ft 8 1/2 in
Trucks: Krupp-MaK high-speed bolsterless
Wheel diameter: 40 in
Wheelbase: 43 ft 2 1/2 in
Length: 69 ft
Width: 10 ft
Propulsion: GE 7FDL16

The GE AC4400CW is a 4,400 horsepower diesel-electric locomotive that was built by GE Transportation Systems between 1993 and 2004. It is similar to the Dash 9-44CW, but features AC traction motors instead of DC, with a separate inverter per motor. 2,598 examples of this locomotive were produced for North American railroads. As a result of more stringent emissions requirements that came into effect in the United States on January 1, 2005, the AC4400CW has been replaced by the GE ES44AC. As of 2005, every Class I railroad with the exceptions of Norfolk Southern and Canadian National owns at least one AC4400CW. These units quickly gained a reputation as powerful freight haulers, especially in heavy-haul applications. The AC4400CW was the first GE locomotive to offer an optional self-steering truck design, intended to increase adhesion and reduce wear on the railhead. CSX ordered many of its AC4400CW locomotives with 20,000 pounds extra weight to increase tractive effort. These same units were also modified in 2006-2007 with a "high tractive effort" software upgrade. CSX has re-designated these modified units from CW44AC to CW44AH. Union Pacific ordered many of their AC4400CW engine's with Computerized Tractive Effort software, giving them the designation of AC4400CW-CTE. This software was carried on from their AC4400s and is now standard on Union Pacific ES44ACs.

GE AC6000CW C-C (1995)

The AC6000CW is a 6,000 horsepower diesel electric locomotive that is built by GE Transportation. This locomotive, along with the EMD SD90MAC, is the most powerful single-engined diesel locomotive in the world, surpassed in power by only the dual-engine EMD DDA40X. The AC6000CW was designed at the height of a horsepower race between the two major locomotive manufacturers, Electro-Motive Diesel of London, Ontario and GE Transportation of Erie, Pennsylvania in the early to mid 1990s. The goal was 6,000 horsepower. GE partnered with Deutz-MWM of Germany in 1994 to design and construct the 7HDL engine for the locomotives. The first locomotive constructed was the "Green Machine" GE 6000, the nickname due to the green paint scheme. The first production models were also built in 1995: CSX Transportation 600-602, and Union Pacific Railroad 7000-7009. After testing was completed by GE, they were released to their respective owners in late 1996. Union Pacific Railroad 7391, an example of the 106 "Convertibles" built for Union Pacific Railroad with the 7FDL engine itself. The initial locomotives suffered from various mechanical problems with the most severe being the engine itself. There were major vibration problems which were addressed by increasing the engine mass to alter the resonant frequency. This in turn caused problems with the twin turbochargers. These problems caused GE to push back full production of the new model until 1998. Changes such as stiffer materials and increased engine wall thickness (to increase mass) were in place at full production.

GE built 106 AC6000CWs for Union Pacific with the older, proven 7FDL engine, rated for 4,400 horsepower. The AC6000CW ended production in 2001.

SPECIFICATIONS
Power type: Diesel-electric
Build date: 1995–2001
AAR wheel arr.: C-C
UIC classification: Co'Co'
Gauge: 4 ft 8 1/2 in
Wheel diameter: 42 in
Length: 76 ft
Width: 10 ft 3 in
Axle load: 72,000 lbs
Locomotive weight: 211.5 short tons
Prime mover: GE 7FDL16, 7HDL16, GEVO-16 (rebuilds)
Engine type: 45° V16, four stroke cycle
Aspiration: Twin turbocharger, model 7S1408D
Cylinders: 16
Maximum speed: 75 mph
Power output: 6,250 horsepower

GE Dash 9 Co-Co (1995)

SPECIFICATIONS
Type: Diesel-electric
Gauge: 4 ft 8 1/2 in
Trucks: GE HiAd
Length: 73 ft 2 in
Fuel capacity: 4,600 gall 5,300 gall-C44-9W
Propulsion: GE 7FDL16
Engine type: 45° V16, four stroke cycle
Aspiration: Turbocharged
Alternator: GE
Traction motors: GE
Transmission: Alternator, silicon diode rectifiers, DC traction motors
Power output: 4,000 hp (C40-9, C40-9W, BB40-9W), 4,400 hp (C44-9W), 5,100 hp (C-38AChe)

The Dash 9 Series is a line of diesel locomotives built by GE Transportation Systems. It replaced the Dash 8 Series in the mid-1990s, and was superseded by the Evolution Series in the mid-2000s. The Dash 9 Series is an improved version of the Dash 8 Series. Like that earlier Series, it has a microprocessor-equipped engine control unit, and a modular system of construction of the vehicle body.

All models of the Dash 9 Series are powered by a 16-cylinder, turbocharged, GE 7FDL 4-stroke diesel engine, with electronic fuel injection and split cooling.

Dash 9 Series locomotives also ride on Hiad high adhesion trucks, with low weight transfer characteristics and microprocessor controlled wheelslip.

The C-40 9 version of the Dash 9 was manufactured between January and March 1995. All 125 examples of the model are owned by the Norfolk Southern Railway. It is the only model in the Dash 9 Series to feature the standard cab design. All units have rooftop-mounted air conditioner units which gives them their distinct 'top hatted' look.

All 1,090 units of the C-40 9W were built for the Norfolk Southern Railway, and were road numbered 8889 to 9978. The orders for these units were basically an extension of NS's previous order for the standard cab Dash 9-40C (or C40-9). They were built under the same premise that a lower power rating than the 4,400 hp rating of the Dash 9-44CW would prolong the life of the engine, and use less fuel. However, there is a manual override switch that allows the engineer to run the engine with all 4,400 hp if necessary.

The C44-9W model was in production between 1993 and 2004.

Of all the Dash 9 Series models, this one received by far the greatest number of orders.

Above: Locomotive EFVM BB40-9W No1165 Aroaba Yard - Serra, Espirito Santo – Brazil.

Below: BC rail Dash 9 -44CWL at Superior, Wisconsin.

A total of 1,697 orders for C44-9Ws were received from Burlington Northern / BNSF Railway alone.

Other large orders from North American operators were placed by Canadian National Railway, Atchison, Topeka and Santa Fe Railway, and a number of operators that have since been absorbed by the present day Union Pacific Railroad.

Above: BNSF Railway GE C44-9W No4617, at Commerce, California.

Right: Norfolk Southern No8801 D9-40C heads a long grain train at Marion, OH.

Following page: Mixed BNSF freight train between Kennewick and Wishram, WA. The locomotives are four GE Dash-9 C44-9W.

BNSF
4923
BNSF
5365

5482
5482
BNSF
5482
BNSF

Tri Rail DMUs (2002)

SPECIFICATIONS
Gauge: 4 ft 8 1/2 in
Type: Diesel multiple unit railcar
Propulsion: Detroit Diesel
Power output: 600 hp

In 2003, after receiving a grant from the Florida Department of Transportation, Tri-Rail contracted to purchase two pieces of rolling stock from Colorado Railcar: a self-propelled diesel multiple unit (DMU) prototype control car and unpowered bi-level coach entered regular service with Tri-Rail in October 2006. The new purpose-built railcars are larger than the Bombardier BiLevel Coaches, holding up to 188 passengers, with room for bicycles and luggage. Tri-Rail possessed four DMU control cars and two unpowered trailer cars. One DMU train usually consists of two DMU power cars at each end of a trailer coach (making for two complete DMU+trailer+DMU sets on the system). These cars now reside in the SunRail Rand Yard in Sanford, FL, where they are unused.

Colorado Railcar had originally designed two prototypes, one being a bilevel rail car, the other single level. The self-propelled vehicles can pull two other coaches with their two 600 horsepower Detroit Diesel engines. The single level vehicles can carry up to 92 passengers, 188 for the bilevels. Colorado Railcar had offered non-powered single and bilevel commuter coaches that had a high level of parts commonality with the DMU offerings. Colorado Railcar is now U.S Railcar.

GE ES40DC Diesel-Electric C-C (2003-2014)

SPECIFICATIONS
Power type: Diesel-electric
Builder: General Electric Transportation
AAR wheel arr.: C-C
Gauge: 4 ft 8 1/2 in
Length: 73 ft 2 in
Fuel capacity: 5,000 US gall
Prime mover: GEVO
Engine type: 4-stroke diesel engine
Cylinders: V12
Power output: 4,000 hp

The GE ES40DC was the first engine in the Evolution Series. This was a line of diesel locomotives built to meet the EPA's Tier 2 locomotive emissions standards that took effect in 2005. The first pre-production units were built in 2003. Evolution Series locomotives are equipped with either AC or DC traction motors, depending on the customer's preference. They are all are powered by the GE GEVO engine. The Evolution Series was included as the only post-1972 locomotive to be included in the "10 Locomotives That Changed Railroading" by *Trains Magazine*. Four different Evolution Series models have been produced for the North American market. They are all six axle locomotives and have the C-C or Co-Co wheel arrangement, except for the ES44C4 which has an A1A A1A wheel arrangement. The ES40DC replaced the Dash 9-40CW model in GE's range and has been delivered exclusively to Norfolk Southern Railway.

GE Evolution series Co-Co (2005)

SPECIFICATIONS
Type: Diesel-electric
Gauge: 4 ft 8 1/2 in
Length: 73 ft 2 in
Fuel capacity: 5,000 gall
Propulsion: GEVO
Engine type: 4-stroke diesel engine
Cylinders: V12
Power output: 4,000 hp (ES40DC)
4,400 hp (ES44DC, ES44AC, ES44C4)
Disposition: Almost all still in service.

The Evolution Series is a line of four diesel locomotives built by GE Transportation Systems, designed to meet the U.S. EPA's Tier 2 locomotive emissions standards that took effect in 2005. The first pre-production units were built in 2003. Evolution Series locomotives are equipped with either AC or DC traction motors, depending on the customer's preference. All are powered by the GE GEVO engine.

The models produced for this series are; ES40AC, ES44DC, ES44AC, ES44C4. Operator(s) UP, BNSF, CSX, NS, CN, CP, KCS, KCSM, GECX, FXE, IAIS, SVTX, CREX, FEC

The Evolution Series was named as one of the "10 Locomotives That Changed Railroading" by industry publication Trains Magazine. It was the only locomotive introduced after 1972 to be included in that list. Disposition: almost all still in service.

GE ES44AC Diesel-Electric (2007)

SPECIFICATIONS
Type: Diesel-electric
Gauge: 4 ft 8 1/2 in
Length: 73 ft 2 in
Fuel capacity: 5,000 gall
Propulsion: GEVO
Engine type: 4-stroke diesel engine
Cylinders: V12
Power output: 4,000 hp (ES40DC)
4,400 hp (ES44DC, ES44AC, ES44C4)
Disposition: Almost all still in service.

The ES44AC is the third variant in General Electric's Evolution Series. It had 4400 horsepower and AC traction. The engine replaced AC4400CW model in the range. These locomotives have been ordered by Union Pacific Railroad, BNSF Railway, CSX Transportation, Norfolk Southern Railway, Kansas City Southern Railway, Kansas City Southern de Mexico, Ferromex, and Canadian Pacific Railway. Iowa Interstate Railroad ordered 14 ES44ACs to handle an expected traffic growth of 25%-30%. CSX began receiving an order of 200 ES44ACs in December 2007. In September 2008, Norfolk Southern purchased 24 ES44Acs to be used in long haul coal trains. Canadian National announced in January 2012 that 35 ES44ACs were on order for 2012-2013 delivery to be used primarily on the grades of its ex-BC Rail main line. Citirail/CREX acquired 100 ES44ACs for lease service. They are painted silver with blue and yellow nose striping and blue numbers. As of August 2014, most of these units are leased to BNSF Railway.

GE ES44DC Diesel-Electric C-C (2008)

SPECIFICATIONS
Power type: Diesel-electric
AAR wheel arr.: C-C
Gauge: 4 ft 8 1/2 in
Length: 73 ft 2 in
Fuel capacity: 5,000 US gall
Prime mover: GEVO
Engine type: 4-stroke diesel engine
Cylinders: V12
Power output: 4,400 hp

General electric's Evolution Series locomotives are visually similar to the Dash 9 and the AC4400CW locomotives, although small differences are evident. The most significant differences are the radiator section at the rear of the locomotive is larger and the wings on top of the radiator section are also larger. Unlike any previous GE locomotive the grills under the radiator are at two different angles. The radiator is larger to the necessity of greater cooling capacity in the locomotive in order to reduce emissions. The other major difference between the Evolution Series locomotives and older models are the vents below the radiators, which are larger than those on previous GE locomotives. The ES44DC is a member of the Evolution Series. It replaces the Dash 9-44CW model in General Electric's range. The main users of the model are the BNSF Railway, CSX Transportation, and the Canadian National Railway.

GE ES 44 C4 A1A-A1A (2009)

SPECIFICATIONS
Power type: Diesel-electric
AAR wheel arr.: C-C
Gauge: 4 ft 8 1/2 in
Length: 73 ft 2 in
Fuel capacity: 5,000 US gall
Prime mover: GEVO
Engine type: 4-stroke diesel engine
Cylinders: V12
Power output: 4,400 hp

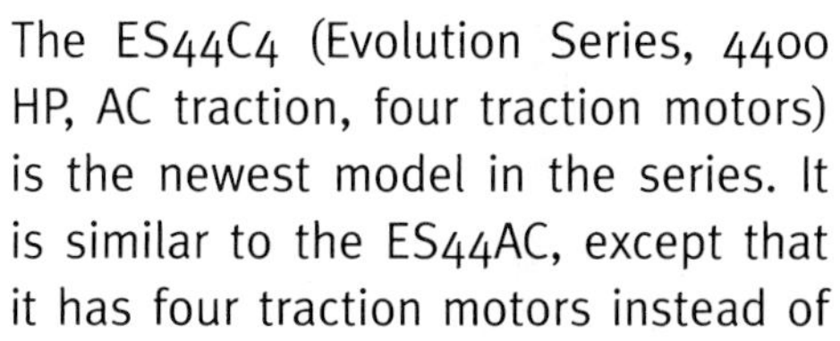

The ES44C4 (Evolution Series, 4400 HP, AC traction, four traction motors) is the newest model in the series. It is similar to the ES44AC, except that it has four traction motors instead of six. The center axle of each truck is unpowered, giving an A1A-A1A wheel arrangement. BNSF Railway is the launch customer for this model, ordering an initial batch of 25 units numbered 6600-6624. BNSF was the first buyer of this model and has over 700 units. On January 30, 2014 Florida East Coast Railway announced that they would buy twenty-four ES44C4s, to be numbered 800-823, for heavy haul service and intermodal traffic. All are to be delivered by the end of 2014 to beat the EPA's deadline on exhaust-emissions standards for new-built Tier-3 locomotives. A unique feature of these units is that there is a variable traction control system in their computer systems. One of the differences between an ES44AC and an ES44C4 are the air cylinders and linkages on the truck sideframes of the ES44C4; these are part of the traction control system. Every time a variable in grade, traction, or wheel slip occurs, the computer will adjust the pressure in these cylinders to maintain sufficient adhesion, by varying the weight on the drive axles.

Electric Power: 1895-Present

Nos 1-3 Bo Bo B&O (1895)

SPECIFICATIONS
Type: Main line electric locomotive.
Gauge: 4 ft 8 1/2 in
Propulsion: Direct current at 675V fed via a rigid overhead conductor to four gearless motors of 360 hp each.
Weight: 192,000 lb
Max. axleload: 48,488 lb
Overall length: 27 ft 1/2 in
Tractive effort: 45,000 lb
Max speed: 60 mph

The world's first main line electrification was carried out on a section of the Baltimore & Ohio RR as it ran through the city of Baltimore and in particular through the 1 1/2 mile Howard Street tunnel, adjacent to a new main passenger station at Mount Royal. The tunnel was on a gradient of 1-in-l 25 and trouble with smoke and steam was a problem. The solution adopted was an electric locomotive designed by General Electric of Schenectady, New York State.

An innovative decision for B&O, the locomotives were over nine times heavier and nine times more powerful than their nearest rivals.

Gearless motors were used, but not mounted direct on the axle, although concentric with it. Torque was transmitted to the wheels through rubber blocks; this flexible drive was yet another feature many years ahead of its time. Each four-wheeled tractor unit was mechanically quite separate, although two were permanently coupled to form one locomotive. There were three double locomotives in all.

This set up worked reasonably well and the locomotive had no problems hauling 1,800 ton trains up the gradient. This load included the train's steam engine, which did no work in the tunnel. Trouble was encountered with corrosion of the unusual conductor arrangements; a brass shuttle ran along a Z-section overhead rail, the shuttle being connected to the locomotive by a one-sided tilted pantograph which was replaced by a conventional third rail mounted outside the running rails in 1902.These locomotives stopped work in 1912, but one was laid aside for many years—in fact, until B&O's centennial "Fair of the Iron Horse" in 1927, at which it was exhibited after which it was unceremoniously scrapped. Thus this important pioneer of electric powered railroads was not preserved for posterity.

NYC S-Motor 2-do-2 (1904)

SPECIFICATIONS
Gauge: 4 ft 8 1/2 in
Leading wheel diameter: 36 in
Driver diameter: 44 in
Length: S-1, S-2: 39 ft 0 in, S-2a 37 ft 0 in, S-3 43 ft 2 in
Weight on drivers: S-1, S-2: 148,000 lb, S-2a: 140,000 lb, S-3: 150,000 lb
Weight: S-1, S-2: 228,000 lb, S-2a: 224,000 lb, S-3: 242,000 lb
Electric system(s): 660 Volts DC
Current supply: Third rail
Traction motors: 4 x GE 84 550 hp
Transmission: Resistance controlled DC current supplied to gearless DC traction motors mounted directly on the axles.
Maximum speed: 60 mph
Power output: 2,200 hp Starting, 1,695 hp Continuous
Tractive effort: 37,000 lb
Max speed: 60 mph

S-Motor was the class designation given by the New York Central to its Alco-GE built S-1, S-2, S-2a and S-3 electric locomotives. The S-Motors hold the distinction of being the world's first mass-produced main line electric locomotives with the prototype No6000 being constructed in 1904. The S-Motors would serve alone until the more powerful T-motors began to arrive in 1913, eventually displacing them from main line passenger duties. From that point the class was assigned

Right: S-Motor No100 is awaiting restoration in a forested flood plain south of Albany, NY. The locomotive belongs to the Mohawk Chapter (NRHS) who received it from Amtrak and Conrail sometime around 1980.
S-Motor No100 is the prototype for the first class of New York Central electrics and the S-Motors were amongst the earliest class of independent main line electric locomotives to be in service.

to shorter commuter trains and deadhead rolling stock between Grand Central Terminal and Mott Haven coach yard. Some examples, including the prototype later renumbered #100, would serve in this capacity through the Penn Central merger in 1968, only being retired in the 1970s as long distance passenger traffic to Grand Central dried up.

Above: No113 was built by Alco-General Electric in 1906 as New York Central & Hudson River Class T-2 No3413. It was renumbered No3213 in 1908 and then, in 1909, was rebuilt with 4 wheel leading trucks as a Class S-2. It was renumbered No1113 in 1917 and then as No113 in 1936. Designed to operate on the New York Central's electrified track out of Grand Central Station in New York City, thirty-two S-2s were rostered by the railroad and the last one was only retired in 1981.The NYC donated No113 to the St Louis Museum of Transportation in 1963.

Left: A very rare photograph of S-Motor unit in service in the 1960s.

EP-1 Bo Bo (1906)

SPECIFICATIONS
Type: Electric passenger locomotive.
Gauge: 4ft 8 1/2 in
Propulsion: Alternating current at 11,000V 25Hz fed through transformer to four 240hp bogie-mounted gear-less traction motors with spring drive; alternative supply at 660V dc from third rail.
Current supply: two dc pickup shoes on each side of each bogie, a small dc pantograph and two ac pantographs.
Weight: 204,000 lb
Max. axleload: 51,000 lb
Overall length: 37 ft 6 1/2 in
Tractive effort: 42,000 lb
Max. speed: 65 mph

By the turn of the 20th century, the advantages of ac current for electrical generation and transmission were well established, and early in the new century Westinghouse marketed a high-voltage single-phase ac traction system.

General Electric established an early lead in the supply of dc traction equipment in the United States, but in 1895 its main rival in the heavy electrical industry, Westinghouse, began a long association with the Baldwin Locomotive Works. In that year an experimental locomotive was built at GE's East Pittsburgh Works, with mechanical components by Baldwin. This locomotive was used for experimental work on both dc and ac traction equipment.

The first application was in 1905, when a 41-mile line between Indianapolis and Rustville was electrified at 3,300 Volts.

The New York, New Haven & Hartford Railroad had electrified various branches in New England from 1895 onwards, using dc with third-rail current collection. In 1903 the law which compelled the New York Central to electrify its lines into Grand Central Terminal also applied to the New Haven, whose suburban trains used that station. The New Haven developed an even more elaborate scheme than the NYC. Rather than face the administrative and financial problems of a change from electric to steam working outside the area covered by the ban, the road decided to electrify to the limit of suburban territory at Stamford, 33 miles from Grand Central, making it the longest main-line electrification so far undertaken. Having taken the plunge the railroad then looked for a system of electrification which would be suitable for further extension along its very busy main line to New Haven, and even Boston.

In 1905 GE and Westinghouse both submitted schemes for dc and ac systems. Despite their limited experience of ac traction, the New Haven commissioned Westinghouse to carry out the electrification using ac at 11,000V 25Hz with locomotive haulage. Design of the locomotives was difficult, for not only had teething problems in ac traction to be solved, but the locos had also to be able to work on 660V dc over the 12 miles of New York Central electrified line by which the New Haven gained access to Grand Central. The solution was a Bo-Bo design with commutator motors which could operate on both ac and dc, the ac being transformed to 660V. The motor casings were mounted on the bogies above the axles, and the armatures on hollow quills through which passed through the axles. The quills were supported in rigid bearings, and they were connected to the axles through spring connections. There

Above: A New Haven EP-1 electric locomotive, circa 1907. Note the small DC pantograph between the two larger AC pantographs.

was sufficient clearance in the quills for the axles to move vertically on their springs. The weight of the motor, including the armature, was thus fully spring borne, whereas in the GE bi-polar motors the weight of the armature was on the axle.

The size of the locomotives was suited to average trainloads, which included hauling expresses of 250 tonnes at up to 60 mph but to allow the locomotives to work in twos and threes on heavier trains, multiple-unit control was fitted. Current collection provision. was generous, with two dc pickup shoes on each side of each bogie, a small dc pantograph and two ac pantographs.

A total of 37 had been delivered by mid-1907. Some initial difficulties were encountered, particularly with nosing of the bogies at high speed. This was solved by fitting a pair of guiding wheels at the outer ends of the bogies, making the wheel arrangement 1-Bo + Bo-1. After some electrical problems had been solved, the locomotives settled down to long and successful careers. By 1924 they had accumulated an average of 1 1/4 million miles much of it at high speed. At times they exceeded 60mph and on test a figure of 89mph was recorded. A further batch of six locomotives was built in 1908.

These locomotives were notable in being the first main-line ac units, as well as the first dual-voltage machines. Despite these innovations, several of them, including the prototype, were still in main line service in 1947.

MP-54 Pennsylvania Railroad Bo-Bo (1908)

SPECIFICATIONS
Entered service: 1908-1972(LIRR), 1915-1981(PRR) refurbished 1950
Number built: PRR 481, LIRR 923, PRSL 18
Operators: Pennsylvania Railroad, Long Island Rail Road, Pennsylvania-Reading Seashore Lines, Penn Central Railroad Conrail, New Jersey Department of Transportation, SEPTA
Car body construction: carbon steel
Car length: 64 ft 5 3/4 in
Width: 9 ft 11 1/2 in
Height: 14 ft 6 in
Floor height: 4 ft 0 in
Doors: 2, end vestibule
Maximum speed: 65mph
Power output: 400 hp (MP-54E1/2), 736 hp (MP54E3), 450 hp (MP54E5), 508 hp (MP54E6)[1]
Electric system(s): 650 V DC third rail and 11,000V 25 Hz AC catenary
Current collection method: contact shoe, pantograph
Braking system(s): Pneumatic
Safety system(s): Cab signaling, Automatic Train Control
Coupling system: AAR

The Pennsylvania Railroad's MP-54 class of electric multiple unit cars was their first and largest class of this type of car. Manufactured severally by the Pennsylvania Railroad, American Car and Foundry Company, Pressed Steel Car Company, Standard Steel Car Company. The class was initially constructed as an unpowered locomotive hauled coach for suburban operations, but were designed with the capacity to be rebuilt into self-propelled MU as electrification plans were realized. The first of these self-propelled cars were placed in service with the PRR subsidiary, the Long Island Rail Road with DC propulsion in 1908 and soon spread to the Philadelphia-based network of low frequency AC electrified suburban lines in 1915. The cars came to be used throughout the railroad's electrified network from Washington, DC to New York City and Harrisburg, Pennsylvania. They became a commuting tradition during their long years of service in several major cities,and were known as red cars ,remaining in service with the PRR until the Penn Central merger in 1968 at which point they were already being marked for replacement by new technology railcars such as the Budd M1 and Pioneer III.

DD-1 2-Bo+B0-2 PRR (1909)

SPECIFICATIONS
Type: Electric passenger locomotive.
Gauge: 4 ft 8 1/2 in
Propulsion: Direct current at 600V fed to two l,065hp motors, each driving two main axles by means of a jackshaft and connecting rods.
Current collection: third rail or by overhead conductors and miniature pantographs at places where third rail was impractical
Weight: 199,000 lb adhesive, 319,000 lb total.
Max. axleload: 50,750 lb
Length: 64 ft 1 in
Tractive effort: 49,400 lb
Max. speed: 80 mph

The DD-1 s were a landmark in electric locomotive design, with exceptionally high power and unusual reliability for their day, but at the same time their design was conservative. Simplicity of design and the flexibility of the "double 4-4-0" chassis contributed greatly to their success. With 72in driving wheels, each half of the unit resembled the chassis of an express 4-4-0 steam locomotive. This similarity to steam design was apparent in a number of early electric engines, but unusually for such designs the Pennsylvania's was highly successful. The first two rod-drive units appeared in 1909-10, the individual half-units being numbered from 3996 to 3999. They were followed in 1910-11 by a further 31, numbered from 3932 to 3949 and from 3952 to 3995. It was capable of 80mph and there was no appreciable clanking of the rods. The Pennsylvania classified its steam locomotives by a letter, denoting the wheel arrangement, followed by a serial number, letter D denoted 4-4-0. For electrics the road used the same system, the letter being doubled when appropriate, so that the 2-B+ B-2, being a double 4-4-0, was a DD. The main production batch of 31 units was classified DD-1, and the two prototypes odd DD.

The Pennsylvania Railroad gained entry to New York City and its new Pennsylvania Station by single-track tunnels, two under the Hudson River and four under the East River, and for the operation of these tunnels electrification was essential. The third-rail system at 650V was chosen, and between 1903 and 1905 three experimental four-axle locomotives were built, two B-Bs at PRR Altoona shops and a 2-B by Baldwin. The motors of both types of locomotive drove through quills, an early version of the drive which was to be used a quarter of a century later on the GG-1 electrics.

An important change in design from the experimental locomotive was in the drive to the axles, which incorporated a jackshaft mounted in bearings in the main frame of the locomotive, with connecting rods from the motor to the jack-shaft and from the jackshaft to the driving wheels. The technical problems with the quill drive were left for solution at a later date.

. Maintenance costs were very low, and were helped by the design of the body. The whole casing could be removed in one unit to give access to the motors and control equipment.

This feature was repeated in all subsequent PRR electric designs. A small pantograph was fitted to allow overhead current collection on complicated trackwork.

The DD-1 locomotives worked all the express passenger services on this section of line until 1924 when newer types began to appear, but they continued to share the work until 1933 when overhead electrification reached Manhattan Transfer from Trenton, and the remaining section into Pennsylvania Station was converted to overhead. Third-rail current collection was retained between Pennsylvania Station and Sunnyside, because the Long Island Rail Road used this system.

After 1933 DD s continued to work empty trains between Pennsylvania Station and Sunny-side for many years. After the arrival of newer power in 1924, 23 of the DD- Is were transferred to the Long Island Rail Road, and these remained in service until 1949-51.

Class T-Motor Bo-Bo+Bo Bo NYC Railroad (1913)

SPECIFICATIONS
Type: Express passenger electric locomotive.
Gauge: 4 ft 8 1/2 in
Propulsion: Direct current at 660V collected from under-contact third rail supplying eight 330hp gearless traction motors mounted on the bogie frames, with armatures on the axles.
Weight: 230,000 lb
Max. axleload: 28,730 lb
Overall length: 55 ft 2 in
Tractive effort: 69,000 lb
Max. speed: 75 mph

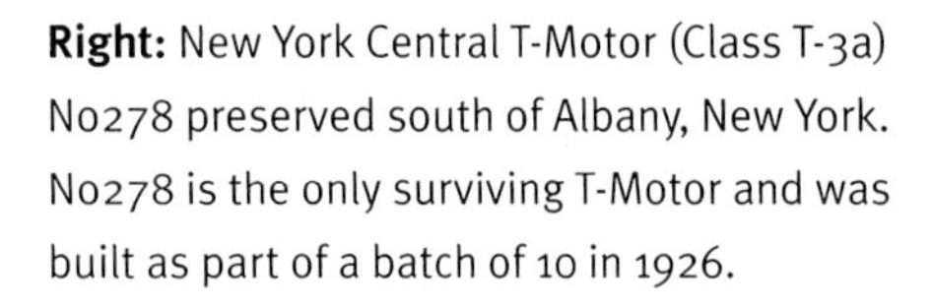

Right: New York Central T-Motor (Class T-3a) No278 preserved south of Albany, New York. No278 is the only surviving T-Motor and was built as part of a batch of 10 in 1926.

In 1913 the New York Central Railroad completed its Grand Central terminal, and in the same year it also completed an extension of electrification along the Hudson River main line to Harmon, 33 miles from Grand Central. The heaviest expresses would now have some 20 miles of fast running with electric haulage, and a new type of locomotive was therefore commissioned, which would be both more powerful than the S class 2-Do-2 and also less damaging to the track at high speed. The first locomotive was completed in March 1918; this was designated Class T-la. Nine more, classified T-1b, were delivered later in the year. Sub-class T-2a of 10 units came in 1914, a further 10 designated T-2b in 1917, and a final batch of 10 in 1926; these were T-3a. Successive batches differed mainly in the size of train-heating boiler and its fuel supplies, but the T-2s and T-3s were 20in longer than the T-ls. The bodies of successive batches were a bit longer, with the end overhanging slightly less.

The most important change was adoption of an articulated layout in which every axle was motored, but at the same time the whole chassis was more flexible than in the earlier class. There were two sub-frames, each with two four-wheeled trucks pivoted to it. The trucks were connected by arms, and the sub-frames were hinged together at their inner ends and carried the couplers at the outer ends. The body rested on two pivots, one on each sub-frame, and one pivot had some end play to allow for changes in geometry on curves. The whole assembly was therefore flexible, but with restraining forces on all its elements to discourage the build-up of oscillations. The wheel arrangement was Bo-Bo + Bo-Bo, and it was equivalent to two Bo-Bo locomotives hinged together.

Nominal power of the original motors fitted to Class S was 550hp but in Class T the motor size was reduced to 330hp This enabled the weight of the motors to be reduced, with a corresponding reduction in the forces on the track and with an improvement in riding. The new class was therefore allowed 75mph later reduced to 70 compared with 60mph on Class S. The motors were of the same bi-polar type as before, with almost flat pole faces to allow for the vertical movement of the axle (and armature) on its springs.

Another change from Class S was that the motors had forced ventilation, which involved some complications of ducting to the bogies. As in the earlier units, in addition to pick-up shoes for the third rail, a small pantograph was fitted. This was used on complicated track layouts where there were long gaps in the third rail, and overhead wires were installed locally. There was an oil-fired train heating boiler, with supplies of water and oil.

With a continuous rating of 2,610hp at 48mph they were powerful locomotives for their day, and they were highly successful. They continued to work single-headed all express trains out of Grand Central until 1955, when some locomotives surplus from a discontinued electrification in Cleveland began to displace them. They could handle trains up to 980 tons at 60mph.

These locomotives established the practicability of the all-adhesion machine for high-speed work, and they showed that it was possible to avoid the heavy and complicated rod drives of many of their contemporaries.

EF-1 2 Bo-Bo+Bo-Bo-2 (1915)

The Chicago, Milwaukee, St. Paul and Pacific Railroad (Milwaukee Road) class EF-1 comprised 42 boxcab electric locomotives built by the American Locomotive Company Alco in 1915. Electrical components were from General Electric. The locomotives were composed of two half-units semi-permanently coupled back-to-back, and numbered as one unit with “A” and ‘“B” suffixes. As built, 30 locomotives were assigned to freight service, classified as EF-1 and numbered 10200–10229. The design was highly successful, replacing a much larger number of steam locomotives, cutting costs and improving schedules.

SPECIFICATIONS
Type: Electric freight locomotives
Gauge: 4 ft 8 1/2 in
Weight: 576,000 lb
Power output: Continuous 3,340 hp,
One hour 4,100 hp
Retired: January 1951–June 1974

In 1919, with the arrival of a newer generation of passenger power, the EP-1 locomotives were converted to EF-1 freight locomotives, and renumbered 10230–10241. In this role, they served until the 1950s, when the arrival of the Little Joe locomotives began to replace them in freight service.

Below: As No10200 was hauled west from the Alco works in Schenectady, New York, more than 60,000 people came to stations along the way to see it. Its electrical components were supplied by General Electric, and the locomotive consists of two half units numbered as one, with “A” and “B” suffixes, coupled back to back using couplers and a ball and socket joint. Thirty were assigned to freight service classified as EF-1 road numbers 10200-10229).

C. M. & St. P.
10203B

E50A

ITC B-Class No1575 (1918)

SPECIFICATIONS
Gauge: 4 ft 8 1/2 in
Propulsion: four GE 69-C motors
Weight: 120,000 lb
Tractive effort: 30,000 lb

Built at the Illinois Traction System's Decatur, IL, workshops in 1918, this electric freight locomotive was powered by four GE 69-C motors. In 1937, Illinois Traction was renamed the Illinois Terminal Railroad, which it is commonly known as today. It operated 550 miles of track serving passenger and freight business in central and southern Illinois. The Illinois Terminal Railroad was a heavy duty interurban electric railroad with extensive passenger and freight business in central and southern Illinois from 1896 to 1982. When Depression era Illinois Traction was in financial distress and had to reorganize, the Illinois Terminal name was adopted to reflect the line's primary money making role as a freight interchange link to major steam railroads at its terminal ends, Peoria, Danville, and St. Louis. In the 1950s, with the final dominance of the automobile, ITR's passenger service became hopelessly unprofitable. This was even after IT had purchased three expensive electric multiple car streamlined train sets ("Streamliners") from St. Louis Car Company designed somewhat upon the North Shore Line's Electroliners. These were capable of decent speeds on ITR's well-maintained open country roadbed, but had to negotiate tight streetcar-style curves in the numerous towns along the line; moreover, they suffered an abnormal amount of failures, unlike the Electroliners upon which they were based. Interurban passenger service slowly was reduced, and it ended in 1956. Freight operation continued but was hobbled by the same tight street running in some towns requiring very sharp radius turns. In 1986, ITR was absorbed by a consortium of connecting railroads.

The only remaining locomotive No1575 was sold to the St. Louis Car Company Division of General Steel Industries in 1953, who donated it to the museum in 1963.

Left: No300, one of three streamliners of Illinois Terminal Railroad in the early 1950s as the company bought these only a few years before they stopped passenger service in 1956.

Class EP-2 Bi Polar (1919)

In 1917, following the tremendous success of the 1915 electrification of the Mountain Division, the Milwaukee Road decided to proceed with electrifying its Coast Division. As part of this project it ordered five new electric locomotives from General Electric for $200,000 each. They were delivered in 1919, factory serial numbers 6978-6982, and were designated Class EP-2 by the railroad. Their design was radically different from the boxcab locomotives previously provided by GE for the initial electrification of the Mountain Division two years earlier. The Milwaukee Road was the only railroad to order this design of locomotive from GE.

They were often known as Bipolars, which referred

Above: This is one of five electric locomotives ordered from GE by the Milwaukee Road in 1919 for passenger services on the newly electrified Coast Division. Costing $200,000 each, they were very different from the earlier EP-1 GE had built for service on MILW's Mountain Division in 1915. The five units (Nos10250-10254) realized immediate savings over steam locomotives, however, as they ran long distances without requiring servicing and hauled trains up grades that had required double-headed steam engines. Their distinctively modern design also made them the most famous of MILW's electric locomotives. In 1939, the units were renumbered NoE-1 - E-5 and, in 1953, were heavily rebuilt, including adding the rounded front ends (until then, these were flat).

SPECIFICATIONS
Type: Electric locomotive
AAR wheel arr.: 1B+D+D+B1
Gauge: 4 ft 8 1/2 in
Length: 76 ft 0 in between coupler pulling faces
Axle load: 38,500 lb
Weight on drivers: 457,000 lb
Weight: 530,000 lb
Electric system(s): 3,000 Volts DC
Current Supply: Dual pantograph
Transmission: Twelve 370 hp gearless traction motors mounted directly on the axles
Maximum speed: 70 mph
Power output: 4,440 hp
Tractive effort: 116,000 lb

to the bipolar electric motors they used. Among the most distinctive and powerful electric locomotives of their time, they epitomized the modernization of the Milwaukee Road and came to symbolize the railroad during their nearly 40 years of use, and remaining an enduring image of mainline electrification were rebuilt in 1953.

The most remarkable mechanical improvement was arguably the traction motors used on the new locomotives. They were known as bipolar motors because each of the locomotive's 12 motors had only two field poles, mounted directly to the locomotive frame beside the axle. The motor armature was mounted directly on the axle providing an entirely gearless design. This design was almost entirely noiseless as it eliminated not only gear noise but also the whine of higher-RPM electric motors typically used in standard nose mounted applications. The layout of the bipolars was unusual as well. The locomotive carbody consisted of three sections. A small center section contained a boiler for heating passenger cars, while the larger end sections contained the locomotive's electrical equipment and operator cabs in distinctive round-topped hoods. The locomotive's frame was split into four sections, hinged at the joints. There were twelve sets of driving wheels, plus a single idler axle at each end, for a 1B+D+D+B1 wheel arrangement. All buffering forces were transmitted through the locomotive frame.

Between 1958 and 1960 all five locomotives were gradually retired, by which time they had received the Union Pacific-inspired yellow and gray passenger paint scheme. In 1962 all except for E2 were towed to Seattle and scrapped.

Locomotive E2 was donated to the Museum of Transportation in St. Louis, Missouri in 1962 and moved there that year. It has remained on static display ever since, and has been fully restored as it would have looked after its 1953 rebuild.

The five EP-2s, numbered 10250-10254, No10251 shown here prior to renumbering in 1939,were placed into regular service in 1919 on the Coast Division. The Milwaukee Road saw immediate cost savings over the steam locomotives previously in use, as the Bipolars could run from Tacoma to Othello without stopping for servicing and could haul trains up grades that had required double-heading steam engines. The Bipolars operated on the Coast Division from 1919 to 1953, for most of that period without any serious rebuilding. In 1939 they were renumbered E1-E5. In 1953 all five of the EP-2s, which were 35 years old and worn out from heavy wartime service, were heavily rebuilt by the Milwaukee Road at a cost of about $40,000 per locomotive.

Pacific Electric (1920s)

SPECIFICATIONS
Reporting mark: PE
Locale: Greater Los Angeles Area
Dates of operation: 1901–1961
Track gauge: 4 ft 8 1/2 in
Headquarters: Los Angeles, CA

Pacific Electric, also known as the Red Car system, was a privately owned mass transit system in Southern California consisting of electrically powered streetcars, light rail, and buses and was the largest electric railway system in the world in the 1920s. Organized around the city centers of Los Angeles and San Bernardino, it connected cities in Los Angeles County, Orange County, San Bernardino County and Riverside County.

The system shared dual gauge track with the 3 ft 6 in narrow gauge Los Angeles Railway, "Yellow Car," or "LARy" system on Main Street in downtown Los Angeles (directly in front of the busy 6th and Main terminal), on 4th Street, and along Hawthorne Boulevard south of downtown LA toward the cities of Hawthorne, Gardena, and Torrance.

No455 CNS&M Steeple Cab loco (1927)

SPECIFICATIONS
Type: Electric rail car
Gauge: 4 ft 8 1/2 in

No455 was built in 1927 by GE and spent its life on the Chicago, Northshore & Milwaukee RR. .No455 had extra batteries to avoid loss of power when losing the overhead supply during switching duties. It was scrapped in 1963.

A steeplecab is a style or design of electric locomotive; the term is rarely if ever used for other forms of power. The name originated in North America and has been used in Britain as well as the alternative camelback. A steeplecab design has a central (or nearly central) driving cab area which may include a full-height area in between for electrical equipment. On both ends, connected to the full-height cab areas, lower (usually sloping) "noses" contain other equipment, especially noisy equipment such as air compressors not desired within the cab area. When overhead lines are used for power transmission, the cab roof usually supports the equipment to collect the power (either by pantograph(s) bow collector(s) or trolley pole(s)), although on some early designs (such as the North Eastern Railways Electric number 1 -- later known as an ES-1) a bow collector might be mounted on one of the bonnets (or nose hoods) instead. The steeplecab design was especially popular for electric switcher locomotives, and on electric locomotives ordered for interurban and industrial lines. It offers a large degree of crash protection for the crew combined with good visibility.Disadvantages include reduced room for bulky electrical equipment compared to other designs.The overall design pattern of a central crew area with lower and/or narrower equipment hoods on each end has been repeated many times, although the lack of equipment space has meant it has largely died out in recent years.

Below: A Milwaukee Road class ES-2, an example of a larger steeplecab switcher for service on an electrified heavy-duty railroad.

Car No. 749 (1928)

Built in 1928 at the Pullman Company's south side Chicago factory, Car No749 was acquired by the Illinois Railway Museum from scrappers after the abandonment of Chicago, North Shore & Milwaukee Railway service in 1963.

The 749 restoration is the first complete restoration of an interurban car attempted by the museum.

On June 19, 2010, the Illinois Railway Museum in Union Illinois celebrated the return to service of Chicago, North Shore & Milwaukee Railway Car No. 749, following completion of a 23 year restoration. Eighty-two years young and looking as good as the day it was new, Car No. 749 beckons to riders who want a taste of classic electric rail transportation. This historic interurban car served the Chicago to Milwaukee route—downtown to downtown—for more than 30 years.

Above: No749 takes to the tracks again after its 23-year renovation.

Below: No409 at the 2014 Trolley Pageant at the Illinois Railroad Museum.

P-5 Class 2-co-2 PRR (1931)

SPECIFICATIONS
Type: Mixed-traffic electric locomotive
Leading wheel diameter: 36 in
Driver diameter: 72 in
Trailing wheel diameter: 36 in
Wheelbase: 49 ft 10 in
Length: 62 ft
Width: 10 ft 6 in, 10 ft 8.25 in (P-5a (modified))[4]
Height: 15 ft
Axle load: 74,000 lb 77,800 lb (P-5b) [3], 77,000 lb (P-5a (modified)) [4]
Weight on driving wheels: 220,000 lb (P-5, P-5a) [2], 444,700 lb, (201.7 t) (P-5b, all wheels driven) [3], 229,000 lb (104 t) (P-5a (modified)) [4]
Weight: 392,000 lb (P-5, P-5a)[2], 444,700 lb (P-5b)[3], 394,000 lb (P-5a (modified))[4]
Electric system(s): 11 kV AC @ 25 Hz
Current collection method: Overhead pantograph
Traction motors: 6 x 625 hp (466 kW) AC motors;[2][4] plus 4 x 375 hp (280 kW) motors on the trucks on P-5b[3]
Transmission: AC current fed via transformer tap changers to paired motors geared (25:97) to quill drives on each driving axle;[2][4] plus single motors geared to driving axles on end trucks on P5b (gear ratio: 17:50)[3]

The Pennsylvania Railroad's class P-5 comprised 92 mixed-traffic electric locomotives constructed 1931–1935 jointly by the PRR- 13 made at their Altoona workshops, Baldwin/Westinghouse -54 and General Electric-25 .They were road numbered 4700–4791. The first P-5s were built with box cabs. A fatal grade crossing accident on the New York Division confirmed train crews' concerns about safety when the crew were killed after colliding with a truckload of apples. A redesign was undertaken, giving the locomotives a central steeple cab, raised higher, with narrower-topped, streamlined noses to the locomotive to enable the crew to see forward. The final 28 locomotives were built to this design, which was not given a separate class designation since it was mechanically and electrically identical; they were called class P-5a.This design modification was also adopted by the GG-1.

Although the original intention was that they work many passenger trains, the success of the GG-1 locomotives meant that the P-5 class was mostly used on freight. A single survivor, prototype No4700, is at the Museum of Transportation in St Louis, Missouri.

Left: This is one of two prototype P-5 electric locomotives built by the Pennsy in 1931.Ninety more P-5s were produced from 1932 to 1935 designated P-5a, by Westinghouse (fifty-four units), General Electric (twenty-five) and PRR at its Altoona shops (ten).

GG-1 PRR (1934)

SPECIFICATIONS
Type: Heavy-duty express passenger electric locomotive.
Gauge: 4 ft 8 1/2 in
Propulsion: Medium-frequency alternating current at 15,000V 25Hz fed via overhead step-down transformer to twelve 410hp traction motors, each pair driving a main axle through gearing and quill-type flexible drive
Current supply: overhead catenary.
Weight: 303,000lb adhesive, 477,000lb total.
Max. axleload: 50,500lb
Length: 79ft 6in
Tractive effort: 70,700lb
Max. speed: 100mph

Since 1928, PRR had been pursuing a long-considered plan to work its principal lines electrically. The statistics were huge; $175 million in scarce depression money was needed to electrify 800 route-miles and 2,800 track-miles on which 830 passenger and 60 freight trains operated daily.

Regarding the current supply the medium-frequency single-phase ac system with overhead catenary was adopted. The reason lay in the fact that the dc third-rail system used in New York City was not suitable for long-distance operations and, moreover, since 1913 the PRR had been gaining experience working its Philadelphia suburban services using overhead wires on 25Hz ac. It required a corporation of colossal stature to keep this costly scheme going through the depression years, but by 1934 impending completion of electrification from New York to Washington meant a need for some really powerful express passenger motive power. A prototype locomotive was built with a steamlined casing by the famous industrial designer Raymond Loewy. Between

1935 and 1943, 139 of these GG-ls were built and remained in service until 1982.

Some of the GG-l s were constructed in-house at PRR's Altoona shops, others by

Baldwin or by General Electric. Electrical equipment was supplied by both GE and Westinghouse. The design relied on well tried and tested precedents such as the arrangement of twin single-phase motors, the form of drive, and many other systems were essentially the same as had been in use for 20 years on the New Haven.

Low traffic levels during the depression years meant that the physical upheaval of electrification was easily introduced while its completion (there was an extension to Harrisburg, Pennsylvania, in 1939) coincided with the start of the greatest passenger traffic boom ever known caused by World War II. The peak was reached on Christmas Eve 1944 when over 175,000 long-distance passengers used the Pennsylvania Station in New York. It was true that anything that had wheels was used to carry them, but coaches old and new could be rostered in immense trains which the GG-l s had no problem at all in moving to schedule over a route which led to most US cities from Florida to Illinois. In numerical terms, a GG- rated at 4,930hp on a continuous basis, could safely deliver 8,500hp for a short period. This was ideal for quick recovery from stops and checks. In this respect one GG-l was the totally reliable equivalent of three or four diesel units .With the demise of the Pennsylvania Railroad in the post-war years the GG-1 fleet passed piecemeal to the later owners of the railroad, or parts of it—Penn Central, Conrail, Amtrak, and the New Jersey Department of Transportation.

Conrail also recently de-electrified the parts of the ex-Pennsylvania lines it inherited on the bankruptcy of Penn Central and so it too had no use for the locos. Sadly, on October 28, 1982, the last GG-1 was withdrawn from service by the New Jersey DoT which had a few still in service. Fortunately for rail fans no fewer than fifteen production locomotives and the prototype are preserved in museums. None are operational, due to difficulties with running their transformers.

Electroliner C,NS &M (1941)

SPECIFICATIONS
Type: High-speed articulated electric interurban train.
Gauge: 4 ft 8 1/2 in in
Propulsion: Direct current at 550V (600/650V post-World War II) fed to eight 125 hp Westinghouse nose-suspended traction motors geared to the driving axles of all except the third of the five Commonwealth cast steel bogies.
Current collection: Via trolley wire and poles (or 600V on third rail on the Loop)
Weight: 171,030 lb adhesive, 210,500 lb total.
Max. axleload: 21,380 lb
Overall length: 155 ft 4 in
Max. speed: 85 mph

Above: The legendary Chicago, North Shore & Milwaukee Railroad "Electroliner" trains consisted of five articulated cars of which the outer two pairs are depicted here. They ran until 1963.

The Electroliners were a pair of streamlined four-coach electric multiple unit interurban passenger train sets operated by the Chicago North Shore and Milwaukee Railroad between Chicago, Illinois, and Milwaukee, Wisconsin. They were built by St. Louis Car Company in 1941. Each set carried two numbers, 801-802 and 803-804.

They represented America's early move toward establishment of a substantial network of electric driven trains which heralded later moves toward greener trains. Unfortunately by that time the network have depleted rapidly due to the competition from the motor car. One of the best examples of this was the C,NS&M RR's North Shore Line where travellers started their journeys at selected stops on the famous central loop of Chicago's elevated railway—which meant that trains had to be flexible enough to turn street corners on 90ft radius curves. This was achieved by making the cars articulated as well as rather short. Only a few minutes later they would have to be rolling along at 85mph on the North Shore's excellent main line tracks. In Milwaukee, the trains made their final approach to the city center terminal on street-car tracks, with all that that involves in control at crawling speeds. It was a superb feat of design to build rolling stock that was able to suit both such a high-speed as well as such a low-speed environment. The St Louis Car Company cars seated 146 and boasted a tavern-lounge car. The two "Electroliner" trainsets were scheduled to make the 88-mile journey from Chicago to Milwaukee and return five times daily from February 9, 1941 until the flexibility of the motor car finally won out. The last full day on which the North Shore line operated was January 20, 1963.They are preserved at the Illinois Railroad Museum.

Class W-1 Bo-Do-Do-Bo GNR (1947)

SPECIFICATIONS
Type: Mixed-traffic electric locomotive for mountain grades.
Gauge: 4 ft 8 1/2 in
Current collection: Overhead catenary
Propulsion: Alternating current at 11,500V, fed to two motor-generator sets supplying direct current to twelve 275 hp nose-suspended traction motors geared to the axles.
Weight: 527,000 lb
Max. axleload: 43,917 lb
Length: 101 ft 0 in
Tractive effort: 180,000 lb
Max. speed: 65 mph

The Great Northern Railway's class W-1 comprised two electric locomotives designed for mixed traffic over the mountains. The locomotives were used on the 73-mile electrified portion of the railroad, from Wenatchee, Washington to Skykomish, Washington, including the Cascade Tunnel which is the longest Railroad tunnel in America at 7.8 miles.

The 3,300 horsepower W-1 motor-generator locomotives were built at General Electric's Erie works in 1947. Road numbered 5018 and 5019, they were the largest single-unit electric locomotives used in North America.

Nos 5018 and 5019 were retired in 1956, with No5019 being scrapped in 1959. No5018 was sold to the Union Pacific, who used its body and running gear as part of an unsuccessful experimental coal burning gas turbine-electric locomotive X-80.

Below left: Only two of the huge W-l class electrics were built; 3,300hp was a very modest output for such bulky units.

Below: GN No5019 in the Cascades region.

Little Joe 2-Do-Do-2 (1947)

SPECIFICATIONS
Type: Heavy duty mixed traffic electric locomotive.
Gauge: 5 ft 8 1/2 in
Current collection: Overhead catenaries
Propulsion: Direct current at 3,000 Volts fed to eight 470 hp traction motors geared to the main axles.
Weight: 440,800 lb adhesive, 535,572 lb total.
Max. axleload: 55,100 lb
Tractive effort: 110,750 lb
Overall length: 88 ft 10 in
Max. speed: 70 mph

Before the Cold War between West and East began in the late-1940s, General Electric had begun work on an order for 20 giant 3,300V dc 2-Do-Do-2 electric locomotives. These were intended for the Soviet Union, then involved in the first stages of implementing Lenin's own pet project for railroad electrification on an immense nationwide scale. When the time came for delivery, however, an embargo existed and for a time the locomotives which had been completed languished at the GE works. Later, the whole order was completed and offered for sale, the later units having been finished to standard gauge 4ft 8 1/2in instead of the 5ft gauge used in the Soviet Union.

The obvious buyer was Chicago, Milwaukee, St Paul & Pacific which used the same voltage and current and had not acquired any new locomotives for over 20 years. After a trial of a completed unit, this railroad in 1948 made an offer to purchase all 20, but the sum offered was considered too low. Before long, three were bought by the interurban-cum-freight railroad Chicago, South Shore & South Bend. Eventually in 1950, the remainder—now blessed with the name Little Joes after the ruler of the country for which they were built—did reach the Milwaukee's electrified divisions. During a distinguished 24-year career amongst the mountains of the north-west, the Little Joes already provided with regenerative braking, were modified to run in multiple even with diesel-electrics. By the mid-1960s, they were sole electric power to run regularly over the electrified sections, normally acting as helpers in the mountains. Alas, in 1972 the Milwaukee ceased using electric traction and the Little Joes were retired. The three which went to the South Shore were modified to run on that system's 1,500Volt DC current and lived long lives, although all have now been retired. No803 being preserved in the Illinois Railroad Museum.

Above: 2-Do-Do-2 "Little Joe' originally destined for the Soviet Union waits for an assignment under the 3,000v catenaries of the Milwaukee Road's electrification.

Left: Chicago; Milwaukee St Paul & Pacific "Little Joe" No.E20 at Deer Lodge, Montana, in 1952.

Silverliner series I-IV (1963-present)

In 1958 the PRR placed an order with the Philadelphia-based Budd Company for 6 Pioneer III MU cars in two subclasses to test out various options. Numbered 150 to 155 the even-numbered cars had fabricated truck frames and disc brakes, while the odd-numbered cars had cast steel truck frames and tread brakes. The PRR initially had hopes to MU cars such as the Pioneer IIIs in intercity service along its electrified routes and the cars were split between long distance and suburban duties. However as testing went on

Above: Penn Central operated Silverliner II coupled to a Silverliner III at Chestnut Hill West in 1974.

Right: SEPTA No145, a Silverliner IV, makes an eastbound station stop at Paoli, Pennsylvania.

Above: SEPTA Silverliner III making a stop at the Cornwell's Heights station on the Trenton Line in 2010.

SPECIFICATIONS
In service: SL-I: 1958-1990, SL-II: 1963-2012, SL-III: 1967-2012, SL-IV: 1973-present
Manufacturer: SL-I/II: Budd Company, SL-III: St. Louis Car Company, SL-IV: Avco General Electric
Number built: SL-I: 6, SL-II: 59, SL-III: 20, SL-IV: 232
Formation: SL-I/II/III: Single unit, SL-IV: Married Pair and Single Unit
Operator: Pennsylvania Railroad, Reading Railroad, Penn Central Railroad, Conrail (under SEPTA), SEPTA, US DOT
Line(s) served: SEPTA Regional Rail
Car body construction: Stainless steel
Doors: SL-I/II/III/IV 2 end doors w/ traps
Maximum speed: 100 mph
Traction system: SL-II/III/IV: Transformed alternating current fed initially through Ignitron and later Silicon-controlled rectifiers to phase angle DC motor controller.
Electric system(s): Catenary 11-13.5 kV 25 Hz AC
Current collection method: Pantograph
Bogies: SL-I/II: Budd Pioneer SL-II/IV/: General Steel GSI 70
Braking system(s): Pneumatic, Dynamic (SL-IV only)
Coupling system: WABCO N2

they were soon limited to suburban service in the Philadelphia area when a full-scale production order of 38 PRR "Silverliner" cars were delivered in 1963. A total of 38 cars were purchased for the PRR with the remaining 17 going to the Reading. While some referred to the new vehicles as "PSIC Cars," the modern stainless steel body shells quickly defined the fleet and the name "Silverliner" was soon adopted referring to the train's shiny stainless steel body shell compared with the painted (or rusting) carbon steel railcars in service by the Pennsylvania and Reading Railroads in the past.

Jersey Arrow (1968)

SPECIFICATIONS
Gauge: 4 ft 8 1/2 in
Body construction: Stainless steel
Car length: 85 ft
Width: 9 ft 11 1/2 in
Doors: 2 end doors w/ traps 1 middle door high level only
Maximum speed: 100 mph
Traction system: Transformed alternating current fed through either Ignitron (Arrow I/II) or Silicon-controlled (Arrow III) rectifiers to phase angle DC motor controller. Arrow III fleet converted to AC traction 1992-95.
Power output: Arrow III rebuilt (pair): 1,150 hp, Arrow III rebuilt (single): 750 hp
Electric system(s): 12 kV 25 Hz AC Catenary (Arrow I/II/III), 12 kV 60 Hz AC Catenary (Arrow II/III), 25 kV 60 Hz AC Catenary (Arrow II/III)
Current collection method: Pantograph
Bogies: General Steel GSI 70
Braking system(s): Pneumatic, dynamic
Coupling system: WABCO Model N-2

The Jersey Arrow is a type of electric multiple unit (EMU) railcar developed for the Pennsylvania Railroad, and used through successive commuter operators in New Jersey, through to New Jersey Transit. Three models were built but only the third is in use today.

The first series of Arrows (known as PRR MP85s) were built in 1968-69 by the St. Louis Car Company; 35 were built and purchased by the New Jersey Department of Transportation. These cars were initially numbered 100-134. These cars were built with higher capacity 3-2 seating which caused grumbling by the passengers at that time.

In 1974, General Electric produced 70 Arrow II cars in the married pair format, classed MA-1G. These cars were built in GE's Erie (PA) shops with car shells from Avco. The Arrow IIs were numbered 534-603. They were purchased specifically to replace the ancient PRR MP54s, which were slowly phased out of New Jersey service in late 1977. The Arrow IIs briefly returned to service on the Newark Division, but were ultimately reassigned to the Hoboken division for the rest of their service lives. In 1997, the decision was made to retire them due to rotting floors and holes in the roofs. Most were scrapped in 2001.

The Arrow IIIs were built between 1977 and 1978 by General Electric in the same fashion as the Arrow IIs. They consist of 200 cars built as married pairs (1334–1533) and 30 single cars (1304–1333). These cars were initially ordered as part of a plan to rehabilitate the NJDOT (Later NJ Transit's) Hoboken division, converting the 3,000 volt DC system to a 25 Kv 60 Hz AC system. The Arrow IIIs were given a mid-life overhaul between 1992 and 1995 by ABB. The rebuild replaced the original DC propulsion system with a new solid state AC system that also included higher power traction motors with a total of about 375 hp per two axle truck.

Right: New Jersey Transit train led by Arrow III No1327 pulls into the Far Hills station.

Amtrak AEM-7 Bo-Bo (1978)

The AEM-7 is a twin-cab B-B electric locomotive that is used in the United States on the Northeast Corridor between Washington DC and Boston and the Keystone Corridor between Philadelphia and Harrisburg in Pennsylvania. They were built by Electro-Motive Division from 1978 to 1988. In the Boston Mechanical Department of Amtrak they are known as "Meatballs" and in the Washington Mechanical Department they are known as ASEAs since some of their major parts and components were designed in Sweden by ASEA which merged with Brown Boveri in 1988 forming ABB. They are also referred to as "toasters" by railfans, owing to their boxy appearance. There are two versions of the AEM-7 as of 1999: the original AEM-7DC which has DC propulsion equipment and the newer, modified AEM-7AC which uses AC propulsion equipment.

SPECIFICATIONS
Type: Electric
Designer: Allmänna Svenska Elektriska Aktiebolaget (ASEA)
Builder: General Motors Electro-Motive Division
Total produced: 65
Gauge: 4 ft 8 1/2 in
Length: 51 ft
Weight: 202,000lb
Electric system(s): 11-13.5 kV 25 Hz AC Catenary, 11-13.5 kV 60 Hz AC Catenary, 25 kV 60 Hz AC Catenary
Current collection method: Dual pantographs
Maximum speed: 125 mph for Northeast Regional, 110 mph for long-distance trains
Power output: AEM-7DC and AC: 5,100 kilowatts (6,800 horsepower) maximum at rail; 4,320 kilowatts (5,790 horsepower) continuous at rail
Starting Tractive Effort: AEM-7DC: 53,924 lbf ; AEM-7AC: 51,700 lbf to 43 mph
Continuous Tractive Effort: AEM-7DC: 30,000 lb @ 77 mph; AEM-7AC: 42,500 lb @ 60 mph
Operator(s): Amtrak, MARC and SEPTA
Number(s): AMT 901-902, 904-910, 912, 914-921, 923-929, 931-932, 934-953; MARC 4900-4903; SEPTA 2301-2307
Safety system(s): Advanced Civil Speed Enforcement System

Above left: AEM-7 No923 pushes the Keystone Service train to Harrisburg, PA.

Left: AEM-7 locomotive No901 wearing the Phase III livery introduced in 1979; photographed in the 1980s.

Above: Amtrak Train 670 with locomotive AEM-7 No948 at Elizabethtown, PA Station in 2012.

ABB ALP-44 Bo-Bo (1989)

SPECIFICATIONS
Type: Electric
Total produced: 33
Gauge: 4 ft 8 1/2 in
Electric system(s): 12.5 kV 25 Hz AC Catenary, 12.5 kV 60 Hz AC Catenary, 25 kV 60 Hz AC Catenary
Current collection method: pantograph
Maximum speed: 125 mph
Power output: Max: 7,000 hp; Continuous: 5,790 hp
Starting tractive effort: 52,000 lb
Train brakes: Direct Release air brakes

The ABB ALP-44 is an electric locomotive which was built by Asea Brown Boveri (Sweden) between 1989 and 1997 for the New Jersey Transit and SEPTA railway lines. As of 2014, only SEPTA still operates the ALP-44 in revenue service, as New Jersey Transit has retired its fleet. The ALP-44 was originally ordered for New Jersey Transit's electric lines, with fifteen units, designated ALP-44O ("O" denotes Original) and numbered 4400 through 4414, delivered in 1990, with prototypes 4400 and 4401 in late 1989. An option order for five more units, designated ALP-44E ("E" denotes Extended) and numbered 4415 through 4419, were delivered in 1995. The final order for 12 ALP-44M units ("M" denotes Microprocessor) numbered 4420 through 4431 were delivered in 1996 and into early 1997 for the new Midtown Direct service.

Below: New Jersey Transport ALP-44 No4408 pushes an eastbound train into Elizabeth station.

Acela Express (2000)

SPECIFICATIONS
Manufacturer: Bombardier, Alstom
Formation: 8 cars (2 x power car; 6 x passenger car)
Capacity: 304 (44 first class; 260 business class)
Car body construction: Stainless Steel
Train length: 665 ft 8 3/4 in
Car length: 69 ft 7 in (Power car), 87 ft 5 in (Passenger Car)
Width: 10 ft 5 in (Power car), 10 ft 4.5 in (Passenger Car)
Doors: Single Leaf Sliding Plug Doors
Passenger Cars: 4 Intermediate, 2 End
Wheel diameter: 40 in (Power Car), 36 in (Passenger Car)
Maximum speed: 165 mph (150 mph max operating speed)
Weight: 1,246,000 lb (Trainset), 204,000 lb (Power Car), 142,000 lb (End Cars; Business and First), 139,000 lb (Intermediate Business Cars), 137,000 lb (Bistro Car)
Axle load: 51,000 lb (Power Car), 35,500 lb (End Cars; Business and First), 34,750 lb (Intermediate Business Cars), 34,250 lb (Bistro Car)
Traction system: Alstom GTO Inverters and 3-Phase Asynchronous AC Traction Motors
Power output: 1,540 hp (Per Motor), 6,200 hp (Per Power Car)
Tractive effort: Starting 51,000 lb (Per power car)
Power supply: 2850 V DC (PWM Rectified) Voltage Regulated from mains re-inverted to three-phase, frequency and voltage controlled AC waveform.
Electric system(s): Catenary, 25 kV 60 Hz AC, 12 kV 60 Hz AC, 12 kV 25 Hz AC
Current collection method: Pantograph, 2 per power car
Braking system(s): Dynamic and Regenerative (Power Cars), Electro-Pneumatic Disk and Tread(Trainset)

The Acela Express is Amtrak's high-speed rail service along the Northeast Corridor (NEC) in the Northeast United States between Washington, D.C., and Boston via 14 intermediate stops including Baltimore, Philadelphia, and New York City. Acela

Express trains are the fastest trainsets in the Americas; the highest speed they attain is 150 mph in revenue service. Acela trains use tilting technology, which lowers lateral centrifugal forces, allowing the train to travel at higher speeds on the sharply curved NEC without disturbing passengers. Because Acela shares its track with normal commuter and freight trains,aging infrastructure in many segments of NEC trackage limits the express to an average speed in regular service much lower than its maximum speed.

For example between Boston and Washington, Acela covers 454 miles in 7 hours an average speed of 65 mph.

All 20 built are still in service.

Left: The Amtrak Acela Express train at New Haven Union Station.

Below: HHP-8 engine in Boston South Station. The Acela branding was only used for a short time on Regional trains like this one.

MLV New Jersey Transit (2006)

SPECIFICATIONS
Gauge: 4 ft 8 1/2 in
Capacity: 127 (cab car), 132 (trailer car with restroom), 142 (standard trailer car)
Operator: NJ Transit, Agence Metropolitaine de Transport of Montreal
Car body construction: Riveted or welded aluminum body on a steel frame
Car length: 85 ft
Width: 10 ft 0 in
Height: 14 ft 6 in
Floor height: 4 ft 3 in
Doors: Pneumatically-operated doors
Maximum speed: 100 mph
Weight: 139,250 lb (cab car)
134,880 lb (trailer car with restroom)
132,990 lb (standard trailer car)
Power supply: 480 V AC, 60 Hz, 3-phase
Braking system(s): Pneumatic Disc and Shoe

The Bombardier MultiLevel Vehicle (MLV) is a bi-level passenger rail car manufactured by Bombardier at La Pocatière, Quebec and Plattsburgh, New York for use on commuter rail lines. It started service in 2006 and is still being produced as of the present day. 643 units have been built and are all still in service. The coaches have a two-by-two seating arrangement and more knee and leg room than single level coaches. The seats are also bigger and it has 15-30% more seating than on single level coaches. The intermediate levels have 5 inward-facing flip-up seats on each side, for wheelchairs or bicycles. On cab cars, a large equipment locker behind the cab replaces one row of seats. There are large side doors at intermediate levels for high-platform loading, and end doors, except at the cab position on cab cars. The end doors of NJ Transit coaches have stepwell traps, allowing these doors to be used for both high and low-platform loading.

Above: NJ Transit Multi-level coach at Millburn.

Left: New Jersey Transit MLV train powered by a Bombardier ALP-46.

Silverliner V (2009)

The Silverliner V is an electric railcar designed and built by Hyundai Rotem for the SEPTA Regional Rail system and is now being offered to other commuter rail providers with Denver's RTD being the second purchaser. n 2010 Denver's Regional Transportation District selected the Silverliner V for its new commuter rail line. A total of 66 cars were purchased in the married pair configuration for a total of $300 million. The first four cars were delivered to Denver on December 3, 2014 with service to start in 2016. Differences between the RTD and SEPTA cars include, support for only 25kv 60Hz AC electrification, four high level doors per side, less powerful traction motors, full-width cabs and bells.

This is the fifth generation railcar in the Silverliner family of single level EMUs. SEPTA had ordered a total of 120 of the cars at a cost of $274 million with the first cars arriving in the United States on 28 February 2010 from South Korea, where they were manufactured by Hyundai Rotem. The cars are built in South Korea and final assembly takes place in South Philadelphia.The cars entered revenue service on 29 October 2010 and the last of the 120 cars arrived on property for testing in February, 2013.

Cars 735, 736, 871 and 872 are owned by Delaware for use on the Wilmington/Newark Line.

SPECIFICATIONS
Number under construction: 62
Number built: Total 124 (SEPTA: 120, RTD: 4)
Number in service: 120
Fleet numbers: SEPTA: 701-738, 801-882; RTD: 4000-4065
Capacity: SEPTA: 107 Single Car, 109 Married Pair Car; RTD: 91 per car
Operator: SEPTA; Denver RTD
Line(s) served: SEPTA Regional Rail; Denver RTD Commuter Rail
Car body construction: Stainless steel
Car length: 120 ft
Width: 10 ft
Height: 15 ft
Doors: quarter point, SEPTA: 3 per side, 2 with traps, RTD: 4 per side
Maximum speed: 110 mph
Operating speed: 100 mph SEPTA, 79 mph RTD
Weight: 146,600 lb
Traction system: Mitsubishi Electric AC Induction motor, VVVF inverter[2][4]
Electric system(s): SEPTA: 12.5 kV 25 Hz AC Catenary; 12.5 kV 60 Hz AC Catenary SEPTA and RTD: 25 kV 60 Hz AC Catenary
Current collection method: Pantograph
AAR wheel arrangement: B-B
Bogies Bolsterless: GSI 70
Braking system(s): Pneumatic, one outboard disc, one tread per wheel.
Coupling system: WABCO Model N-2
Gauge: 4 ft 8 1/2 in

ALP-46A New Jersey Transit Bo-Bo (2014)

The ALP-46 is an electric locomotive built in Germany by Bombardier between 2001 and 2002 based on the German Class 101 for use in the United States. New Jersey Transit (NJT) is the only railroad to operate this locomotive model. They can be found all over the electrified NJT system, but are primarily used for service to and from Penn Station in New York City. In February 2008, NJT ordered twenty seven 125 mph top speed ALP-46A locomotives from Bombardier, which were intended to haul Bombardier MultiLevel Coaches. The estimated value of the order was $230 million. In June 2009 NJT took up an option for a further nine locomotives, and spare parts, at a cost of $72 million. Final delivery of the order took place in April 2011 and all locomotives were in operation May 2011.

SPECIFICATIONS
Type: Electric
Gauge: 4 ft 8 1/2 in
Length: 64 ft
Width: 9 ft 8 in
Axle load: 46A: 50,706 lb
Weight: 202,822 lb (91,999 kg)
Electric system(s): 12.5 kV 25 Hz AC Catenary; 12.5 kV 60 Hz AC Catenary; 25 kV 60 Hz AC Catenary
Current collection method: Pantograph
Head end power: 480 V AC, 60 Hz, 3 phase.
Maximum speed: 124 mph
Power output at rail: 7,500 hp
Tractive effort: Starting: 71,000 lb, Continuous: 54,000 lb at 53 mph
Locomotive brakeforce dynamic: 34,000 lb

Right: New Jersey Transit ALP-46 4600 in Summit, New Jersey. Both the ALP-46 and ALP-46A are have been used to haul New Jersey Transit's Comet IIM, III, IV, V, and Multilevel fleet.The ALP-46 was also used to pull Amfleet consists on Amtrak's Clocker service in its final days of operation.

4609
NJ TRANSIT

4600
4600
NJ TRANSIT
NJ TRANSIT

Amtrak Cities Sprinter Bo-Bo (2012)

In October 2010, Amtrak ordered 70 ACS-64 locomotives at a cost of US$466 million, to be delivered beginning in February 2013.The order was the second part of Amtrak's company-wide fleet-replacement program, after an order for 130 Viewliner II passenger cars was placed in July 2010.

The Amtrak Cities Sprinter, or ACS-64, is an electric locomotive designed by Siemens Mobility for Amtrak. The locomotives are to operate on the Northeast Corridor (NEC) and the Keystone Corridor, replacing the railroad's existing fleet of AEM-7 and HHP-8 locomotives. The design is based on the EuroSprinter and the Vectron platforms, which Siemens sells in Europe and Asia.Significant structural changes to the design were made to comply with American crashworthiness requirements, including the addition of crumple zones and anti-climbing features as well as structural strengthening of the cab, resulting in a heavier locomotive than the previous models.The body is a monocoque structure with integral frames and sidewalls.

The first ACS-64 entered service in February 2014; deliveries lasted until 2015.

Left: ACS-64 No600 leads southbound Northeast Regional No171 through Readville on its first revenue run on February 7, 2014.

SPECIFICATIONS
Type: Electric
Gauge: 4 ft 8 1/2 in
Trucks: Siemens model SF4
Wheel diameter: 43.98 in
Minimum curve: 249 ft 4.1 in
Wheelbase: 32 ft 5.8 in
Length: 66 ft 8 in
Width: 9 ft 9.5 in
Height: 12 ft 6 in excluding pantograph
Axle load: 54,250 lb
Weight on drivers: 100%
Weight: 215,537 lb
Electric system(s): 12 kV 25 Hz AC Catenary, 12.5 kV 60 Hz AC Catenary, 25 kV 60 Hz AC Catenary
Current collection method: Pantograph
Traction motors: 3-phase, AC, Fully Suspended, Siemens built (Norwood, Ohio)
Head end power: 1,000 kW (1,300 hp) 3-phase, 60 Hz, 480 VAC, 1000 kVA
Transmission: Pinion Hollow Shaft Drive w/ Partially Suspended Gearboxes
Multiple working: Yes
Maximum speed: 125 mph Service, 135 mph Design
Power output: 8,600 hp Maximum (Short-Time), 6,700 hp Continuous
Tractive effort: Starting:72,000 lb Continuous: 6,700 hp, 63,000 lb@40 mph, 20,000 lb @125 mph Continuous: 8,600 hp, 61,000 lb@53.5 mph , 26,000 lb@125 mph
Factor of adhesion: 2.99 (33.4%)
Locomotive brake: Regenerative braking, NYAB Electro-Pneumatic Cheek Mounted Disk Brakes
Locomotive brakeforce: 6,700 hp Max.
Train brakes: Electro-Pneumatic

Index

D

E

O

P

Q

R

S

Acknowledgments

The publishers would like to express their thanks to the following:

JP Bell Photography
Steve Barry, Railfan
Mark Karvon Art Studios
Richard Grant, RgusRail
Bob Vogel, RR Picture Archives